Entomology at the Land Grant University

NUMBER NINE:
Texas A&M University Agriculture Series
C. Allan Jones, General Editor

Entomology at the Land Grant University

Perspectives from the Texas A&M University Department Centenary

Edited by
Kevin M. Heinz
Raymond E. Frisbie
& Carlos E. Bográn

Texas A&M University Press
COLLEGE STATION

First edition

The paper used in this book meets the minimum requirements
of the American National Standard for Permanence
of Paper for Printed Library Materials, z39.48–1984.
Binding materials have been chosen for durability.
∞

Library of Congress Cataloging-in-Publication Data

Entomology at the land grant university : Perspectives from the Texas A&M
University department centenary / edited by Kevin M. Heinz,
Raymond E. Frisbie, and Carlos E. Bográn.
p. cm.
ISBN 1-58544-432-4 (cloth : alk. paper)
1. Entomology—Study and teaching—History. 2. State universities and
colleges—History. I. Heinz, Kevin M. II. Frisbie, R. E. (Raymond E.)
III. Bográn, Carlos Enrique, 1968–
QL468.5.E58 2005
595.7'07—dc22
2005000440

To all former, current, and future students
of Texas A&M University, Department of Entomology

Contents

Preface

Entomology departments, which form a component of all Colleges of Agriculture within Land Grant institutions in the United States, have acted as either catalysts for change or as resultant examples of these changes in agricultural science and technology. Departments of Entomology also serve a unique function within Colleges of Agriculture by contributing significantly to the teaching, research, and service missions of their institutions. Within the context of changing demographics, land-use patterns, environmental issues, agriculture, and the life sciences, departments of Entomology are frequently cited as models to providing leadership to the missions of universities, the economies of private industry, and the overall improvement of human society. In 1999, the Department of Entomology at Texas A&M University celebrated its 100th anniversary as an organized unit on the College Station campus. This celebration represented a time for both reflection of past accomplishments and planning for future challenges. To accomplish these tasks, the Department sponsored a series of mini-symposia to study entomological issues of academic, scientific, economic, and social importance. A product of these symposia is this edited volume in which symposia participants have contributed chapters representing a perspective on central themes.

Several issues characterized by Entomology Departments and important to the Land Grand University System are covered in this text. Key elements within Land Grant University departments are how they programmatically address teaching the general public at the undergraduate and graduate levels; retain responsibilities to the citizenry of their state that range across traditional agricultural and newly emerging urban clientele; obtain a balanced, innovative, and productive research program; and develop a clear vision of the future and a realistic path to making that vision become reality. By comparison with many other academic departments associated with life sciences, the physical sciences, or humanities, Entomology departments represent a mixture of professionals with highly varied responsibilities. A large umbrella in terms of philosophy, administration, and execution is needed to accomplish such a diversity of tasks. Is having such diversity a benefit or detriment to Entomology? How do Entomology departments excel in each of the areas? How are the conflicting demands resolved? How does the clientele served perceive the various areas of specialization? Important perspectives to these issues, which transcend all orders of organization, are also addressed.

The delivery of science-based knowledge to the public is a charge of Cooperative Extension Service, and in this role, Extension Entomology will continue to provide an invaluable service to our country. However, modern Cooperative Extension Services face many challenges in providing leadership in education on important issues, retaining flexibility to respond to ever-changing issues, and continue delivering services and programs in areas of strength. Addressed are how Extension handles controversial topics in agriculture, retains value in an increasingly diversified world, aligns technology transfer with the needs of end-users, and executes its initiative to provide education about key issues.

Due to their small size, short generation time, high reproductive capacity, and genetic diversity, insects have been model systems for the advancement of science. In addition, global policy is to have Integrated Pest Management (IPM) be practiced on the majority of the agricultural lands. Entomology has historically been the leader in developing IPM programs using a variety of approaches that include biological control, host plant resistance, pest-reducing cultural practices, chemical control, and other methodologies. IPM finds its scientific basis in applied ecology; however, this connection between ecology and pest management has been taken for granted. Pest problem-solving based purely on empiricism is limited, and only through the study, development, and implementation of ecological principles, as they relate to pest populations, can we better forecast how introductions of new pests or modifications of production practices influence IPM. These topics are discussed in a section addressing the field of entomology as a model for IPM research and for the advancement of science.

Entomology departments in U.S. universities originated from the need to improve food and fiber production. Traditional Entomology students prepared for careers in agriculture and related fields, and the curriculum offered by Entomology departments was framed to meet the needs of those students. However, there has been a significant, rapid, and fundamental shift in the student population in many U.S. institutions over the past fifteen years. The pool of students interested in taking entomology courses at Land Grant institutions has changed. Larger numbers of students are coming from urban or suburban areas and fewer students have agricultural backgrounds. Today, students are interested in programs in molecular, cell, and developmental biology; health professional programs; environmental biology; conservation biology; and urban entomology in addition to traditional agriculture programs. As a result, Entomology has followed a diversity of paths as it has moved into the mainstream of human society. Four different perspectives on this subject are offered in the section on Entomology as a science for all society. It is fascinating to identify similarities between pest management and military strategies and how portrayals of insects in movies have molded public perception of these animals. As evidenced by these examples, insects are not just for entomologists any more; they provide a foundation for educating the public about science. Departments frequently seek to establish or strengthen formal and financial ties with industry. The benefits derived from such ties with private industry, and the benefits eventually relayed to the public at large are addressed.

Finally, the text hopes to challenge the spatial and temporal scales on which Entomology may be explored. Insects have no regard for international boundaries. Today's pest problem in one country or region is tomorrow's pest problem in another country or region. International collaboration among entomologists presents a model by which basic and fundamental research should be conducted in an arena where research dollars are scarce and the problems large. Globally important questions relevant to human health, biodiversity, and agricultural sustainability are addressed, as are views to the future of entomological research in terms of process, researchable issues, and utilization of new techniques for advancing the science and making contributions toward the solution of real world problems.

Acknowledgments

This book is the outcome of a series of symposia held at Texas A&M University during the spring and fall semesters 1999. The symposia were part of a yearlong celebration of the 100th anniversary of the Department of Entomology. The success of the events surrounding the celebration was made possible by the work and contributions of several individuals, institutions, and corporations. We would like to thank the administrative support staff at the Department of Entomology, especially Victoria Bienski, Lela Cape, Laura Nelson, and Sherry Harmon. Their hard work, dedication, and special attention to details throughout the development of the centennial celebration are greatly appreciated. The many positive comments and praises received regarding the organization and professionalism of the centennial celebration were due to their efforts, thank you.

We would also like to recognize the generous contributions of Texas A&M University Agriculture Program, Texas A&M University Center for Advanced Invertebrate Molecular Sciences, Abbot Laboratories, American Cyanamid, Cheminova, Cotton Foundation, Cotton Incorporated, Delta and Pine Land Company, Dupont Corporation, National Cotton Council, National Grain Sorghum Producers, Novartis Corporation, Texas Corn Producers, Texas Cotton Ginners, Texas Cotton Producers, Texas Grain Sorghum Board, Texas Wheat Producers Board, and the Texas Pest Management Association. Without their support, the events leading to this book would not have been possible. We look forward to their continued support in the next 100 years.

Part I

Entomology: Fulfilling the Land Grant University Mission

Chapter 1

Contributions of Entomology to the Mission of the Twenty-First-Century Land Grant University

RICHARD L. JONES
Agricultural Experiment Station
University of Florida–IFAS

The major challenge facing a twenty-first-century Department of Entomology in a Land Grant University (LGU) is to remain relevant in a changing environment. To remain relevant as departments we must continuously adapt. To do this successfully, we must know where our science is going, where the LGU is going, and where society in general is going. All of these entities are constantly coping with a rapidly changing environment; therefore, change is critical to our success. The Department of Entomology, and for that matter any department, will be successful to the extent to which it continually adapts to contribute to the LGU success. Consequently, we should look at factors impacting future directions of the LGU. This chapter discusses new directions of the LGU, new directions and issues in agriculture and natural resources, new directions for a Department of Entomology, and a few points about planning.

NEW DIRECTIONS OF THE LAND GRANT UNIVERSITY

The key to any success is performance. Beyond that, such performance must be in areas important to society. As I stated, our success as a department should be keyed to the success of the LGU. The directions taken by the twenty-first-century LGU will be driven largely by accountability: accountability to our overriding mission to serve the public. Over the past fifteen to twenty years, the LGU has undergone much criticism. This criticism comes from our clientele and from our political leadership. It has focused on our undergraduate program and specifically, on low graduation rates, poor attention to student needs, an obsession with research and its resultant scramble for grant dollars, a faculty that is invisible, and an attitude of arrogance among faculty. We are fortunate that the traditional Land Grant component of the LGU that I will refer to broadly as agriculture is often cited as an exception, and in fact, a good example. There are several reasons behind this. For one, our departments have faculty teaching in the classroom. In addition, we pursue problem-solving research and we boast a customer-friendly extension program. However, even though we have been pointed out as examples for the rest of the LGU to follow, we must not become complacent.

The roots of criticism against the LGU can be attributed largely to a loss of contact and communication with our clientele. To some extent, the LGU has forgotten for whom it works, and our attitudes often *are* arrogant. A recent study commissioned by the Kellogg Foundation titled "Returning to Our Roots: The Engaged Institution" (NASULGC, 1999), addresses these issues. The theme of this report is that we must re-engage with our clientele. An important feature of this study is a seven-part test to assess our level of engagement. The seven parts of the test include responsiveness, respect for partners, academic neutrality (objectivity as information source), accessibility, integration (of teaching, research, and extension functions), coordination (across units), and resource partnerships. Along with the commission, I believe these seven issues are key indicators of successful Land Grant universities of the future. In fact, they were the strengths upon which the LGU system was built. They will be key indicators of performance for a Department of Entomology or any other unit of the university in the future.

The seven issues addressed by the Kellogg Foundation report, dealing with contact and communication with our clientele, are equally applicable within a university, within a department, and between our partners and us. If we analyze our failures over recent years, most would fall under one or more of the seven characteristics. For example, how often do we hear from our clientele that we are not responsive or that we are not accessible? On the other hand, how often do we have organized listening sessions or joint planning sessions with our clientele? How often do we, usually as individuals, ridicule the opinions of the public, the press, or politicians? How objective are our recommendations now that we profit from intellectual property ownership, both as an institution, as units, and as individuals? Recommendations on plant cultivars now take on a different light. Are we accessible? Institutions are confusing, and as it has often been said, clientele have problems and the LGU has departments. We must be user friendly. Are teaching, research, and extension optimally integrated? Do we have faculty sending in proposals to do the same thing from multiple departments or from multiple units? Coordination is important. Partnerships will be essential—partnerships with the corporate world as well as partnerships with the rest of our clientele. However, partnerships must not compromise our mission to serve the public good. The successful department of tomorrow will aggressively address all of these issues to maximize its performance.

Another important component of performance, along with emphasis areas and clientele, is that of quality. All LGUs strive to be in the top ten or the top five. At the University of Florida we pride ourselves as a member of the Association of American Universities (AAU), but not resting on our laurels, we are pressing hard to be among the top institutions within AAU. This means increasing graduate education, increasing grant success, increasing merit scholars, and increasing graduation rates. These are worthwhile endeavors for they mean that taxpayers are getting a greater return on their investment. But a drive for quality based entirely on these criteria can be a tradeoff with accessibility and perhaps with academic neutrality. Striving for increased grant funds can be counterproductive to our mission, especially in the Experiment Station, and it can also interfere with our partnerships. We must identify quality factors that address all seven traits of the engaged institution.

Perhaps the most important quality aspect of any organization is its human resource base. The most important quality decisions at the LGU are the hiring and tenure of faculty. Maintenance of strong faculty development programs after employment is also important. As a rule, the LGU does a poor job in this area and it must be improved. In addition, quality and any other desirable trait must be rewarded. Once the department has decided the direction it must take, it must reward those employees who contribute the greatest to move the department in that direction.

Measurements of quality in research are particularly difficult to demonstrate to the pubic. Grant resources are inputs that can be helpful in assessing quality when all other things are equal—which they rarely are. Quality is really assessed best by peers, but even then it is hard to completely remove the human element. Of course the best measure of quality is the outcome. How has the research impacted society? Citation indexes are often suggested and often used to assess quality. To some extent they are being used at the University of Florida. But citation indexes are only useful within narrowly defined parameters. A scientist's publication on corn will get cited a lot more than the same research conducted on strawberries, but we must work on strawberries anyway. Research on fundamental methods will be cited more often than research on applications.

Another aspect of performance is productivity. We are constantly asked: how can the LGU be more productive? How can we run faster? This is a special challenge for the LGU. Attempts to deal with perceived less-than-optimum productivity of faculty by legislative bodies has resulted in some undesirable consequences. At the University of Florida, components of research and service have been virtually eliminated from the effort allocation of academic faculty. The purpose of this is to increase the effective number of faculty Full-Time Equivalents (FTEs) in the classroom, but it was done at a cost to the other two missions of the LGU.

Measurements of productivity are a special challenge with research and extension. If one ignores quality, outputs are easy to measure—publications, patents, contacts, and such, but outcomes are more difficult. Outcomes are the answer to the question, "So what?" Outcomes serve the public good. They are the real purpose of our work and it is the outcome in which the clientele are interested. What impact has the release of a new variety had on the economic well-being of the industry, or what is the impact of a new rotation or a new tillage method? Of what benefit were these new outputs to the consumer? How has the adoption of Integrated Pest Management (IPM) practices enhanced public well-being? The cost of making these assessments across the board is extensive, but we will have to find a way. Concerns about performance and accountability have led the University of Florida to publish quarterly "Measuring University of Florida Performance." Examples of topics over the past several months include efficiency, research productivity, improvement, graduate growth, and undergraduate program enhancements. Within these categories the University of Florida is compared to peer institutions. Such parameters can serve also as guideposts for departmental success.

FACTORS IMPACTING CHANGE IN OUR TRADITIONAL LGU PROGRAMS

Agriculture is changing. It is becoming larger, more integrated, and much more technological. Environmental integrity has become more important. The corporation is more deeply involved in production research such as variety development and even extension education, activities that were once a hallmark of the LGU but now are being done in the private sector. These changes in the agricultural sector are impacting the Florida Agricultural Experiment Station (AES) and the Cooperative Extension Service. In response to these changes the AES can pursue three avenues with respect to research programs. We can either collaborate with private industry, compete head to head with industry, or we can do something different. I am not sure we can compete head to head. Compared to private industry, our resources are limited. Some major corporations have as many as 300 or more scientists working on one or two crops, for example. But there are other opportunities and agriculture is more and more a global system. The only way for the United States to compete is with new production technology. We will play a greater role in this area. In addition, good production technology, plus good management, plus skilled labor can lead to success, even when labor costs are high.

I believe we will see more collaborative arrangements with the private industry, and if done properly this can be a mutually beneficial arrangement. These arrangements range from sponsored research agreements to industry-funded centers. But we must remember that our first commitment is to our clientele, and we must work for their best interest. Many of the challenges facing agriculture and natural resources management will have solutions based on knowledge that can't be packaged and sold. In fact, this knowledge probably represents our greatest opportunity. Thurow (1993) says the future is not in product technologies but in process technologies. I believe this applies to agriculture. As an example, Thurow cites electronics in Japan. While the United States concentrated on development of new products, Japan concentrated on the development of new processes to manufacture those products. Consequently, Japan was able to pay higher wages and still make the products to sell at a lower price than we could in the United States. There is an important lesson here for U.S. agriculture.

Among the biggest challenges to agriculture are the environmental challenges. At the University of Florida we conduct research and offer educational programs in the development and use of the best management practices. For example, how to grow sugar cane in South Florida and keep phosphorus in the run-off below 5 ppb? The solution here will be knowledge and management (process technology) with minimal opportunity for a use of new products. Much research work needs to be done, so lots of opportunities exist. The same is true for animal waste management and integrated pest management. In fact, to effectively implement many of our biological control programs, the producer will need to be involved in management, or very specialized businesses will be required, or in some cases, government involvement will be needed. But knowledge and management will be key and more important than product.

Food safety is another critical issue in want of research solutions that may not be addressed by private enterprise. Again, process technologies will be the driving forces leading to improvement. Agriculture and natural resources are biology-based systems. We live in a biology-dominated era, and as biologists (and entomologists) we are in an excellent position to contribute. Our AES mission has broadened beyond agriculture, and this is clearly reflected by the large number of LGU programs adding life sciences, or environment, or both to their name. Our research, extension, and teaching programs will continue to expand beyond traditional production issues to address basic questions in biology and to solve problems in agriculture and natural-resource management. This will include such topics as water conservation, natural ecosystems, species diversity, and the human dimension of our interactions with the environment.

WHAT ABOUT ENTOMOLOGY?

The application of many previous statements to a Department of Entomology is clear. First, I would like to address the popular administration question of: Why do we have so many entomologists? This question is often raised because of invalid comparisons with other departments. It is noteworthy that faculty in a discipline only get summed when the discipline is coincident with a departmental name. The most common comparison of entomology is with plant pathology. But plant pathology is a specialization of microbiology. The fair comparison of numbers of entomologists is with microbiologists. The fair comparison of number of plant pathologists is with plant entomologists (a heretofore unrecognized subdiscipline). An illustration at the University of Florida is as follows: Statewide we have about seventy-two entomologists, twenty-six of whom work on plants. We have forty-one plant pathologists. The remaining forty-six entomologists work in areas of medical entomology, livestock entomology, apiculture, household/structures, and the basic entomological sciences (ecology, physiology, systematics, etc.). To continue with the analogy, microbiologists involved in medical issues work for the health sciences whereas medical entomologists are usually attached to Entomology departments. The same is true for veterinary microbiologists and livestock entomologists, and for food microbiologists and apiculturists. The "basic" subdisciplines of microbiology are found in the Department of Microbiology and Cell Science (a faculty of seventeen at the University of Florida). Regardless of subdisciplines, we entomologists call ourselves entomologists, and we are usually aggregated in one department and hence, very visible. Decisions should not be based on total numbers of entomologists, but based on the amount of work needed to be done and the opportunity to make an impact.

The analogies between the number of entomologists and microbiologists can also be used to illustrate the diversity of entomological programs and their interface with society and its institutions. Diversity is a strength of a Department of Entomology because it provides multiple opportunities to contribute to the future of the LGU. I also believe that within agriculture, entomologists are the best all-around biologists. For example, few other disciplines in Colleges of Agriculture have faculty so broadly trained in the basic biological sciences including biology, ecology, evolutionary theory, zoogeography, and animal behavior.

One of the major challenges of our Department of Entomology continues to be pest control. I believe there will continue to be immense pressure to eliminate the use of conventional pesticides on our commodities, especially foods, both at pre- and post-harvest. Many of these pesticides are used to control insects, and the development of control alternatives will be an area of increased need.

Entomology has traditionally provided leadership in IPM programs, but a step further will be required for the future. We need better tools for the IPM toolbox. We must press forward with research into nonchemical methods such as cultural control, host plant resistance, and biological control. There are many opportunities in the more fundamental aspects of insect science, genetics and biochemistry to name two. Other focus areas include a growing urban entomology field, biodiversity, ecotourism, natural systems and ecology, and wetland conservation. We must also look at collaborative opportunities with the social sciences. E. O. Wilson (1998), in his recent book *Consilience* points out the need to communicate with the social sciences and to reach understanding levels that can lead to common goals. Much of science is now butting heads with social policy, and this is an interaction facing immense communication gaps. We must establish new linkages within the LGU. Scientific progress is made at disciplinary interfaces, and although interdepartmental research is to be pursued, the need to interface is often found within departments as well. Parrella et al. (1998) provides an excellent overview of the ways one Department of Entomology is positioning itself to meet the Land Grant mission in the next millennium. Entomology has a potential for interactions with other disciplines. We have many opportunities for joint programs with education, the health sciences, architecture, landscape architecture, and human ecology.

In addition to pest control, one of the traditional challenges of the Department of Entomology is graduate education. To remain competitive in the LGU system, we must have strong graduate programs. At a minimum, we must maintain graduate programs at their current level and perhaps even promote the growth of these programs. Most of our graduate programs in agriculture have remained constant in numbers for the past decade or more. To overcome this lack of growth we must address the two major challenges: financial support and job placement after graduation. We must direct additional resources into financial support of graduate students, and we cannot allow research efforts to shift entirely to post-doctoral scientists for productivity reasons. We should look for help in the private sector in this regard. In addition, we must be innovative in training for job placement. Traditionally, we have trained Ph.D.s for research. That market has been saturated for years in many disciplines. We should think more about professional degrees. Our industries are now more technologically oriented than in the past, and consequently industry needs more extensively trained employees. At the University of Florida, we have just approved a Doctor of Plant Medicine program. We know there is a risk in this decision but also know there is a market, and if our first graduates prove successful the program should grow.

Departments of Entomology must also be innovative at the undergraduate level. As the LGU strives to teach students such skills as critical thinking, writing, and teamwork, in addition to the basic biological sciences, entomology programs should be proactive in the provision of courses. Dollars will flow to credit hours. At the University of Florida, a "bank" has been established and money flows out based on credit hours. Successful undergraduate programs will assess the job markets and design innovative curricula to meet the market expectations. The design of curricula should have strong clientele input. The College of Agriculture at the University of Florida benefits from a strong pre-med/pre-vet program in the Department of Microbiology and Cell Science. Such professional school preparation could also be attained in a well-designed curriculum within the Department of Entomology.

The availability of technologies for distance education offers new teaching opportunities. The University of Florida is presently collaborating with community colleges and other state universities to implement new University of Florida degree offerings at four distant locations. Community colleges and state universities provide the first two years of education or upper division supporting courses and we provide core disciplinary courses taught through distance education or by on-site scientists. These types of programs include, for example, a degree offering in urban entomology at Ft. Lauderdale.

Successful Entomology departments will address some of the key indicators noted earlier. As director of the Florida Agricultural Experiment Station, I look for a strong fundamental program that competes effectively and creatively for support funds and that is on the cutting edge of the science. I expect strong applied programs that address the needs of the clientele. I expect faculty to realize for whom they work, the clientele. A good new-faculty orientation program can accomplish this. I also expect faculty members to develop teamwork, collegiality, and to work for the good of the greater program. I expect responsiveness to the clientele. I am disturbed by calls indicating that Professor *Y* won't return calls, or won't address an industry problem; or about a department head who is not willing to address new issues without new faculty or dollars. I am much more inclined to award new positions and support to programs that strive to meet new clientele needs without excuses. It is often said that we cannot be everything to everyone, but instead we must assess our resources and focus our efforts. While this makes a lot of sense, in reality I believe specialization can be a recipe for disaster. The moment we start restricting our accessibility or responsiveness to any portion of our clientele, we will be planting seeds of discontent, and this discontent can take years to mend. As a public institution, we must be very judicious about what we decide not to do. Although we may not be able to be everything to everybody, we must continue to be something to all of our clientele.

SETTING THE NEW DIRECTIONS

Strategic planning is essential for the survival and success of the twenty-first-century Department of Entomology. The most important aspect of strategic planning is strategic thinking, a component that is often under-emphasized. Planning structures must be implemented that foster extensive discussion and debate about (1) the strengths and weaknesses of the department as a unit, (2) the changing needs of society and institutional directions and their impact, and (3) the actionable imperatives that the department must undertake. To be successful we must continue to assess our environment and act accordingly. We should make it a practice to look at futuristic ideas in articles and newsletters to position ourselves for upcoming changes. It is of interest to note that there are people who still think the Internet is a passing fad. Many opportunities can be seen in the recent article by Brown et al. (1998), analyzing the impact of the world's increasing population on our water supply, animals, crops, waste, forestry, biodiversity, and climate change, among others. A strategic plan must encompass a mission, a shared vision, and a set of shared values. In the book, *Built to Last* (Collins and Porras, 1998), a set of shared values was found to be the most consistent attribute of successful U.S. companies. It is also interesting to note that it didn't matter much what the values were, just that the companies had them. Common departmental goals are essential. No department can succeed as a collection of "cats." The most successful industries have learned this and practice it faithfully.

It is also important to remember that a strategic plan is dynamic. It should be regularly reexamined and refined. The plan should address programmatic goals as well as organizational goals. In other words, it should identify research, extension, and teaching imperatives, but it should also identify a strategy to succeed within the institution, to finance its operations, and to match the goals and expectations of the institution. To implement a strategic plan involves change. LGUs have taken a beating over the word change. We are perceived to be antichange. Faculty are perceived to be set in their ways, and the institution is often viewed as an unmanageable ship. This has not made sense to me because faculty, especially research faculty and their discoveries and inventions, are the instigators of change. We certainly are anxious to change to the latest equipment or the most up-to-date computers. To me, the problem is not change but failure to identify the objective of change. We are often encouraged to "change" for the sake of change and taught that we must change "to keep from being left behind." We must take the emphasis away from change itself and place it on the goal. First, identify where we want to be, and then design the change necessary to get there. Sometimes, our concern with change is a fear to be worse off after the change than we were before. It is common to read that for most of us, our aversion to loss is twice our desire to gain. If we get everyone on board and with a clear destination, then perhaps the journey will not be so controversial. In summary, the challenge to excel in the twenty-first century can be met by identifying and meeting the needs of our clientele and our institutions and by doing this better and more economically than our competitors.

REFERENCES CITED

Brown, L. R., G. Gardner, and B. Halweil. 1998. *Beyond Malthus: Sixteen Dimensions of the Population Problem.* Worldwatch Paper #143. Washington, D.C.: Worldwatch Institute.

Collins, J. C., and J. I. Porras. 1994. *Built to Last.* New York: Harper Business.

NASULGC (National Association of State Universities and Land-Grant Colleges), 1999. *Returning to Our Roots: The Engaged Institution.* Washington, D.C.: NASULGC Office of Public Affairs and Kellogg Commission on the Future of State and Land-Grant Universities.

Parrella, M. P., J. Granett, S. S. Duffey, and D. D. Kitterman. 1998. Exploiting the "Entomology Advantage." *American Entomologist* 44: 198–201.

Thurow, L. 1993. *Head to Head.* New York: Warner Books.

Wilson, E. O. 1998. *Consilience.* New York: Alfred A. Knopf.

Chapter 2

Entomology

A Major Component of Pest Management at the University of California in the Twenty-First Century

SEYMOUR D. VAN GUNDY
College of Natural and Agricultural Sciences and
Department of Nematology
University of California–Riverside

NICK TOSCANO
Department of Entomology
University of California–Riverside

Let me set the stage for my comments by saying that my administrative experience at one of the three land grant institutions of the University of California occurred five to ten years ago when the University of California was going through one of its worst financial crises since the Great Depression. As Dean of the College of Natural and Agricultural Sciences and Director of the Citrus Research Center and Agricultural Experiment Station from 1988 to 1993, my administration had to provide leadership for three consecutive 5 percent budget cuts as a direct result of a loss of $433 million in cuts to the University of California's permanent state budget. My coauthor was associate dean for Integrated Pest Management and Extension during this same period and continues as an entomology specialist in the Department of Entomology. These problems are not unlike those experienced at other land grant institutions during this period. Thus, some of our comments may be jaded by these past experiences; however, we will try to present some of the changes and trends that have or are taking place in the University of California that will influence the pest sciences into the twenty-first century. Hopefully, these observations will be helpful and thought provoking to new faculty and graduate students at Texas A&M University and other land grant institutions.

UNIVERSITY OF CALIFORNIA

The University of California became a land grant institution in 1907 with the Berkeley campus. Today there are three land grant institutions in the University of California (UC) system: UC Berkeley (UCB), UC Davis (UCD) and UC Riverside (UCR). Together they are the largest public university in the land grant system. The Agricultural Experiment Station and the Cooperative Extension Service are under an umbrella organization housed in the Office of the Vice President of the Division of Agriculture and Natural Resources (DANR). The Agricultural Experiment Station is comprised of about 650 academic appointees across more than fifty departments in four colleges and schools. The Cooperative Extension Service numbers approximately 150 specialists and 280 advisors. The research and extension programs of the Division are generally classifiable under three subject matter area sciences: Human, Environment, and Agriculture. These are divided into Animal, Avian, Aquaculture and Veterinary Sciences, Economics and Public Policy, Food and Nutrition, Human and Community Development, Natural Resources, Plant Sciences, Pest Management, and Basic Science in Support of Agriculture. Although many of the research and extension resources (including personnel) have been decentralized to the individual campuses, the Office of the Vice President plays a role in the development and administration of twenty-two special initiatives and special projects (fifteen state funded and seven federally funded). Three important state funded projects pertinent for today's seminar are the Integrated Pest Management (IPM) Initiative of 1980, the Sustainable Agriculture Research and Education Program (SAREP) Initiative of 1986, and the Integrated Hardwood Range Management Program of 1986.

CALIFORNIA AGRICULTURE

California is the leading agricultural state in the nation with over 250 different crops; the state leads the nation in fifty-eight crops. Total farm receipts have grown steadily, reaching $25 billion in 1997. Assuming a 3 to 1 multiplier in economic activity, agriculture contributes almost $75 billion to the state's economy. Only the electronics industry exceeds agriculture in contribution to the economy of California. This agriculture production is challenged daily by the introduction of a new pest/disease every sixty days. It has been estimated that California loses $1.2 billion per year from exotic species. Consequently, because of pests and diseases the university's pest management disciplinary departments have been or are now among the top ranked departments in the country. In 1980 the Division of Agriculture and Natural Resources developed a new budget initiative in Integrated Pest Management to develop and promote the use of environmentally and economically sound pest management strategies for California. This program has flourished as a special legislative initiative ever since and has provided strong continuing support to the pest management sciences within the UC system. In the California context, pest management will continue to be important to agriculture for the foreseeable future.

MEGATRENDS LEADING TO THE TWENTY-FIRST CENTURY

Budgetary Trends/New Initiatives

Part of the University's budget reduction was implemented through three early retirement plans for both faculty and staff. For example, on the University of California–Riverside campus there was the loss of twenty-two faculty through early retirement in the Agricultural Experiment Station. Many positions in the UC system have been directed into new areas or have not been replaced for budgetary reasons. College undergraduate enrollments were on the downturn and there was even talk of closing the Riverside campus. Today the University of California is experiencing large increases in student enrollments and budgets to accompany the increase in students. However, there has not been a parallel increase in funding for the Agricultural Experiment Station. Consequently, there is greater emphasis being placed on teaching for new appointments, with a concomitant decrease in research associated with the Agricultural Experiment Station.

As the University of California embraces the twenty-first century it has prompted the administration to a reexamination of its missions and budgetary initiatives. The new educational mission strategy is to serve increasing numbers of undergraduate students. The number of graduate students (particularly out-of-state students) is declining due to increasing tuition and fees with an insufficient increase in support dollars. All campuses are currently struggling with how to increase internal support for out-of-state students. Unlike in the past century, agriculture now competes with other priority disciplines for finite resources. The service/outreach mission involves more extensive collaborations/partnerships with industry, with K–12 institutions, and with other institutions of higher education. California's higher education system consists of numerous two-year community colleges, and four-year institutions that include the California State College System and the University of California system. A Joint Policy Council on Agriculture and Higher Education is exploring collaboration between California's colleges and universities (Baker and Gomes, 1997). They predict that collaboration will become more commonplace and accepted over the next five to ten years as demands on higher education's teaching, research, and outreach programs grow.

In addition, the new century poses a daunting challenge as it strives to serve all the geographic regions, socioeconomic groups, and ethnic populations within the state as well as extending its economic reach to global issues. California's financial recovery is now being driven by growth in knowledge-based industries such as biotechnology, multimedia, telecommunications, software and computers, and new environmental technologies and not by the previous traditional sectors of aerospace and defense. New resources for university facilities/research will take the form of special initiatives. These are some of the major parameters that the three land grant campuses of the University of California will operate under in the twenty-first century.

More Strategic Planning

In its planning for the twenty-first century each of the three land grant campuses has developed new mission statements. These visionary plans acknowledge the need to address the current socioeconomic and ethnic diversity of the state. As a result, the land grant university will become more important in the future as society contends with increasingly complex agricultural issues and deals with global market ramifications of agricultural production.

Increased Teaching

The budgets, academic priorities, and student backgrounds of the land grant institutions have clearly changed in the last ten years. Permanent budgets at the University of California are tightly linked to increases in undergraduate students. This is further magnified by the recent action of the Regents of the University of California to accept the top 4 percent of all high school students. The university is already planning a tenth campus in Merced to open in the twenty-first century. Thus, many new faculty positions and replacements will be justified on teaching priorities rather than research priorities. In the past most agricultural research positions were appointed in the traditional land grant formula: the Experiment Station as the primary research appointments with only a smaller fraction of their salary from the teaching budget primarily to cover graduate instruction. Today the new replacement positions are more heavily supported from the teaching budget. This transition has been difficult because research faculty have been reluctant to take on more teaching responsibilities while striving to retain world-class research programs. Increased teaching loads have reduced the efforts of faculty to concentrate on field research and their contributions to agricultural experiment.

Today, larger numbers of undergraduate students are coming to the University of California with more urban or suburban backgrounds than with agricultural backgrounds. However, this should not be a problem from a pest management perspective because more land grant universities are increasing their efforts in urban pest management. Certainly, insect, nematode, and bacteria/fungi/virus models are important to studying biological principles and have broad application to pest management.

Increased Faculty/Program Realignment

Many new land grant administrators are opting to realign faculty into new departments and/or into new interdisciplinary programs, or in some instances to phase out disciplinary departments. A good example is the "Berkeley Experience," which resulted in the reorganization of highly ranked pest management programs at UCB in 1992 (Weinholt, 1996). The emergence of modern molecular biology and the decision to reduce applied agricultural research provided the impetus for this reorganization. The merger of several agriculturally oriented departments (including Entomology) into one single large department called Environmental Science Policy and Management (ESPM) was unprecedented in the UC system, and faculty were offered the opportunity of early retirement or to transfer to Davis or Riverside. This action resulted in what we think are major losses to California agriculture, to the individual disciplines and to graduate instruction in these disciplines. However, the Division of Insect Biology on the Berkeley campus has regained its status within ESPM and contributes broadly to the biological sciences using arthropods as model systems. The question for the twenty-first century is, will the emphasis on molecular biology occur at the expense of more traditional agricultural fields of study? UCR formed Graduate Research Umbrellas (GRUs) to enhance and maintain top quality research and education in agriculture. The GRUs are independent administrative structures linking individuals with related research and teaching interests from different departmental Ph.D. programs and allowing them to function under a single structure rather than through disciplinary departments. Each GRU is responsible for the recruitment and selection of new graduate students. GRUs may eventually become Interdepartmental Graduate Programs with their own Ph.D. granting program. The objectives of the GRUs are to increase program visibility of the biological sciences, to increase the scientific breadth of academic training of graduate students, and to increase the depth of research foci. Similar models exist on the Davis and Berkeley campuses.

Increased Research Partnerships

The new direction of biological sciences into molecular biology are also having a major impact on the traditional disciplinary pest management sciences research programs. As colleges of agriculture and their departments seek new sources of extramural funding, often outside the traditional sources, conflicts often arise between the academic freedom of the university and the political agendas of private and other nongovernmental organizations. Two such examples have received considerable attention at the flagship UCB campus during the last year. The Novartis special-funded partnership program in the College of Natural Resources at UCB is one example of a new working agreement between the private sector and the university. While the administration looks on this as budgetary innovation, many faculty still think it sets a deep and dangerous trend toward the blurring of conflict of interests and selling of academic freedom within the university (Berring, 1999). Another example is the formation of a private company (Lawyers Economic Consulting Group) by UCB faculty whose expertise was the result of public-funded programs. These programs are not unique as many land grant universities are participating in and actively promoting private research parks to encourage partnerships and market university research as a mechanism of increasing research budgets. In the twenty-first century the private sector will play an ever-increasing role in setting research priorities and even in some instances get built in to budgets and facilities. Berring (1999) suggests one of the faculty's tasks in the twenty-first century should be to reconsider what are the universities' central values and question the appropriateness of these new directions in the land grant university.

Increased Internationalization

The internationalization of activities of colleges and universities offering programs in food, agriculture, and natural resources is being promoted by the National Association of State Universities and Land-Grant Colleges (NASULGC). Many land grant institutions have strong programs in responsible contributions to a global economy, a global demand for food, and implementation of governmental policies that impact at both the national and international level. Within the last year UCD and UCR have recognized the importance of international activities and have created new vice chancellor–level positions for international programs and outreach to make university activities a more important part of the lives of Californians. The goals are also to promote faculty involvement in programs in the agricultural, social and natural sciences worldwide. The concept of "global university" will be strengthened in the twenty-first century.

Decreased Service to Agriculture

In the past, most research faculty were involved with service to the agricultural industry. Currently, faculty who increase their teaching are unable to spend as much time in service to their clientele groups. This reduction in service has often been covered by extension specialists who are also expected to do more mission-oriented research. This new role for extension specialists suggests that there may be less emphasis on service to the agricultural industry. Since 1988, the appointment of many new cooperative extension specialists has taken the form of joint research or teaching appointments with extension (Table 2.1). Fewer than half of the extension specialists are 100 percent Cooperative Extension (CE) appointments.

Cooperative Extension has also suffered losses in personnel. For example in Southern California the number of CE advisors has decreased 13.3 percent

Table 2.1. **Changes in Extension Specialist Appointments at UCR (CE = Cooperative Extension, OR = Mission Oriented Research, I&R = Instruction and Fundamental Research).**

Year	*Total FTE*	*100% CE*	*10–60% OR*	*15–20% I&R*
1988–89	22	20	2	0
1998–99	24	10	11	3

Table 2.2. **Changes in Faculty and Staff FTE between 1988 and 1998.**

Faculty FTE	*Botany and Plant Science*	*Entomology*	*Nematology*	*Plant Pathology*	*% Change*
1988–89	37.5	36.0	8.5	19.0	—
1998–99	27.4	34.6	5.8	17.5	−15.5%
Staff FTE					
1988–89	47.5	58.0	11.5	31.0	—
1998–99	36.0	39.7	3.1	21.4	−32.2%

in the last ten years. Much more program funding for advisors is through extramural sources than through county, state, and federal appropriations.

Rebuilding Pest Management Sciences

Within Agriculture and Natural Resources more faculty specialize in Integrated Pest Management; this consists of 25 percent of the Agricultural Experiment Station scientists and extension specialists and about 15 percent of the County Extension advisors. However, with the changing parameters and priorities of the university all of the pest management disciplines on the campuses are now struggling to maintain their strength, breath, balance, and identity. As an example, the UCR campus has large contingents of pest management scientists (85 FTE). It has pioneered in biological control and the concept of economic thresholds. It has had strong Plant Pathology and Nematology Departments.

From 1988 to 1998 at UCR there was a 15.5 percent reduction in FTE in the Integrated Pest Management disciplines (Entomology, Plant Pathology, Nematology and Botany, and Plant Sciences). In addition there was a major reduction of 32.2 percent in supporting staff (secretaries and laboratory technicians; Table 2.2). Prior to 1988, nearly every faculty member had a staff research associate and some secretarial support from the permanent budget. Today, there are few staff research associates, and there is almost no secretarial support for individual faculty. Presumably, computers have replaced the secretarial support, and post-doctorates on extramural funds have replaced the state-budgeted staff research positions. These losses in Experiment Station research funds will probably never be totally replaced through the permanent budgetary process. Thus, the emphasis to increase the Agricultural Experiment Station budget into the twenty-first century at the University of Cali-

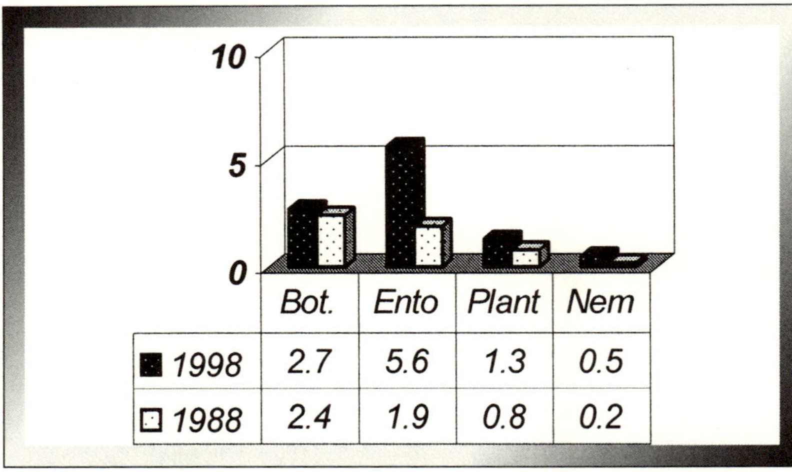

Figure 2.1. Increase in extramural awards between 1988 and 1998 in millions of dollars (U.S. $).

fornia will be oriented toward more teaching of undergraduates, through extramural funding, partnerships with nongovernmental organizations, and through new legislative initiatives. For example, the extramural funding for pest management at UCR increased by 90 percent from 1988 to 1998 (Figure 2.1). We believe that pest management is and will continue to be the foundation for new budget initiatives in California into the twenty-first century.

ENTOMOLOGY AT UCR

The Entomology Department has six core research areas: biological control, Integrated Pest Management, chemical ecology/behavior, systematics, urban/medical/veterinary, and physiology/toxicology. One of the advantages of the Entomology Department over most other departments in the United States and abroad is its size of 36 FTE. In the last ten years, however, the faculty FTE of the Department of Entomology has decreased by 10 percent, thus threatening its balance and breath. It is further threatened by the possible realignment of five positions in toxicology to a new department of Neuroscience, Cell Biology, and Environmental Toxicology. The department has identified six new positions to be added as a result of enrollment growth in the biological sciences and possible realignment of positions. Five of the new replacement faculty would be in the area of molecular biology and one in biological control. Extramural funding for the department has increased by 195 percent between 1988 and 1998. The department is uniquely situated to combat urban and agricultural pests. As stated previously, a new pest is introduced into California every sixty days. Some important pests that are a threat to the state's agricultural industry and to urban society include the Mediterranean fruit fly, the fire ant, the Africanized honey bee, the Formosan termite, the Eucalyptus longhorn borer, and numerous whiteflies. UCR has put forth several new initiatives: the Center for Exotic Pest Research with $5 million in research funds to expand IPM, a new Insectary and Quarantine Facility including BL level of quarantine with 16,685 usable square feet, and a new Entomology Building that will increase their total usable square feet by 10,000. Obviously, all of these new program initiatives are contingent on recognition and support by the university administration, agricultural industry, scientific community, state and federal lawmakers, and the general public. The Department of Entomology at UCR has fared well through the budget cuts of the last ten years and appears to be in excellent shape to enter the next millennium.

REFERENCES CITED

Baker, Warren, and W. R. Gomes. 1997. Finding Common Ground. A Report of the California Joint Policy Council on Agriculture and Higher Education, 1–48. Oakland: California Joint Policy Council on Agriculture and Higher Education.

Berring, Robert, 1999. Is Berkeley Off Course? *California Monthly* 109 (4): 19–20.

Weinholt, A. L. 1996. Plant Pathology at Crossroads. *Annual Review of Phytopathology* 34: 1–11.

Chapter 3

The Texas A&M University Department of Entomology

A Model of Commitment to the Land Grant University Mission

EDWARD A. HILER
College of Agriculture and Life Sciences
Texas Agricultural Experiment Station and Texas Cooperative Extension
The Texas A&M University System

Thank you for inviting me to join with you in commemorating the 100th anniversary of the founding of the Department of Entomology at Texas A&M University. We are here to celebrate the many achievements of those 100 years—the education of generations of young people, scientific advances in the discipline of entomology, and the development of new methods and technologies that made American agriculture the most productive in the world. In celebrating these achievements we must also acknowledge that they would not have been possible without the foresight of those who created the land grant university system and its guiding philosophy of serving the people. So today I want to talk briefly about the land grant system both historically and in today's context. I'll share my view of how the Texas A&M Department of Entomology is a model disciplinary department in terms of fulfilling the land grant mission. Then, I'll look ahead to the future of agriculture and the role entomology will play.

Texas, and entomology, had a special part in establishing the nation's land grant system. The federal Morrill Act of 1862 established land grant colleges in every state, and Texas A&M was founded in 1876 as a result. That was followed by the founding of state Agricultural Experiment Stations through the Hatch Act of 1887, with the Texas Agricultural Experiment Station established in 1888. So the land grant system now had two key components—teaching and research. But if it was to succeed, there needed to be a way to get information and education from the colleges to the people. That's where Texas came in. One of the first demonstrations in cooperation with farmers took place in Terrell, Texas, when Seaman A. Knapp demonstrated methods for boll weevil control in 1904. This method of transferring technology was so successful that it became the basis for forming the Cooperative Extension Service through the Smith-Lever Act of 1914. So the boll weevil was not only the centerpiece insect for the founding of this department but also for the founding of a national system of adult and youth education.

The roots of this department are deeply embedded in the land grant university mission, and that is surely its defining quality and greatest strength. Being part of a land grant university means we are philosophically committed to providing opportunity for high-quality education for all, to world-class research, and to outreach education that helps people solve problems and improve their lives or livelihoods. Simply stated, we are about serving people.

Today, I see the Department of Entomology at Texas A&M serving as a model of how the land grant mission is fulfilled by faculty in a disciplinary area. That model begins with the outstanding work you have done and continue to do in the three primary functions of teaching, research, and extension. The Department of Entomology is recognized for the quality of its undergraduate and graduate teaching programs, as well as for having one of the largest programs in the country. But you have not rested on reputation alone and have continued to seek improvements. In recent years, expectations at Texas A&M have changed in terms of teaching quality and number of students taught. While this department has a distinguished reputation for quality teaching, the faculty has had to face some of these issues. The result has been a series of teaching retreats and faculty meetings to continually revise the undergraduate and graduate curricula to assure that it meets contemporary needs and expectations. The faculty has responded heroically over the past four years by introducing new classes and expanding teaching outside of the department. This has increased student contacts and reduced the cost per student contact hour. More importantly, teaching faculty have been better utilized, and a broader range of students are now being exposed to the entomological sciences.

World-class research and leadership are hallmarks of this department. Dr. Perry Adkisson and Dr. Max Summers are both members of the National Academy of Sciences. Perry's distinguished career as father of Integrated Pest Management and winner of the World Food Prize has brought great honor and recognition to the department. Several faculty members have or are serving as presidents of national professional societies. Extensive publications in refereed journals point to the quality research being produced by faculty members.

As a faculty, your research interests range from the very basic to the applied and from the molecular level to the landscape. While the discovery of knowledge is imperative in a research university, the application of that knowledge to meet societal needs is essential and fulfills the land grant mission. I encourage you to maintain your balance and seek ways to bridge fundamental studies and approaches in your basic research programs with those that are more mission-driven. The diversity of your research programs is an inherent strength, and as a faculty you will want to seek new opportunities to capitalize on its strength. In addition, several subdisciplinary areas in the department are strong. Examine these areas, focus on them, and come up with new and innovative ways to link these areas to strengthen your overall research program.

The Department of Entomology serves significant needs of the people across this state through a strong, robust, and aggressive extension program. To quote Seaman A. Knapp, the father of extension, "What a man hears, he may doubt; what he sees, he may possibly doubt; but what he does, he cannot possibly doubt." Through extension entomology and related Integrated Pest Management (IPM) programs you have invited farmers to *do.* The economic success of these programs and the impact they have had on agricultural producers is well documented and something that we point to with pride in showing what the Agriculture Program does for the people of Texas.

Extension entomology programs have also done a very credible job of addressing environmental concerns. Pesticides affect all Americans and are an environmental issue worldwide. Through IPM programs, you have found the appropriate middle ground on which to balance the economic and environmental concerns of a diverse public. You have not only discussed environmental problems but have also been in a unique position to address them.

While work in the teaching, research, and extension areas is basic for a department to fulfill its land grant mission, it's really only the beginning. How those functions are put together and delivered to the people to meet their needs is equally important. The Department of Entomology has set an excellent standard in three areas. First, the department has established and supported a truly statewide scope for its program. The entomology department has its base in College Station, but due to regional demands faculty members are distributed widely across the state at Texas A&M Research and Extension

Centers. You are unique among departments in that you have county extension agents designated to work in IPM. This grass-roots support at the county level places the department in close contact with the agricultural and urban constituency of the state. Few departments have such a network of professionals as a support base. In fact, the statewide reputation of this department is largely reflected through its off-campus faculty.

In addition to its statewide scope, the department has been exemplary in integrating functions to build synergy and address increasingly complex problems. A primary rationale for forming the Agriculture Program nine years ago was to foster integration of, and collaboration among, the functions of teaching, research, and extension so that we could better serve the people of Texas. It is clear that the entomology department has made a great effort in this area, which has created positive results. In entomology, off-campus research and extension faculty now participate on graduate student committees, guest lecture in classes, and collaborate in research and extension programs. I know that campus-based entomology faculty are working closely with extension faculty and user groups to share their research and in turn gain insights of problems faced. This type of integration and collaboration strengthens the department and makes it more efficient and effective. It's what the land grant university should be about and what our legislators, the taxpayers, and our clientele want us to do.

A third way in which the department sets an example is by listening to clientele and citizen groups. For a department to remain competitive, the work of its faculty must be on the cutting edge of teaching, research, and extension programming. That means being aware of local, state, and national attitudes and policy changes. The Department of Entomology has done an outstanding job of continually assessing clientele priorities, whether the clientele are students, research agencies, farmers, ranchers, or urban residents. You have been blessed with a series of strong departmental leaders who have had their pulse on society's needs and have been able to capitalize on opportunities for the department in response to these needs. Through county-level personal contact and statewide surveys, the department has continually asked clientele for their opinions on how to improve its programs. As with peer and student review of your research and teaching programs, you have openly courted reviews from your extension customers to improve your programs to meet clientele needs.

Although we may not think of it this way, peer review is also a part of listening to clientele as it brings the knowledge of our disciplinary peers to bear on evaluating strengths and weaknesses as a means of improvement.

The Department of Entomology will be the first department in the College of Agriculture and Life Sciences to undergo a doctoral program review. You were selected for this review because you are recognized as having one of the better graduate programs in the college. As you prepare for your doctoral program review, I would urge you as individual faculty members to engage fully in the process. This review provides you with a unique opportunity to conduct a self-study of how you find yourselves in the area of graduate education and where you want to be as we head into the next century. Individual participation and action is vital to fully reflect the true potential of your department.

The Department of Entomology's strong teaching, research, and extension programs, along with its ability to maintain a statewide scope, build synergy among functions, and listen to clientele and citizens make it a model in many ways for what an academic department should be today. However, what of the coming years? We all know that the twenty-first century will pose significant challenges for agriculture and for the land grant university system. This past January, I had the opportunity to make a presentation on the future of agriculture at the Southern Association of Agricultural Sciences meeting. I believe today, as I stated then, that there are five "drivers of change" that will significantly and forever change the face of agriculture. I'm going to briefly talk about those drivers of change as the context for discussing the future of the Department of Entomology, our college, and the land grant university system.

The first driver of change is, and will continue to be, information technology. The era we live in today was named the Information Age almost thirty years ago. The central tool of the information age is obviously the computer. Computer technology has allowed expansion in the availability of information on a scale unseen since the invention of the Gutenberg printing press. Computers allowed the amalgamation of data and computations to take place at increasingly faster speeds. We quickly learned, however, that data alone aren't much good unless they can be used as information. Information is the arrangement of data into meaningful patterns. And while the Information Age was initiated by the marriage of computers and telecommunication, it is still in a growth phase and we have no idea of its limits. Some argue that the next age is the Knowledge Age. Knowledge, or the application and productive use of information, requires a new technology platform that allows us to integrate and use existing technologies easily.

Clearly, we must successfully move from the Information Age to the Knowledge Age in all aspects of our work. The department's Knowledge Engineering Laboratory is a good example of the integration of simulation technology, heuristics, and research-based information to assist agriculture in the decision-making process. You are pioneers in this area. Your extensive use of database management to array and organize information places you squarely in the front of the infomatics era. Your capability to use simulation models and related computer applications to study and integrate complex biological systems will provide fuel for the Knowledge Age. Your commitment to World Wide Web technology to distribute information in creative and innovative ways speaks to your ability to design new learning environments to the benefit of students and the general public. Your work in spatially referenced databases through Geographic Information Systems merged with simulation and other computer technologies has allowed landscape-level analysis and problem solving. Much of the work done and ongoing in entomology directly relates to precision agriculture. All of these things position the department to move into the Knowledge Age.

The second, and perhaps most powerful, drivers of change in agriculture are the life sciences and biotechnology, which are rapidly building a bio-based economy. The next century may be termed the Biotech Century. *Business Week* made this declaration last year in an article asserting that biology will replace physics as the dominant science. Public attention has been stimulated by the amazing work being done on the frontiers of genetic research and biotechnology that promises, among other things, the map of the human genome within the next decade.

Agriculture will need a multi-billion-dollar infusion of capital into research to hasten the evolution of biotechnology in the twenty-first century. And this research must go beyond the traditional food, feed, and fiber approaches. If that investment is made, the results will be phenomenal. There will be tremendous new opportunities for agricultural producers as biotechnology creates options for a far more diverse set of products from pharmaceuticals to fuels. Genetic alteration of crops and animals will play a vital role in increasing food and fiber productivity and improving food nutritional properties while limiting environmental stress. Benefits to agriculture will be enormous:

1. Increased production is possible through increased plant and animal yields and efficiencies, and through decreased losses from insects, diseases, and weeds.

2. Genetic engineering will provide improved nutrition by providing consumers with new and healthier food choices.

3. Genetic engineering will protect the environment by allowing production of food and fiber on less land.

4. It will also forestall deforestation and make it possible to produce foods and fibers that require none, or few, of the additives that have potential to harm water, air, and environmental quality.

5. Genetic engineering tools will be even more useful in producing pharmaceuticals and advancing diagnostics of human, animal, and plant diseases.

Entomology has secured a place for itself in the life sciences and biotechnology revolution underway. Your faculty has been involved in the development of pharmaceuticals and diagnostics that are products of biotechnology. You have traditionally been involved in the development of insect resistant crops. You see the advantages that biotechnology can bring to you not only in host resistance but also in biological control and IPM. Many of you are using the tools of biotechnology to unravel the basic, biochemical, and physiological processes of insect systems. However, as with all disciplines, your work in this area represents the very beginning of the possible opportunities in the life sciences. To better position the entire university for a prominent role in the life sciences Texas A&M University has appointed a Life Sciences Task Force, and the Agriculture Program has developed a committee to identify priority areas in the life sciences. The Department of Entomology is expected to play a key role in both those groups.

The third driver of change is the global economy. We are into a global market economy that has dramatically influenced American agricultural production and will likely do more so in the future. International Food Policy Research Institute studies suggest that the global economic demand for food will be matched by global supply at constant or slightly falling real prices during the next twenty years. However, long-term projections are subject to many uncertainties. Developments in China, India, Eastern Europe, and the former Soviet Union and sub-Saharan Africa will be particularly important for the future of the global food situation. Developing countries as a group will more than double net imports of grain between now and 2020, and policies of traditional exporting countries, such as the United States, will play an important role in meeting these import demands.

There will be greater opportunities if we can cooperate with, rather than fight, the competition. For example, there is already vegetable and field crop production on both sides of the Mexican border. The next step is cooperation with Mexican producers to develop more efficient ways to move product to market so the retail food sector has a steady supply. This is really the Wal-Mart model of just-in-time, efficient consumer response. With some 200 food retailers across the nation and continued consolidation in that sector, these kinds of arrangements could be both workable and profitable for producers.

The entomology department has established an excellent track record in terms of cooperating internationally in the codevelopment of research and educational programs. Your faculty and former students are present virtually on every continent in the world sharing technology that will build stronger bonds with the United States in a global economy. Multinational trade agreements, such as GATT and NAFTA, while opening the flow of goods, also introduce exotic pests such as the Russian wheat aphid and the Formosan termite into the United States. Your continuing battles with foreign insect introductions will present both challenges and opportunities. We will rely on entomology as a strong participant in international agricultural programs in the future as we move toward a more global economy.

Demographics, including both a growing population and consumer power, is the fourth driver of change in agriculture. World population is expected to double to more than 10 billion people by the year 2050. The issues are not just feeding the hungry but also meeting the demand for higher quality and more nutritious food worldwide. As Norman Borlaug, father of the Green Revolution has said, the world has the technology—either available or well advanced in the research pipeline—to feed a population of 10 billion people. The larger and more difficult question in feeding the world will be as much in the realm of culture and politics as in agriculture and science.

The expansion of the U.S. population has directly affected prime farmland. Even as we remain an urbanized country in the decades ahead, Americans' desire to live outside central cities will continue to consume farmland. According to the American Farmland Trusts report, "Farming on the Edge," America's best farmland is being needlessly destroyed by scattershot urban development. Between 1982 and 1992, 4.3 million acres of prime and unique farmland were overrun, nearly fifty acres every hour of every day. Of the top twelve most threatened agricultural areas in the nation, two are in Texas: the Blackland Prairie and the Lower Rio Grande Plain.

Agriculture is moving from a production, supply-driven system to a consumer, demand-driven system. The changes in agriculture we see most readily today are often described as industrialization and are associated with capital, technology, production, and efficiency—all elements of production or supply-driven change. As we move into the next century, demand, consumption, and demographics will be even greater forces to be reckoned with.

Your work in entomology reflects the reality of demographic shifts and the role consumers play in both agricultural and urban programs. You struggle daily in designing IPM systems that work at the urban/agricultural interface. As the city moves to the country there is more concern about designing effective, multitactic IPM systems that will protect crops and livestock, yet not impact the urban population. Complaints about pesticide applications, odors, and flies have skyrocketed simply because the city has moved to the country and does not like the smell, taste, or touch of agriculture. Your urban programs are directly in touch with consumers as you strive to protect citizens from insect pests that attack man and his possessions. Your IPM in the Schools program is a national model in this regard. Consumers are demanding pure food without pesticide or insect contamination. That means new and advanced IPM systems must be designed to meet the consuming public's needs in the future. You will be challenged to continue broad and creative thinking regarding environmentally safe IPM programs that provide needed crop protection as we move into the twenty-first century.

The fifth driver of change for the next century is public policy, and probably the greatest public policy issue facing us for the next decade is restructuring agriculture. We have continued to see small and mid-sized farming operations go out of business. In the southern states, the number of farms fell by 13.5 percent between 1982 and 1992. With drought and other natural disasters as well as low prices, such as we are seeing in most crop and livestock commodities today, that percentage will obviously increase.

The agricultural sector is in the midst of a massive restructuring with large corporate farms and vertical integration as a fact of life. Restructuring has been dramatically illustrated in the poultry industry, and now the pork industry, with farmers as contractors to large corporations. The expansion of multinational chemical/pharmaceutical companies in buying up seed companies at an ever-increasing rate is a clear illustration of the vertical integration of agribusiness. Prescription agriculture, where farmers buy their seed and produce crops under close and strict rules to be sold to a contract buyer, may be the wave of the future in agriculture.

Entomology faculty work closely with many of these large companies such as Dupont, Monsanto, and Novartis in the area of crop protection. You have a strong and active program in host plant resistance. You have redirected faculty positions into crop biotechnology and insect transgenic systems. You are seeking ways to form partnerships with colleagues from other disciplines to strengthen your programs, but also you actively seek partners in the biotechnology/seed industry. What impresses me most is the balanced approach this department has taken toward optimum use of conventional breeding programs while taking advantage of the new biotechnologies. That to me signals your willingness to remain on the cutting edge of your science and deal rapidly with changing technologies in the industry.

There are other public policy issues that Entomology will continue to confront, particularly changes regarding insecticides. The Food Quality Protection Act (FQPA) will reshape the type and availability of insecticides over the next decade. This department has responded rapidly to work with the agricultural industry to determine the impact of this policy on Texas. You have once again involved farmers and ranchers in predicting impacts of FQPA and in providing alternatives. The Texas Pest Management Association, a non-profit, farmer-operated IPM association, is an outstanding example of a public/private partnership you turn to with great frequency to deal with public policy issues that impact the management of pests. The variety and complexity of public policy issues will increase in the years to come, affecting our work in numerous ways.

As you can see, the entomology department has already demonstrated its ability to operate in the most important arenas of change in agriculture for the years to come. And while I certainly cannot predict the impacts of these watershed changes for the long term—thirty or fifty or 100 years from now—I remain highly optimistic about the future of this department, Texas A&M University, and our land grant system. For as long as we never lose sight of our heritage and mission of serving the people, the future is ours to make. Congratulations to each of you in the Department of Entomology on this 100 years of accomplishment. May there be 100 years more.

Part II

The Multiple Roles of Entomology Departments: Past, Present, and Future

Chapter 4

The Three-Function Department of Entomology

Is It Really Suited or Ideally Suited for the Next Century?

DONALD A. RUTZ
Department of Entomology
Cornell University

There undoubtedly are many challenges facing departments of entomology today. Budgetary constraints, departments becoming smaller in size, attempting to do more with less, and mergers with other departments are just a few of these challenges. In this chapter I will present my views as to how I believe we can effectively deal with these challenges while maintaining our identity as departments of entomology with a successful integration of teaching, extension, and research. I will also discuss the benefits and challenges of the three-function department of entomology in dealing with the numerous audiences that have come to rely on our departments as sources of credible and unbiased information.

DEPARTMENT OF ENTOMOLOGY–ITHACA

In 1874 John Henry Comstock founded our Department of Entomology at Cornell, the first department of its kind in the United States. Comstock assembled a group of fine scholars including Bradley, Forbes, Johannsen, Matheson, Needham, and Slingerland. Their influence is still evident in the department's Comstock Memorial Library, which is among America's foremost collections of entomological literature, and in the department's insect collection, which houses approximately six million specimens including many designated as types.

Cornell's Department of Entomology at Ithaca has responsibility in the areas of teaching, research, and extension. It is a unit of the New York State College of Agriculture & Life Sciences and the Cornell University Agricultural Experiment Station. Among entomology departments in the United States, our department is unique in that it is a part of a university that is both an endowed and a land grant institution. It is also unusual in that it has equal strengths in basic and applied aspects of entomology. We currently have seventeen faculty. We have three joint appointees and twelve adjunct professors. The major research areas within the department are in morphology and systematics, pest management (especially of forage, greenhouse, vegetable, and livestock insects), biological control, ecology, evolutionary biology and behavior, medical and veterinary entomology, pathology, apiculture, toxicology, and physiology. The department's *teaching program* offers training for undergraduate and graduate students. Our Field of Entomology has fifty-two graduate students. Undergraduate majors number approximately twenty.

Our *extension program* applies the results of research to alleviate insect problems in New York State, throughout the Northeast and across the United States. This effort reflects the ever-changing needs of our agricultural industry, our recreational resources, and our urban dwellers. In addition to the domestic programs, the department contributes to the college's program in International Agriculture. We also have a sister department at Geneva. The Geneva Department of Entomology has thirteen faculty with responsibilities in both research and extension. If research, teaching, and extension are to be successfully integrated under one roof, it is critical that the department, the college, and the university embark on a concerted effort to make certain that such an integration is indeed achievable and then make every effort to see it through to fruition.

INTEGRATION PROCESS

Recruitment of New Faculty Members

The recruitment of new faculty is the cornerstone of the successful integration of research, teaching, and extension under one roof. There is a continual need for strategic, long-range planning in our departments with a reevaluation every two to three years. We need to constantly consider where we wish our department to be in the next ten to twenty to thirty years. This must involve frank, open discussion and must include a clear understanding of economic realities. We must anticipate the entomological/agricultural needs in our state for we are in the College of Agriculture and Life Sciences. We also must anticipate the needs of the general public as well, along with the impact of urbanization in agricultural areas, which is a problem not unique to New York but is occurring throughout the country.

One of the challenges we are faced with is how do we fill positions that historically have had three-way (research, teaching, and extension) splits. At Cornell the days of three-way splits for assistant professor positions are over. The only exception to this would be endowed professorships. At Cornell our assistant professor positions for the most part are all two-way splits, as either research/teaching positions or as research/extension positions. If we are working to fill an existing position that was originally a three-way split, we are now faced with the challenge as to how the teaching effort is to be maintained if the position is now research/extension. Or how will the extension effort be maintained if the position is now research and teaching. Again, we need to take a long, hard look at all of these positions to determine if indeed they are critical to our anticipated departmental needs over the next ten to twenty to thirty years.

Second, it is critical that position descriptions be written to clearly depict our departmental expectations. This needs to be done not to the detriment of casting a big net to attract all qualified candidates but not so broadly written that potential excellent candidates may not apply. Furthermore, good justification statements need to be prepared to get all members in the department on the same page. Essentially we are all in agreement—here is our position description and this is why this position is needed.

Other groups need to be involved in this process as well. Depending on the position, faculty from other departments or even from other colleges participate on our position search committees. Graduate students serve on these search committees as well. If we are filling extension positions we involve the position stakeholders. This has worked out exceptionally well for us. For example, when we filled our apiculture research/extension position, five beekeepers were completely involved in the search process as well as in the reappointment decision. In my opinion, stakeholder involvement is critical to assure that our programs/our positions, particularly those involving extension, are indeed on the right track. In addition, salary and start-up dollars for equipment and technical support to facilitate the success of new faculty must be effectively negotiated with college- and university-level administrations.

In summary, the position description and its justification need to be well thought out and well written to be certain that we are attracting the best candidates available. Furthermore, performance evaluations are ultimately based on this position description; that is, how well does this individual meet our expectations for this position? We have all seen position descriptions that are absolutely atrocious. These descriptions beg the question: what do they really want or perhaps what don't they want when position descriptions are so broadly or poorly written? With such a poorly written position description, how can or will this individual's performance be effectively evaluated for reappointment and ultimately for promotion and tenure.

The Selection Process

The selection of candidates for either a short list or for an interview is all too frequently difficult to do from the documentation provided by the candidate. The letter of interest from the candidate is, in my opinion, the most important component of their application documentation. It is critical that the candidate goes through the position description and describes how their qualifications specifically meet our needs, including such things as training, experience, interests, and such. I cannot emphasize this enough. There is absolutely nothing more disappointing than reading a letter of interest from a candidate in which the candidate does not address a teaching component or an extension component in a position. We are then left to conjecture whether the candidate is indeed interested, willing, or able to teach or conduct an extension program. In our department, the interview process involves our faculty, academic staff, graduate students, and stakeholders. We also require that the candidate give two presentations—a research seminar plus an extension presentation or a class lecture depending on the position. When selecting the candidate for the position, it is critical that the candidate is truly the best fit, the best qualified for the position. Again, it is critical to stick to the original position description. If this is not done, certain components of the position may never be addressed. Furthermore, such hire defeats the entire departmental strategic planning process. Recruitment is the most critical component in shaping our departments of entomology of tomorrow. We are in effect making a thirty-year commitment to this position, to this effort. In summary, it is essential that the department hire the individual who is the best qualified to fill the position. One must make sure that all of the needs described in the position description are met. It is critical not to change your mind during the interview process to essentially change the position in a major way.

The issue of spousal hires is also one that is becoming more prevalent across the country. Again, strategic, long-range planning is critical here. If the position is one that has already been decided by the department as a priority, great! However, it is essential that this individual is considered to be the best available to fill the position. This is usually accomplished by conducting a directed search with feedback from the most noted scientists in a particular area. It is the department's responsibility to be certain that we are hiring the best-qualified candidate available. This is in the best long-range interest of the department, the college, and the university.

Faculty Performance Evaluations

As faculty and as college administrators it is our responsibility to do all that we can do to make a faculty hire a truly successful hire. This involves working with the individual through the reappointment process as an assistant professor, then through the promotion and tenure process to associate professor and then on to full professor. In my opinion, the current Cornell system for performance evaluation is outstanding. Within a month or two of being in the department our new assistant professors meet with the associate dean and chair to go over what is specifically expected of the individual in the position. The criteria for reappointment as an assistant professor and what will ultimately be needed for promotion and tenure are all discussed at this time. Essentially, our new assistant professors are provided with information on what our expectations for excellence really are in research, teaching, or extension depending on their position. Furthermore, our new assistant professors also receive documentation from the college administration that outlines what will be needed from them for reappointment after three years and ultimately for promotion and tenure after six years. Our assistant professors are extremely pleased with this early introduction to how their performance will be evaluated. At Cornell, annual reviews of assistant professors are mandatory, and these involve the entire tenured faculty of the department and not a smaller committee as done in some departments. This is the best type of review possible in that all members of the tenured faculty are kept up-to-date with the individual's performance in research, teaching, or extension, and a better understanding and appreciation of the importance of all three functions is fostered.

Assistant professor annual review documentation is provided to all tenured faculty for review. They then provide the chair with their letters of evaluation. An updated C.V., research, teaching, or extension statements, publications, grant activity, and such are all included in this documentation. If there are no problems or issues raised, the chair prepares a departmental letter and meets with the assistant professor to discuss his or her overall performance. If, by chance, the tenured faculty are not all in agreement regarding the individual's performance, I, as chair, prepare the departmental letter of evaluation and share it with the tenured faculty before meeting with the assistant professor to be certain that I have truly captured their concerns in my letter. Thus, our assistant professors know from the very beginning, after year one, how the tenured faculty in the department view their overall performance. This way our assistant professors get no surprises at reappointment or at promotion and tenure. These are undoubtedly very stressful times for our assistant professors and I am certain that this problem is not unique to Cornell but is widespread in departments across the country. It is extremely difficult to convince our assistant professors as to the importance of enjoying their jobs while at the same time striving for satisfactory performance. This has been and continues to be a real personal challenge for me as chair.

Outreach

We continue to hear more about outreach and its importance to our science. Outreach is a critical component of our work as entomologists. Unfortunately, at present, we spend too much of our time preaching to the choir. It is no longer acceptable for us to merely prepare scientific publications and present our work at scientific meetings. It is critical that we as entomologists educate the general public about what we are doing. When we are publishing our research in scientific journals we should at the same time prepare an article for the popular press. We should give presentations to school children, community groups, public health and environmental advocacy groups, alumni associations, and such. All of our faculty, academic staff, and graduate students should be involved in outreach efforts. I have found that our young faculty, our assistant professors, are most anxious to get involved in outreach. However, it is essential that they get credit at reappointment and promotion and tenure time for these outreach efforts. At present, criteria for excellence in outreach have yet to be developed by our college or university administrations at Cornell. I would venture a guess that we are not unique in this regard. Outreach must be considered scholarly activity along with our research, teaching, and extension efforts. It goes without saying that public awareness and public education are very, very important with regard to our land grant university mission. I am pleased with the progress that I am currently seeing at Cornell but we still do have a long, long way to go.

Reappointment (Year Three)

If annual reviews are done correctly and completely, reappointment as an assistant professor should be a very easy process. The documentation that is ultimately needed for promotion and tenure is identical to reappointment documentation, making the overall effort on the part of our assistant professors one of updating what has been developed over the past five years.

Retention-Promotion with Tenure

The primary responsibility for promotion and tenure decisions rests in the department making the recommendation. If a departmental recommendation for promotion and tenure is rejected at either the college or university level, then the department has failed in one or more of its responsibilities in hiring, mentoring, or documenting the activities of excellent faculty, or the college and department have not maintained suitable communication in terms of priorities. Unfortunately, I am aware of situations where recommendations for promotion and tenure are made by departments and college ad hoc committees only to be rejected by the college administration, or are made by the college only to be rejected by the university administration. I would like to make a couple of suggestions to avoid such situations. The first is that college administrators, including the deans and directors, should participate fully in all departmental CSREES reviews. College and university administrators need to be an integral part of these reviews from the very beginning to the very end, and not be merely participants at the very beginning and/or at the very end. By such participation, administrators would learn firsthand more about the department and offer their constructive input as to directions and goals relevant to the college. With this firsthand knowledge, all levels of administration would be better able to effectively explain what is indeed going on in each of their departments to our stakeholders, the general public, legislators, regulators, public health and environmental advocacy groups, and such, throughout our states, regions, and country. I would also suggest that our college and university administrators participate in retreats with our assistant professors. This would provide our administrators with a solid appreciation of the challenges our young faculty face during the early stages of their careers. This would also provide an excellent opportunity for our assistant professors to learn more about the college and university and its expectations for them. All in all this would create a win/win situation for both our assistant professors and our college and university administrators.

FUNCTIONS

Research

All faculty in our department are involved in research. We have no positions with a teaching/extension split. The criteria for determining excellence in research vary considerably from department to department, from college to college, and from university to university. It is essential that each department establish its criteria for excellence in research based on the percentage of effort in the individual's position description. This should include such things as grants, scientific publications, presentations at professional meetings, graduate students mentored, and outreach activities. There also needs to be a solid appreciation of both basic and applied research in our departments. Unfortunately, the basic versus applied research issue still exists, but I believe that it is much less of an issue today than it was a decade ago. The majority of the research programs in our department today have an excellent mix of both basic and applied work.

Teaching

In reality all of the faculty in our Department of Entomology are involved in teaching, even if they do not have a formal teaching component in their position description. Undergraduate and graduate level courses, guest lectures, undergraduate and graduate student training/mentoring/advising are all teaching efforts. Furthermore, extension is indeed teaching. At the undergraduate level, another wonderful advantage of getting a degree from a research university is the opportunity for undergraduate research projects, employment in research programs, and participation in lab meetings. I believe that we are doing a fairly good job in this regard within our College of Agriculture and Life Sciences at Cornell; however, I believe that research opportunities for our undergraduate students should be better publicized and made available university-wide. Departmental, college, and university administrators all need to work together to make this happen. The benefits of undergraduate participation in our research projects are numerous and profound.

We are in effect educating the decision makers of tomorrow, our legislators and regulators at the state and federal level and the general public, on the benefits of the land grant university system, and enabling them to better understand and appreciate our importance to society as a whole, both in the present and for the future.

Extension

For a three-function department to operate fairly and effectively, it is critical that criteria for excellence in extension programming be established/outlined. It is also essential that extension programming be considered scholarly activity at the departmental, college, and university levels. The challenges facing extension in departments of entomology today are, unfortunately, numerous. More extension associate positions are being substituted for faculty positions in departments of entomology across the country. The rapid turnover of off-campus extension educators (county agents and regional specialists) is also a real problem requiring constant training/educating of our extremely important multipliers. Furthermore, extension programming in smaller states continues to decline, which places a significant burden on the extension programs of the larger states. In effect, larger-state programs are becoming regional programs without a concurrent increase in funding. More formalized regionalization efforts/agreements in extension are gravely needed in many parts of the country today. Nationally, our college and university administrators have yet to step up to this Herculean challenge. This is indeed a critical need for us in the three-function department of entomology. I am very concerned about what I see happening or not happening in extension today. Concentration on process, with at times a complete disregard of product, by extension will undoubtedly lead to its demise. We can no longer be everything to everyone. Prioritization efforts at both the state and national levels are greatly needed or extension will ultimately be characterized as mediocre instead of excellent. At present, extension personnel are considered a very credible source of unbiased information. We need to be certain that we maintain this exceptionally high level of credibility.

Audiences

The audiences of our three-function departments of entomology are vast for it is our primary responsibility to educate. Of course, our undergraduate and graduate students are at the top of all of our audience lists. However, depending on the position, our audiences quickly expand beyond our campuses to include extension personnel, growers, industry groups, private consultants, the general public, our federal and state legislators and regulators, public health and environmental advocacy groups, and the media.

SUMMARY

Our three-function departments of entomology have the primary responsibility to increase our knowledge of insects and to apply this knowledge to benefit society. Insects offer unique opportunities for studying biological principles and, as pests of major importance to food, fiber, human, and animal health, they continue to challenge our ingenuity and inventiveness. The recruitment of new faculty is the cornerstone in the successful integration of research, teaching, and extension under one roof. It is also the most critical component in shaping our departments of entomology of tomorrow. We are in effect making a thirty-year commitment to this position, to this effort. It is critical that our efforts in our three-function departments of entomology reflect the ever-changing needs of agriculture, of consumers, of urban dwellers, and of society as a whole. Our audiences are vast, our challenges are numerous, and our successes have and continue to be significant. And yes, our three-function departments of entomology are indeed ideally suited for the next century.

CHAPTER 5

Integration of Entomology Research, Teaching, and Extension

The Past, Present, and Future at the University of Florida

JOHN L. CAPINERA
Department of Entomology and Nematology
University of Florida–Gainesville

Over the past century, land grant institutions have struggled with the need to provide separate federal accounting for the research, teaching, and extension programs, and yet to blend these resources efficiently into an overall program that best meets the needs of America's citizens. Following is a brief history of entomology research, teaching, and extension programs in Florida. Although in some respects unique, it serves to illustrate some of the challenges faced by land grants in the past, and some remaining to be overcome.

ORIGINS OF THE RESEARCH, TEACHING, AND EXTENSION MISSIONS

The first Morrill Act of 1862 creating land grant universities spurred the development of agricultural and applied sciences, including entomology. Few pieces of legislation have had such profound effects on the development of education in the United States. The force behind, but not the author of, the Morrill Act was Jonathan Baldwin Turner. Turner was a New Englander, a rugged individualist with unorthodox views who taught college in Illinois. He had a vision, inspired by the Jeffersonian ideas of education for all young people and cultivation of their morals, so that agriculture, commerce, and manufacturing could prosper for the benefit of every American. He had a plan that called for creation of colleges that would be accessible to the working class, with curricula in practical and vocational subjects that would benefit the working class, and suggested that these institutions could be funded with grants of land from the federal government. The concept of funding institutions with grants of land was not new, but the practical curriculum proposed by Turner was a radical departure from the approach taken by colleges at that time. Traditionally, colleges were sponsored by churches, offered instruction almost exclusively to the wealthy, and emphasized the study of classical literature, history, and such languages as Latin and Greek. Turner, however, believed that the working class needed an education, and that it should be practical. Not surprisingly, the concept of the "industrial university" was not greeted with universal enthusiasm initially.

After about ten years of effort, including development of a broad base of public support, Congress considered a bill authored by Representative Justin Smith Morrill of Vermont to enact legislation that would create universities as envisioned by Turner. After a series of failures a key provision was included—these new universities would include military instruction in their curriculum. Perhaps because the nation was embroiled in civil war, support for creation of the universities was adequate in 1862, and President Abraham Lincoln signed the Morrill Act into law.

This pioneering legislation was followed by the Hatch Act in 1887, which created the Experiment Station system. The Hatch Act resulted from the efforts of the first U.S. commissioner of agriculture, Norman J. Coleman, and some land grant university presidents, but was sponsored in Congress by William Henry Hatch of Missouri. To aid in the development of scientific knowledge about agriculture, federal funds were allocated to each state to support agricultural experiment stations at the land grant universities. Universities and colleges up to this point in time focused solely on teaching, and did not engage in research. Prior to this time, agricultural research had been the domain of inventive individuals, often conducted by people on their own farms, and lacked government support. Also lacking, in many cases, was the scientific approach or scientific rigor, but this was remedied as agricultural research fell under the influence of university faculty. Thus, commencing in the late 1800s, teaching and research were brought together at the land grant universities, and these institutions played the pivotal role in advancing American agriculture.

The second Morrill Act, in 1890, is known mostly for creation of the historically black land grant institutions. This latter act stated that to continue to receive federal funds, each state had to show that skin color was not an admission criterion for students; if it was they had to designate a separate land grant institution for African Americans. Thus was born in the then-segregated South a group of colleges populated by African American students that included agriculture in their curricula, and that received federal funding. These schools became known as the "1890 land grants." Thus, the land grant system was broadened to include a wider sector of America.

Congress created the Cooperative Extension Service with passage of the Smith-Lever Act in 1914. Seaman A. Knapp is usually credited with being the father of Cooperative Extension. The purpose of Cooperative Extension was "to aid in diffusing among the people of the United States useful and practical information on subjects related to agriculture and home economics and to encourage the application of the same." In reality, the concept of providing practical information to farmers was not new and can be traced back at least as far as 1785. However, congressional appropriations gave legitimacy and impetus to the program, greatly expanding the scope of the educational effort. Financing for this educational effort was to be shared by federal, state, and county governments—the genesis of the "cooperative" designation. It is especially interesting to note that the depredations of the boll weevil were particularly important in fostering development of the educational outreach programs as this new pest was devastating agriculture in the South, and farmers could benefit greatly from demonstrations of management practices. The land grant universities were given the lead in establishing the cooperative education network. The approach to be taken in providing information was a marked departure from the classroom environment favored by university faculty. Rather, the emphasis was on "learning by doing," a hands-on approach reflected in on-farm demonstration work, field days, and farmers' institutes. Again, this represented an expansion of the land grant mission, and a very logical one because it took advantage of the educational orientation of the universities and the research base of the experiment stations.

Finally, in recent years there has been designation of Native American colleges as land grant institutions, and designation of colleges in several American territories as land grants. Land grant colleges now exist in all states and territories, and the educational network associated with the land grants extends to all areas of each state by virtue of the many branch research centers and county extension offices. Thus, access to the land grant system has been extended to include virtually every geographic, economic, and social stratum in the United States (Campbell, 1995; NASULGC, 1995).

EVOLUTION OF THE RESEARCH, TEACHING, AND EXTENSION MISSIONS IN FLORIDA

In Florida, entomology started with creation of the Agricultural Experiment Station in 1888. A Department of Entomology was formed in the Experiment Station some time after 1911. Instruction in entomology started almost immediately with creation of the Experiment Station as entomologists offered a class or two to agricultural students. A formal teaching department was not to appear until 1924, with the creation of the Department of Economic Entomology and Plant Pathology in the College of Agriculture. Extension entomology, or at least the precursor of it, also commenced immediately with creation of the Experiment Station, as entomologists worked closely with farmers in protecting their crops, and as entomologists ranged widely over the Florida landscape as they conducted their research studies. Extension entomology did not exist formally until 1943, when a full-time extension apiculturist was hired. As more extensionists were hired they became part of the Cooperative Extension program, but not part of the Experiment Station or the Department of Entomology housed in the College of Agriculture (Tissot et al., 1954; Brogdon, 1976).

Although the original Experiment Station entomologists conducted classes and outreach programs, as Florida's agriculture developed and more entomologists were added to the faculty the processes of teaching, research, and extension became quite separate. New appointees typically received 100 percent appointment in only one of the functions. The research, teaching, and extension efforts operated independently until 1965 when all were united into a single Department of Entomology. However, the separation of functions continued beyond the amalgamation of entomology research, teaching, and extension functions into a single department. The separation of function was exacerbated by physical separation of the groups, with most extension-oriented faculty housed in Newell Hall, the teaching faculty in McCarty Hall, and the research entomologists scattered in several small laboratories. The need to integrate the land grant functions remained a formidable challenge. Various efforts to foster cooperation were attempted. Foremost was the availability of grant funds for projects that demonstrated integration of efforts, particularly research and extension, or research and teaching. However, there remained philosophical, professional, logistical, and personal barriers that inhibited full integration. One obvious potential solution was to make split appointments, to encourage those who were conducting research in an area to provide educational programs in the area of their expertise, whether the audience was college students or people already working in the field.

SPLIT APPOINTMENTS: KEY TO INTEGRATION OF FUNCTIONS

In the 1990s split appointments began to be commonplace in entomology at the University of Florida. It is a trend that is continuing to grow in frequency if not popularity. Part of the impetus came from construction of a new entomology building, which was occupied in 1990. This new facility provides, for the first time, real opportunity for integration by bringing together faculty who previously were isolated physically and intellectually. It also allows extensionists and teachers who formerly lacked laboratory space the opportunity to conduct laboratory research and enhances their ability to sponsor graduate students.

With split appointments, faculty are confronted with a more diverse clientele. With a split research and teaching appointment, researchers might be expected to teach in the area of their research expertise. Similarly, those conducting research might be expected to communicate the findings and significance directly to farmers or other off-campus clientele, forming the basis for a research and extension split. As the distinction between on-campus and off-campus teaching diminishes, the acceptance of three-way appointment splits is increasing. The magnitude of the shift in faculty appointments in Entomology-Nematology at the University of Florida is shown in Table 5.1.

Examination of the data presented in Table 5.1 should show that a marked shift is occurring in faculty responsibilities. Overall, diversification of effort is occurring both at the main campus in Gainesville and at the Research and Education Centers statewide. In Gainesville, everyone now has some teaching responsibilities. Furthermore, many faculty with research and teaching responsibilities now have small extension appointments. At the Research and Education Centers, about 30 percent of the faculty have assumed some formal extension responsibilities.

In many respects, what we are implementing is restoration of faculty responsibilities in a form similar to the early part of the century (Figure 5.1). In the early days of the Experiment Station, subject-matter specialists con-

Table 5.1. **Comparison of Full-Time Equivalent Faculty (FTE) Assigned Teaching, Research, and Extension Responsibilities, and the Distribution of FTE among Faculty, in 1988 and 1998.**

Year	*FTE availability*	*FTE distribution*
1988	9.1 FTE Teaching	mostly invested in 8 faculty
	7.6 FTE Extension	entirely invested in 8 faculty
	18.3 FTE Research (G'ville)	12/18.3 entirely Research
	44.0 FTE Research (Centers)	44/44 entirely Research
1998	5.8 FTE Teaching	spread over 35 faculty
	13.8 FTE Extension	spread over 31 faculty
	19.2 FTE Research (G'ville)	0/19.2 entirely Research
	40.0 FTE Research (Centers)	29/40 entirely Research

ducted research as their primary responsibility, and then provided instruction to either a college student population or to practicing agriculturalists, as needed. As faculty were added, division of responsibility was made on the basis of audience or clientele rather than subject matter. Thus, extension specialists worked with off-campus clients, teaching faculty worked with college students, and research faculty worked mostly with one another. There was an assumption made that faculty would communicate with one another. Specifically, it was assumed that extensionists would relay needs to researchers, and researchers would relay new findings to extensionists and students. Over time, however, the nature of research and extension changed, with research becoming more and more fundamental and extensionists taking responsibility for applied research. There became little common ground between research and extension, resulting in little dialogue. Practicing agriculturalists came to believe that they were disenfranchised by the research program. In some cases, teaching faculty became detached from current developments in the field because they rarely left the campus environment; in other cases they produced courses that emphasized only their relatively narrow research interests, not truly meeting student needs. A reassessment of responsibilities and activities was overdue, and even cursory examination of departmental operations suggested a need to integrate research, teaching, and extension programs.

At the University of Florida, in Entomology-Nematology we now are emphasizing subject-matter specialization rather than clientele-based specialization. Perhaps it was necessary at one time to have one set of faculty who spoke the language of farmers and another who communicated best with college students. Perhaps at one time some faculty were only capable of research, and others were incapable of research. However, the public is more sophisticated now, and with a little work research scientists can learn to communicate effectively with a variety of clientele. More importantly, by spreading educational responsibilities (on-campus and off-campus) among more faculty, we can have the most authoritative people involved with our educational efforts. We can afford to have specialists in a large number of fields who then translate research findings directly into educational programs.

A principal objection to the split appointment approach is that extensionists and researchers need to have the liberty to be mobile, to be unencumbered by teaching schedules. Over time, however, the travels of both extensionists and researchers have been curtailed due to several factors, including better planning, reduced demand, and financial constraints. Most importantly,

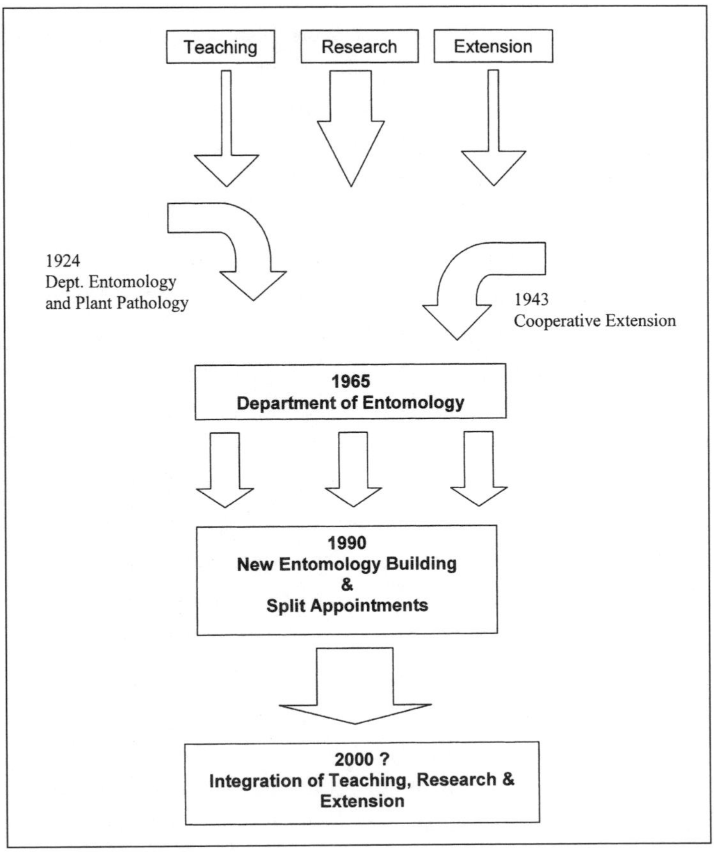

Figure 5.1. Chronology of research, teaching, and extension activities in entomology at the University of Florida.

the availability of electronic technology should reduce the need for on-site visits, especially in extension. Such developments as the use of digital cameras and computers in distance diagnostics, video conferencing, and World Wide Web databases can enhance efficiency of extension efforts. Despite traditional scheduling patterns, we have found that students are receptive to nearly any scheduling we propose, allowing faculty to concentrate teaching into relatively small blocks of time that do not interfere with travel obligations. The institutional bureaucracy is less flexible, but it too can be overcome.

The actual constraint on shifting the research-teaching-extension paradigm is adaptability of existing faculty. Although many faculty can adapt to different audiences, are open to suggestions by clientele, and can juggle several research projects with ease, some are unwilling or unable. As the current generation of faculty retires they will be replaced by flexible and adaptable individuals who readily embrace the entire land grant mission. While not everyone will succeed at all endeavors, and three-way splits will certainly be beyond the capabilities of some, we will be seeking individuals with great technical expertise who can maintain and develop their expertise and simultaneously contribute to education and problem solving for various clientele.

The close of the twentieth century finds us in an increasingly specialized, competitive era. We are expected to be very innovative and current in our research and educational programs. It is increasingly difficult for educators, whether traditional classroom- or extension-oriented, to remain current without direct involvement in research. The knowledge explosion and relentless march of technology demand that teachers devote considerable effort to staying abreast of developments. What better way than to be actively involved in research? After all, it was Seaman Knapp, the father of extension, who preached that the best way to learn was to be actively involved; what better way for extensionists to learn? The credibility of extension has suffered from the lack of technical expertise on the part of some extensionists; this is less the fault of the extensionist than the system that expects technical expertise over too broad an area. We can ill afford to have the land grant university system represented by any less than the most competent extensionists and instructors. We can compensate by reducing the scope of technological expertise for which an individual is responsible by dividing responsibilities more finely and broadening the audience.

The credibility of our research programs has suffered from the lack of researchers' familiarity with the needs of clientele. Direct linkage of researchers with producers or other clientele often is best accomplished through extension activities, and it is these interactions that fine-tune the direction of research programs into relevant, useful expenditures of funds and time. Although the "comfort level" of faculty typically decreases as the breadth of responsibilities increases, there is a clear national trend toward split appointments. Split appointments allow more individuals to be involved in serving the needs of clientele and allow greater technical specialization. Greater technical specialization facilitates high quality programs and competitiveness. Split appointments are one of the key elements in integration of the historic land grant responsibilities of research, teaching, and extension.

REFERENCES CITED

Brogdon, J. E. 1976. Fighting Pests of Crops and Livestock and Promoting Beekeeping. In *Dimensions in History: Recounting Florida Cooperative Extension Service Progress, 1909–76,* ed. J. F. Cooper, 169–77. Gainesville, Fla.: Epsilon Sigma Phi.

Campbell, J. R. 1995. A Historical Perspective of the Land-Grant University System. In *Reclaiming a Lost Heritage: Land-Grant and Other Higher Education Initiatives for the Twenty-First Century.* Ames: Iowa State University Press.

NASULGC (National Association of State Universities and Land-Grant Colleges). 1995. The Land-Grant Tradition. Washington D.C.: NASULGC.

Tissot, A. N., M. Murphey, and R. E. Waites. 1954. A Brief History of Entomology in Florida. *Florida Entomologist* 37: 51–71.

Chapter 6

Strengthening a Department of Entomology through Integrating Teaching, Research, and Extension Functions

RAYMOND E. FRISBIE
Department of Entomology
Texas A&M University–College Station

The Department of Entomology at Texas A&M University has undergone significant self-evaluation and positive change over the last five to ten years. Much of this change is attributable to the faculty addressing entomological needs as a corporate body with greater knowledge of the functional teaching, research and extension responsibilities and opportunities of the department. A departmental review in 1992, public and university priority on undergraduate education, escalating demands by agriculture and urban citizens to provide more rapid solutions to entomological problems, and declining state and federal resources set the stage for substantial reorganization and redirection within the department. The broad and pervasive nature of these needs required the faculty to examine the fundamental goal, purpose, and vision of the department in a comprehensive and integrated way.

Texas A&M University has the only Department of Entomology in the state. Its faculty is located in College Station and across the state at Texas A&M Research and Extension Centers. Additionally, the department provides leadership for extension Integrated Pest Management (IPM) programs with county faculty (extension agents–IPM) involved with agriculture and urban educational programs. The mission of the department is to provide educational and technological advances through teaching, research, and extension. The department also has a regulatory function, which complements its mission, in the Texas Apiary Inspection Service. The faculty agrees that the primary purpose of the department is to provide balanced teaching, research, and extension programs that serve the current and projected entomological needs of the people of Texas, the nation, and the world. The mission and purpose are explicitly stated in the department's Strategic Plan that was developed in 1997. The plan clearly states that the vision of the department is that teaching, research, and extension missions are functionally integrated and of equal importance. The following discussion describes how integration of function is not only relevant to a healthy well-balanced entomology department; it is imperative if the department is going to succeed and prosper.

INTEGRATION IMPERATIVES

Because the faculty is widely distributed geographically, and involved at different levels in teaching, research, and extension programming, integration of these functions and management are of primary importance in terms of scholarship relative to our teaching, research, and extension programs. Integration of functions across the department in order to optimize programming, whether it be undergraduate/graduate instruction, basic or applied research, or conducting extension educational programs, is based on two primary aspects: (1) a clear understanding of each respective faculty member's role and responsibility in teaching, research, or extension by fellow faculty, and (2) a willingness by individual faculty members to work in support of the vision and goals of the department. Without a willingness to understand and share programs with fellow faculty members in terms of active collaboration, a department is only a group of individual faculty internally focused on their own special areas of interest. This is not to say the basis of good entomology departments is not strong individual programs, but it is the willingness of faculty who lead strong individual programs to share and collaborate with others that significantly strengthens a department.

A Planning Process and Environment for Integration

Departments of Entomology should strive for excellence in teaching, research, and extension. Striving for excellence should include interaction and co-activity between these three functional areas. As stated, the department underwent a valuable exercise in developing a strategic plan. One of the main purposes of the plan was to engage the faculty-at-large in terms of sharing with one another their programs and capabilities and then setting future priorities. The strategic plan provides a clear statement of purpose, vision, and goals for the department. The strategic plan has as its foundation the integration of teaching, research, and extension. Specific tasks and actions to be accomplished for research, teaching, and extension, as well the regulatory programs, were identified, reviewed, and agreed upon by the entire faculty.

Teaching

The strategic plan involved both general and precise direction for teaching. As most faculty on the main campus in College Station have the primary teaching responsibility, considerable care was taken in meeting the challenges of teaching accountability, cost per student credit hour, and a curriculum that was contemporary and reflected the job demands of undergraduate and graduate students. Teaching improvement in order to be globally competitive was an overriding theme. The perspective of off-campus research and extension faculty was valuable in terms of graduate student training and recruitment.

The Department of Entomology at Texas A&M, just as other departments of entomology, was undergoing significant change in expectations and direction in teaching. A primary influence on this direction was expectation of the public, legislative, and university administration to expand and improve the quality of undergraduate teaching. Similar plans and ideas regarding this priority of the 1990s were being explored and developed at the University of California–Davis and the University of Florida (Parrella et al., 1995). In the late 1980s and throughout the 1990s, several departments of entomology, across the United States found themselves being merged with other departments for a variety of reasons. Primary reasons were low teaching, in terms of student contact hours, and the low numbers of majors. It was clear that if undergraduate education was not addressed in a meaningful way, status as a department of entomology could be lost.

Historically, the Department of Entomology at Texas A&M was founded and evolved with a central focus to develop and transfer technology in order to solve public entomology problems. Texas A&M's department, like so many departments of entomology, found its focus in research programs that were allied with sound teaching and extension programs. Texas A&M is unique in that it retained its tradition of a strong undergraduate program. Hence, when the priority came to increase undergraduate teaching in terms of hours, the department was well positioned with a faculty dedicated and highly motivated in the area of undergraduate teaching. The development of a non-majors course, "Insects and Human Society," and expanded teaching outside of the department, particularly in the Genetics interdisciplinary program, increased student contact hours. An aggressive communications plan to other departments both within and outside of the College of Agriculture and Life Sciences resulted in meeting university expectation in terms of student credit hours, but more importantly it exposed entomology and entomologists to a much broader student body. Student credit hours in the fall of 1994 stood at 1,332 costing $318 per student credit hour. After expansion of the undergraduate teaching program and an aggressive communications effort, student credit hours had risen sharply to 3,129 and the cost reduced to $145 per student credit hour by the fall of 1998. An additional benefit to expanded teaching was the significant increase in undergraduate entomology majors, thirty-six to seventy, during the period between 1993 and 1997.

These figures illustrate what faculty can accomplish regardless of whether their primary emphasis is teaching, research, or extension. Each functional area committed itself to a common cause. Even though the primary responsibility of undergraduate teaching resides with the College Station faculty, off-campus research and extension faculty were supportive in recruiting undergraduate majors and in willingly participating as guest lecturers in classes.

Research

A basic tenet of the strategic plan was the emphasis on fundamental entomological principles and their applications to problem solving. The faculty chose to promote scientifically rigorous research that developed integrated approaches for managing arthropods. In terms of integration, the faculty set a goal to encourage greater collegiality, dialogue, and cooperation among faculty. Research strengths were focused into Areas of Excellence which included (1) molecular physiology and toxicology, (2) systematics, (3) arthropod ecology, (4) biological control, (5) plant resistance to insects, (6) Integrated Pest Management, and (7) urban, medical, and veterinary entomology. Teams of teaching and research faculty worked together to set goals, actions, and tasks for Areas of Excellence. Graduate students were included and involved in this process. Extension faculty was included on teams for biological control, plant resistance, IPM, and urban, medical, and veterinary entomology Areas of Excellence. The composition of these teams represented a *functional integration* across the department. In addition to an in-depth needs analysis for each Area of Excellence, the teams were asked to consider the linkage or relationship of their Area of Excellence with other Areas of Excellence; thus, an additional dimension of *subdisciplinary integration* across research areas was incorporated in the development of the strategic plan.

Extension

Extension entomology programs have historically been and continue to be closely allied and integrated with research programs that deal with solutions of entomological problems in agriculture, natural resources, and the urban environment. The primary responsibility of extension faculty is to develop comprehensive educational programs for the public. The development of these programs requires broad understanding and direct interaction with research programs. The extension faculty, inclusive of extension agents–IPM, are actively involved in the research Areas of Excellence indicated earlier, particularly in IPM; however, they maintain a degree of understanding across all Areas of Excellence. This knowledge is gained through several planned venues. The Entomology Science Conference agenda is developed each year to highlight specific Areas of Excellence such as biological control, molecular biology, plant resistance, medical/veterinary entomology, and others. Biannual retreats of extension faculty have included research reviews in urban entomology, systematics, biological control, and such.

Extension entomologists are teachers. Their audience in the domain of public education ranges from youth to adult education. Extension entomologists have a keen interest in curriculum development and graduate education. They participate in graduate training and guest lecture in several courses in the department. Extension entomologists acquired professorial rank in the early 1990s. In order to meet promotion criteria set by the department, they, like all faculty, must demonstrate scholarship in their programs. Scholarship infers peer recognition and acceptance that usually is exhibited through publication in peer-recognized scientific journals. Extension faculty who are successful in promotion take advantage of active collaboration with research counterparts as well as taking leadership in research activities that directly support their extension programs. Hence, collaboration with research faculty not only strengthens extension programs but also offers tangible benefits to the individual seeking promotion.

Centers and Laboratories

Centers and Laboratories offer another way to integrate the teaching, research, and extension functions of the department. The Center for Urban and Structural Entomology, for example, serves as a point of integration for research, extension, and teaching programs in urban entomology. Insects that affect man and his near-environment in an urban setting provide a focal point for faculty to develop integrated programs. Similarly, the Biological Control Laboratory was designed to involve research and extension faculty as well as

USDA/ARS and USDA/APHIS scientists working in biological control. This laboratory developed guidelines and activities that foster integration of research and extension education toward the implementation of biological control as an IPM tactic. The sharing of common goals and actively working together has provided a cross-functional synergy to advance biological control in Texas.

Administration

Departmental administration must encourage and foster functional integration across the department with the goal of enhancing and strengthening individual teaching, research, extension, and regulatory programs. A stated task within the Strategic Plan was "to implement an *infrastructure of communication and integration* for teaching, research and extension functions within the department." This is being accomplished in several ways. The committee structure within the department was intentionally designed to include cross-representation from teaching, research, and extension faculty. Additionally, graduate students serve on most committees to assure student input and perspective in the operation of the department. The Faculty Advisory Committee and the Faculty Evaluation Committee are excellent examples of departmental committees that include all functional groups. The Faculty Advisory Committee is elected by the faculty and serves the department by bringing specific issues of opportunity and concern to the department head. Regardless if the issue is teaching, research, or extension, the Faculty Advisory Committee prepares, discusses, and follows up on these issues as a group representing all functions of the department.

To reiterate, the basis of integration is understanding and appreciation of fellow faculty members' programs. The Faculty Evaluation Committee, another elected committee, broadly represents all functions of the department regarding matters of promotion and tenure. This committee, or its subcommittees, also plays a valuable role in mentoring faculty as they proceed through promotion. Each member has the opportunity to understand the demands and needs of their fellow faculty member and their involvement and activities in teaching, research, and extension. As faculty are elected and rotate through both the Faculty Advisory Committee and the Faculty Evaluation Committee, their understanding of fellow faculty members' programs provides opportunity for collaboration and integration. Similarly, other standing committees, special committees, and task forces in the department are structured to enhance communication and understanding, and to foster integration.

BOTTOM LINE

Faculty members are hired and promoted based on strong individual programs. Research programs of high quality have an immediate effect in the classroom through teaching, and they benefit society as discoveries are passed to the public via extension programs. Similarly, the commitment of individual faculty provides the basis for quality teaching and extension programs. Regardless of the faculty member's appointment, the integration of research into teaching and active collaboration with extension not only strengthens the multifunctional mission of the department but also benefits the individual faculty member. Conversely, extension faculty sharing their field knowledge and becoming actively involved in research teams strengthens respective research and extension programs. Training students is a tangible example of where teaching, research, and extension functions are integrated. Classroom instruction, graduate student research, and undergraduate internships all benefit from the integration of teaching, research, and extension. Functional integration of teaching, research, and extension strengthens individual faculty member programs and the department as a whole. The departmental infrastructure and operating environment is specifically designed to foster integration.

REFERENCE CITED

Parrella, M. P., S. S. Duffey, Jeffrey Granett, and D. D. Kitterman. 1995. Departments of Entomology: A Matter of Survival. *American Entomologist* 41: 12–17.

Part III

Development and Delivery of Science-Based Knowledge to the Public

Chapter 7

Science and Stakeholders

New, Traditional, and Nontraditional Ways of Doing Business

FRANK G. ZALOM
Department of Entomology
University of California–Davis

Scientific breakthroughs are occurring with incredible speed. Those of us who work in Entomology can appreciate the powerful new discoveries being made that influence the way we now look at insect populations and their management. Discoveries at the molecular level already permit control of insect pests through the application of molecular tools in the development of designer plants and designer toxins. Molecular biology also permits systematics and evolution to be studied at levels never thought possible only a few years ago. These tools have increased our ability to understand such issues as exotic species invasions and biological diversity. Recent discoveries in physiology and behavior are leading to development of new pesticides, modification of chemical communication, and perhaps even to mechanisms for delaying or avoiding insecticide resistance. New ecological approaches permit true integration to occur at a landscape level rather than at the individual field or organismal levels. Overarching these scientific advances are technologies gleaned from the fields of chemistry, physics, and systems science that can be applied to solving entomological problems.

Entomology is indeed at the brink of a new millennium, one that promises insect populations to be managed reliably and with little environmental risk—but will this actually happen? Moving scientific discoveries from the lab bench to the field, gaining their acceptance by the public, and incorporating them into sustainable pest management systems remains a challenge. The full potential of new discoveries will not be realized without soliciting the understanding, acceptance, and support of affected stakeholders.

DEVELOPING AND APPLYING NEW TECHNOLOGIES—FOR THE RECORD

A Fountain of Knowledge

Throughout history, scientific discovery has been fostered by people who have had the inclination and time to seek answers to questions that were of interest to them. Often, these discoveries occurred in imperial courts, religious settings, and academic institutions where well-educated individuals had an opportunity to explore and satisfy their curiosity, unfettered by the need to generate support for their pursuits. The adaptation of scientific discoveries to address practical problems became a pursuit of the gentry who could apply these to improve the productivity of their land or business.

Until very recently, most people were tied to the land, and farming methods remained relatively unchanged for hundreds, if not thousands of years. It was not until the eighteenth century that innovations were systematically applied to improve agriculture in Europe and the United States, with larger landowners forming groups to discuss their experiences with plant cultivation and selection, animal husbandry, and pest control. These agricultural societies became very popular, and evolved to include a broader base of participants. By the mid-nineteenth century, many of the agricultural societies became state supported and utilized more formal avenues of communication, such as public meetings and printed publications.

The Land Grant Model

The Morrill Act of 1862 established the land grant colleges (land grants) to train students in applications of science-based technology to rural society. The Hatch Act of 1887 established the Agricultural Experiment Stations within the land grant system, freeing time and providing support for scientists at these institutions to systematically apply knowledge to agricultural problem solving and to pursue their scientific curiosity. The Agricultural Experiment Stations became active in reporting the results of their research in bulletins and farm magazines, and land grant instructors organized short courses and traveling schools for farmers. Commonly, local businesses helped to sponsor these courses in an effort to improve the local economy. Railroads even offered free tickets for land grant educators to hold rural education meetings because they saw the potential for increased production and trade. The land grants were successful in teaching farmers and their children about agricultural innovations and in developing new technologies, but many farmers remained skeptical and would not adopt them without seeing the techniques demonstrated on their own farms. These stakeholders required a different educational approach.

One of the first examples of success in on-farm demonstrations was that of Seaman Knapp, a USDA scientist working in Texas (Scheuring, 1988). The crisis caused by severe outbreaks of boll weevil (*Anthonomus grandis* Boheman) in 1904 presented Knapp with an opportunity to demonstrate the benefits of alternative agronomic practices to cotton farmers. Needing help, he employed twenty agents to establish test plots using science-based production practices on demonstration farms in the affected areas. The first resident county agent was hired in Smith County, Texas, in 1906, and was supported by local businessmen and the United States Department of Agriculture (USDA). The Smith-Lever Act of 1914 placed land grant and USDA-affiliated agents in most counties of the United States. The U.S. Cooperative Extension System is unusual in that it is directly associated with land grant universities and their formal teaching and research missions rather than with a regulatory agency as in most other countries.

Postwar Issues and Progress

Food and fiber production increased tremendously in the United States after World War I. Land grants together with the USDA played a significant role in this expansion. Agricultural research was largely responsible for the increase in farm mechanization, the introduction of synthetic fertilizers and pesticides, and the selection and introduction of more productive plant and animal varieties. The widespread use of science-based technologies in agriculture has been beneficial in that fewer people are now required to produce the food and fiber products necessary to sustain the needs of an ever-growing population. In addition, science-based technologies have helped to keep the low cost (as a proportion of income) of food and fiber in the United States, and to keep food supplies relatively stable. However, some nonagricultural interests would argue that these benefits have come at a significant environmental and social cost.

PEST MANAGEMENT DEVELOPMENT AND DELIVERY

The Traditional Partnership

Agricultural research does not occur entirely within publicly supported institutions. Both private industries and growers have a long history of pursuing discovery and adaptive research. Many times, partnerships between industry, growers, and public institutions have been formed to develop new technologies and to promote their adoption. For much of the past century, land grant institutions have assisted private industry to develop commercial products and processes for their production. In return, grower associations and private industry have helped to support land grant institutions financially and politically.

The implementation of synthetic pesticide use as a production tool is a good example of such partnerships between industry and land grant institutions. Pesticide discovery and development occurs both in public institutions and private companies. Their use was effectively promoted by industry, which had a stake in product sales. Land grant and USDA scientists also promoted pesticide use based on their effectiveness and documented ability to increase productivity while reducing variability in production. Growers, working through the political process, secured subsidies and promoted regulations that made pesticide use (and agricultural chemical use in general) more economically attractive.

The Pest Control Industry

Chemical and pharmaceutical companies were well positioned to develop and produce synthetic pesticides, and with a few exceptions companies in this sector became de facto agro-chemical industry leaders. Many chemical companies had established marketing and distribution channels for horticultural oils, sulfur, and other inorganic pesticides that were widely used in agriculture. In addition, similarities regarding product registration existed for pharmaceutical and agricultural products. Unlike pharmaceuticals, however, pesticides could be marketed over the counter through farm supply dealers. Advice on pesticide use was available from the Extension Service but more commonly from salesmen working for farm supply dealers.

Recently, there has been a significant consolidation within the agricultural chemical industry and to some extent, a change in focus in product development to those so-called reduced risk products including biopesticides. In some cases, agro-chemical companies have shifted emphasis and invested considerable amounts of resources on biotechnology as a tool to develop genetically modified seed. New biotechnology-based products can be readily promoted through the existing distribution and marketing channels and will likely be adopted if proven effective and economically competitive with traditional products.

Changing Practices

In the early 1950s, identification of pesticide resistance and pesticide-induced secondary pest outbreaks led some researchers to call for caution in the use of pesticides and to promote the concept of "integrated control." Integrated control was defined as the combination of compatible biological and chemical control tactics (e.g., Michelbacher and Bacon, 1952). The publication of Rachel Carson's *Silent Spring* in 1962 seriously raised public concern over environmental contamination by chlorinated insecticides and led to the placement of restrictions on their use. Public debate over environmental contamination, worker exposure, and pesticide residues on food increased in the 1970s and was accompanied by the passage of various state and federal regulations that placed further restrictions and safeguards on pesticide use. The Food Quality Protection Act of 1996, once thought to be an acceptable compromise between agricultural interests, allied industries, and public advocacy groups is the most recent and perhaps most extensive outcome of this debate.

Trends in the science of entomology, and in Integrated Pest Management (IPM), which encompasses all pest management disciplines in a systems approach, reflect the ongoing changes in pest management practices. Pesticide use first became a political issue in the 1960s, and this directed research efforts into developing alternatives to pesticides as well as quantification of the unintended impacts of pesticide use. Funding for IPM research increased greatly during the 1970s through the U.S. Environmental Protection Agency (EPA) and the USDA-funded Adkisson and Huffaker Projects. These focused attention on pest management for the major cropping systems in the United States and involved a broad range of scientific disciplines. Concurrently, federal funding for Extension Pilot Projects greatly increased IPM implementation and encouraged linkage-development between research and information delivery agencies. Products of the extension IPM efforts include scouting programs, which were the precursors of crop consulting profession in many states, large-scale IPM demonstrations, and various types of how-to publications intended for farmers and the scouts. Acceptance led to new state-funded research and extension programs in California and New York and organizations such as the Texas Pest Management Association. USDA funding for IPM research and education remains level and is allocated through regional IPM grants programs, the Smith-Lever Extension IPM program, and research components of the National Research Initiative. Funding from the EPA in the 1990s targeted partnership programs designed to improve grower adoption of IPM, sustainable, and organic production practices.

SOCIOLOGY AND SCIENCE

Trends

For much of the past century the emphasis of agricultural research and extension has been on greater productivity and reducing production costs. Recently, several trends have been converging which may rapidly change the way agricultural research and extension is conducted. Increasingly, as the proportion of the U.S. population actively participating in agriculture declines and more people reside in urban areas, pressures intensify for public research to address environmental and social issues. Reflecting this new agenda, staffing and programs at land grants and the USDA are in transition, often resulting in reduced emphasis on traditional production issues. Much of the research on these "new" issues clearly does not benefit the farmer economically, but is instead intended to benefit society as a whole by substituting production practices that are considered less environmentally harmful or more socially acceptable.

At the same time, there is a growing public distrust of science and technology. Certain segments openly question the wisdom of addressing problems they perceive as the result of science and technology through a science-based approach. Strong criticism has been mounted against animal research, genetically modified organisms, and development of production technologies perceived to favor large corporate farms (agribusiness). Of course, any research or education involving pesticides is similarly suspect.

Reducing the size of "big government" and demanding accountability in government programs are related trends that have resulted in the erosion of base support for the agricultural research and extension infrastructure. Permanent federal funds for staff and facilities have been declining in real terms for over thirty years. What new additional support has become available is mostly competitively awarded for specific programmatic initiatives leading to a "product" mentality—deliver the product that the funding agency wants, a short-term perspective.

Strained Traditional Partnerships

The partnership between public research, the agricultural industry, and growers has become strained as a result of these trends. Although still an important resource of discovery and development for the agricultural industry, public institutions represent a decreasing source of products. Budget reductions and staff redirections have also resulted in a strained program-delivery infrastructure. Moreover, land grant and USDA involvement in environmental and social issues has resulted in conflict with industry and growers' organizations as their programs become increasingly diverse. Ironically, in spite of changing program emphasis, many advocacy groups still view the relationship between these traditional partners as too cozy.

THERE ARE OPPORTUNITIES, AND THEN THERE ARE OPPORTUNITIES

The future for development and delivery of science-based information by public institutions is excellent in spite of the challenges of recent years. Why? The key is in the content they can provide, and in the structure of the institutions themselves. The land grants and the USDA represent a critical resource for basic innovation—research that is simply too risky to pursue for private industries that must consider the short-term bottom line. They possess unparalleled diversity of expertise, which can be readily tapped for technical information, and represent an unbiased third party for product development and evaluation. Because not all problems need be addressed with products that can be put into a bottle or box and then sold, research that is intended to solve problems with few or no external inputs can only be developed with public support. How best can publicly supported research be delivered to the public?

Via the Private Sector

Delivery of research that results in a commercial product is relatively easy as long as the resulting products have a market. Most public institutions have technology transfer offices to promote commercialization of research products. Once a private company licenses and develops a product, it has an incentive to promote it through established marketing channels.

Via Traditional Extension Methods

The linkage of campus agricultural research programs with county Cooperative Extension staff, farmers, and crop consultants remains one of the primary methods for information delivery in the United States. In most studies on the subject (e.g., Lambur et al., 1989), approximately half of all growers and crop consultants consider county-based extension staff to be a primary information source. Cooperative Extension publications such as manuals, bulletins, and handbooks are the top source of information for over 80 percent of crop consultants (Lambur et al., 1989). Clearly, in addition to meetings and demonstrations, technical information packaged for use by growers and crop consultants is a preferred delivery approach for production information.

Both the American Phytopathological Society and the Entomological Society of America now produce handbooks for pest identification and management. Individual states and regional IPM programs also produce such materials. In California, the Pear Pest Management Manual, produced in 1977 with funding from the USDA Extension IPM program, was so well received by clientele that a whole series of IPM and pesticide safety manuals was written; they have sold well over 100,000 copies.

Via Electronic Media

The potential for electronic dissemination of information was recognized early by scientists involved in IPM development and delivery, probably because of their emphasis on modeling and applications of systems science. Several states developed computer networks in the early 1980s to get IPM models and information out to county extension offices and later, as personal computers became available, to growers and crop consultants via dial-up modems and transient technologies such as videotext (e.g., Zalom and Strand, 1990). In reality, these systems were not widely accessible for users, and were difficult to update and use. However, they provided databases and interactive tools that could be rapidly adapted to be used on the Internet. The World Wide Web presents a tremendous opportunity for delivery of science-based information, and virtually all universities and agencies now have a Web site.

In California, the University of California (UC) Statewide IPM Project was an early promoter of electronic information dissemination, working with scientists to develop and implement models and expert systems on its central computer and later on personal computers. It also developed support databases of weather and pest management guidelines that could be readily updated. These products have been implemented on the program's Web site, which became available late in 1996. The site hosts the university's *Pest Management Guidelines* (over 12,000 pages), over seventy *Pest Notes* for homeowners, and databases of models, meteorology data, pesticide use reports, research grants, and others, and is illustrated with over 14,000 photographs and figures. Its use has doubled every year since its introduction, and was accessed over 2 million times by almost 200,000 users in the past year. Clearly, the Internet holds significant promise for wholesale delivery of science-based information.

Crop Consultants

Some technologies, using a new pesticide or resistant plant variety for example, can be easily implemented as they may simply substitute for another input in a crop production system. Other technologies are more complex, requiring a change in a grower's practice or more intensive monitoring of pests and beneficial insects. Complex technologies can be validated and demonstrated through Cooperative Extension, but adaptation to individual farms may require more extensive experiential work than is possible by public agencies. Growers can do such adaptations themselves. However, crop consultants broadly trained in pest management and crop production offer an excellent opportunity for technology adaptation at the farm level. "Supervised control" of pests by trained crop consultants is not a new idea, and pioneering crop consultants have been in practice for over fifty years (e.g., Michelbacher, 1945). Demonstrating the economic advantages of employing crop consultants was a major focus of the USDA Extension IPM Pilot Projects of the 1970s, and as a result, the crop consultant profession has become established in most parts of the country. In California, state licensing of Pest Control Agents (PCAs) was tied to its extensive regulatory structure in the early 1970s (Kendrick and Stimmann, 1984). Today over 4,000 individuals have passed PCA examinations and are licensed by the state.

Food Certification and Labeling

Although there is extensive scientific evidence to the contrary, the public equates food safety with pesticide residues. A 1990 consumer survey found that 80 percent of respondents were concerned about pesticide residues on food (Food Marketing Institute, 1990). There also exists an interest in "green labeling," which emphasizes that a product was grown in an environmentally sensitive manner. During the past decade, produce certified by a private company (Nutriclean) as having "no detectable residues" has been marketed by several California supermarket chains. Growers pay for the Nutriclean certification service, which is required by the retailers. Organic labeling has also been employed by growers for a number of years, with over forty third-party certification organizations having been established across the United States. In general, organic crops are produced with no synthetically manufactured inputs, and approved inputs are strictly controlled. Organically labeled products are a small but growing segment of the food industry, and in many cases organic growers can command a premium for their products. IPM labeling has been successfully tested by a partnership between a retailer, its processor/suppliers, farmers, and the New York State IPM Program. The producers of the IPM-labeled products must follow specified guidelines. Unlike organic labeling, IPM labeling does not require products to be grown without synthetic inputs, but to be grown in such a way as to result in undetectable or no pesticide residues on food and no environmental contamination. While farmers have yet to receive premium returns with this label, they have been able to promote their sensitivity to consumer concerns. The Integrated Crop Management (ICM) systems developed by the European Union represents another type of product labeling. ICM labels are regulated by the European Union and bring with them certain benefits for producers but also restrict production practices.

Labeling offers an opportunity for growers to promote products grown using technologies that are more environmentally benign than traditional ones. However, unless there is a market incentive, labeling may not become widely adopted. Open communication between growers and consumers must occur that relates the benefits of labeled products in order for market incentives to be achieved. Establishing clear guidelines for IPM use is possible without product labeling, and could offer growers a measure of "how they are doing" in terms of implementing IPM practices. One example of this is a Positive Points System, which was established by grape growers in California to be used as a guide for moving toward increased adoption of environmentally safe practices (Central Coast Vineyard Team, 1998). Unlike labeling programs, the emphasis of the Positive Points System is on setting goals for and demonstrating increased usage of practices.

Community-Based Alliances

The development and delivery of agricultural production information has traditionally relied on "experts." The land grant model of technology development and delivery, the employment of trained crop consultants, and to some extent product development and marketing by the private sector largely rely on a person, the expert, who provides technological innovation and education. Recently, however, there has been an increasing interest in direct involvement of growers and other stakeholders in the development and delivery of agricultural production information. The emphasis of such community-based programs is on local leadership, frequent farmer contact, and collaborative experiential learning. Community-based programs, also known as participatory-extension and farmer-first programs, have been very successful in a number of third-world countries. These programs empower farmers by redefining traditional roles and relationships between stakeholders and value the experience of the participants as much as the experimental results.

In the United States, community-based programs have become widely promoted by federal agencies, especially by the EPA through its partnership programs and the USDA through the Sustainable Agriculture Research and Extension Program. Certain foundations such as the Pew Charitable Trust, and California state agencies such as the Department of Pesticide Regulation, Energy Commission, and Water Resources Control Board have also offered significant financial support. Participants in community-based programs may include growers, pest control advisers, nongovernment organizations (NGOs), researchers, county extension staff, regulatory officials, USDA Resource Conservation and Development and Natural Resource and Conservation Service (NRCS) staff, and private companies. In many cases, NGOs have coordinated community-based programs and have been the primary recipients of the institutional funding used to employ staff who provide monitoring services for growers and coordinate grower meetings and other local outreach efforts. They also help to provide small financial incentives to growers, often through cost share funds from NRCS. In other cases, growers' organizations or private enterprises have been the primary organizers.

In California, there are many examples of community-based programs formed in the past decade. Most of these programs have the primary goal of promoting practices that reduce pesticide use. Some of the most notable examples include:

> BIOS (Biologically Integrated Orchard Systems), a demonstration program designed to help almond and walnut growers reduce the use of synthetic pesticides and fertilizers. The BIOS program is coordinated by the Community Alliance with Family Farmers (CAFF), an NGO with a history of environmental activism. The BIOS model has led to the creation of the BIFS (Biologically Integrated Farming Systems) program administered by the UC Sustainable Agriculture Research and Extension Program. BIFS addresses additional crops, and is funded by the California state legislature, EPA, and the California Department of Pesticide Regulation, DPR (Pence, 1998).
>
> Lodi-Woodbridge Winegrape Commission, formed by a grower referendum in 1991. The Lodi-Woodbridge Commission created an IPM program that is led by a coordinator who holds a Ph.D. and has extensive crop consultant experience (Klonsky et al., 1998). It has become a model for other groups including the Central Coast Winegrape Growers Vineyard Team, Napa Valley Sustainable Winegrowing Group, Sonoma Valley Vintners, and Growers Association and Temecula Winegrowers Association.
>
> Campbell's Soup Company and Sun-Maid Growers Cooperative, which promote IPM through demonstration and educational programs led by IPM coordinators employed by each organization. The Campbell's Soup program (Bolkan and Reinert, 1994) has been funded entirely through the company, while the Sun-Maid program has been funded through a gift from the Pew Charitable Trusts. This model has proven very successful for establishing good communication between growers and handlers of their products in the sensitive area of acceptable quality standards.
>
> Pest Management Alliances, which are funded through competitive proposals submitted to California Department of Pesticide Regulation.

Community-based alliances bring together groups that had not always been in good communication with one another to create organizational structures that address the rapidly changing political and regulatory landscape. As most of these programs are publicly supported, long-term program maintenance may prove to be a challenge as public funding to support these programs inevitably wanes. Providing technical support necessary to support them has already become a challenge as resources of the land grants, particularly in IPM programs and Cooperative Extension, become spread among an ever-increasing number of community programs.

WHO ARE THE STAKEHOLDERS?

Taking the Broad Perspective

In addition to students, of course, farmers and allied agricultural industries have been the traditional clientele for the land grant universities and the USDA. Appropriate methods have been developed for information delivery to these targeted audiences and to focus research efforts on solving production problems. However, recent demographic changes have brought about significant shifts in research priorities and program delivery, and there is now an increased emphasis on addressing social and environmental concerns instead of improving agricultural production. Elected officials, primarily responding to urban and suburban constituencies, are creating a new agenda for agriculture through shifts in public policy. Federal and state agencies are implementing these policy shifts and searching for alternatives that would allow farmers to maintain productivity and profitability while meeting the prevailing spirit of environmental quality and food safety legislation. These shifts emphasize stakeholders' involvement in all aspects of information development and delivery.

Who are some of these stakeholders? They include everyone with a special interest in agricultural, environmental, and consumer issues, as well as businesses that can impact any part of that spectrum—a very diverse group indeed. These groups de facto become the expanded group of possible stakeholders for agricultural research and science-based program delivery. In addition, stakeholders must include those agencies and policymakers responsible for creating and implementing legislation and controlling funding.

Is What We Have Here a Failure to Communicate?

The rapid expansion of the market for agricultural production information, beyond that traditionally served by land grant and USDA programs, has made it difficult for these institutions to adequately respond to the diversity of new expectations from stakeholders. On occasion it has been difficult for some of the involved parties to even comprehend the extent of such diversity. Others cannot accept that a vocal segment of society could mistrust technology or openly question science-based results. These questions have sparked debate on issues of science and society. Clearly, there is a need to redefine approaches and stakeholder groups.

Do we expand clientele to include environmental and consumer groups and conduct research intended primarily to help transition growers to practices that are considered less risky? Do we try to serve traditional clientele by developing and delivering science-based programs, primarily those which enhance productivity and profitability? Do we consider public agencies and politicians to be stakeholders? In reality, these decisions have probably already been made for us. All stakeholders affect and influence land grant universities and USDA policies, and therefore influence their information development and delivery programs. Meaningful communication must be achieved in order to help prioritize research, and improve program delivery.

Reaching an expanded clientele is certainly possible through the use of electronic media and the publication of how-to materials, but these are no substitute for frequent, personal contact with the clients. Personal contacts not only establish relationships but also reinforce messages. Without increased staffing of public institutions, new information delivery models will need to be developed to enhance stakeholder interactions. One wonders if the tone of present debates over pesticide use and the use of genetically modified organisms could have been moderated through better communication. Programs such as IPM and others that seek to draw disciplines and stakeholders together to achieve common goals can play an important part in facilitating communication of science-based information. Perhaps improved communication will allow the promise of science-based technologies to become fully realized.

REFERENCES CITED

Bolkan, H. A., and W. R. Reinert. 1994. Developing and Implementing IPM Strategies to Assist Farmers: An Industry Approach. *Plant Disease* 78: 545–50.

Carson, R. 1962. *Silent Spring.* Boston: Houghton Mifflin.

Central Coast Vineyard Team. 1998. Central Coast Vineyard Team Positive Points System. *Practical Winery and Vineyard* (May–June): 12–27.

Food Marketing Institute. 1990. *Trends: Consumer Attitudes and the Supermarket.* Washington, D.C.: Food Marketing Institute.

Kendrick, J. B., and M. W. Stimmann. 1984. Impacts of Chemicals on Agricultural Production and the Environment. In *Beltsville Symposia in Agricultural Research,* vol. 8, *Agricultural Chemicals of the Future,* ed. J. L. Hilton, 37–42. Totowa, N.Y.: Rowman and Allanheld.

Klonsky, K., F. Zalom, M. Chandler, C. Ohmart, C. Elmore, and L. Tourte. 1998. The Lodi-Woodbridge Winegrape Commission: A Framework for Implementing IPM. *Integrated Pest Management Reviews* 3: 1–3.

Lambur, M. T., R. F. Kazmierczak, and E. G. Rajotte. 1989. Analysis of Private Consulting Firms in Integrated Pest Management. *Bulletin of the Entomological Society of America* 35: 5–11.

Michelbacher, A. E. 1945. The Importance of Ecology in Insect Control. *Journal of Economic Entomology* 38: 129–30.

Michelbacher, A. E., and O. G. Bacon. 1952. Walnut Insect Control in Northern California. *Journal of Economic Entomology* 45: 1020–27.

Pence, R. A. 1998. *Leveling the Learning Fields: An Assessment of the Agricultural Partnership Model and Voluntary Pesticide Use Reduction in BIOS-Merced and BIFS-Lodi.* Davis: Department of Human and Community Development, University of California–Davis.

Scheuring, A. F. 1988. *A Sustaining Comradeship: The Story of University of California Cooperative Extension, 1913–1988.* Oakland: University of California Division of Agriculture and Natural Resources Publications.

Zalom, F. G., and J. F. Strand. 1990. Expectations for Computer Decision Aids in IPM. *AI Applications in Natural Resource Management* 4: 53–58.

Chapter 8

Taking the University to the People

The Changing Role of Extension Entomology

ALLEN E. KNUTSON
Texas Cooperative Extension
Texas A&M Research and Extension Center
Dallas

Teaching, research, and extension, these are the missions of the Department of Entomology. While the nature of teaching and research are well understood and indeed practiced by both students and faculty within the department, extension activities and programs are less familiar to most. This is due in part to the limited exposure to extension of many undergraduate and graduate students in and outside the classroom. Outside the classroom, faculty and graduate students have few opportunities to interact with extension entomologists as most of them are based at Texas A&M Research and Extension Centers, and County Extension Offices across the state. Although extension programs may at times be somewhat invisible within the department, they are aimed directly at educating the public about entomology in general, and providing practical solutions to entomological problems in particular.

Before discussing how the role of extension entomology is changing, and in keeping with the historic theme of the Centennial Celebration, I would first like to briefly review the early history of the Cooperative Extension Service. Texas can claim the birthplace of the Cooperative Extension Service, now recognized nationally and internationally as a premier public agency of technology transfer and adult education. Also, this history is of entomological interest because the invasion of the cotton boll weevil (*Anthonomus grandis* Boheman) into Texas and the southeast, and the associated economic problems, were the catalysts for the creation of the Cooperative Extension Service.

Texas, the Boll Weevil, and the Origins of the Cooperative Extension Service

The Cooperative Extension Service is part of the land grant heritage initiated in 1862 with the passage of the Morrill Act by the U.S. Congress. Each state was provided 100,000 acres to be used to establish a college for the purpose of teaching agriculture and the mechanical arts. Texas A&M was organized as the land grant university for the state of Texas. The architects of the land grant system recognized that lives could be changed for the better through education; they recognized that education is a lifelong process, and that access to knowledge is essential to improving the quality of life. Twenty-five years later, the Hatch Act authorized each state to establish an Agricultural Experiment Station in conjunction with the land grant university to conduct research related to agriculture. Thus, by 1887, the missions of teaching and research had been established at land grant universities in many states, including Texas A&M (Hutchison, 1982). The forces that would bring about the third mission, extension, were soon to converge in east Texas.

At the turn of the century, early settlers had moved into and established farms in east and central Texas where cash income from cotton production provided the economic basis for many communities. Railroads were key to the economic viability of the area as they provided transport of cotton crops to markets, and railroad companies were keenly interested in bettering the plight of farmers along their routes. History records that F. B. McKay, general passenger and freight agent for the Texas Midland Railroad, heard of a new USDA program to establish model farms to show growers the most advanced agricultural practices. Mr. McKay wrote a letter to USDA offices in Washington to request that such a project be started somewhere along the Midland route in Texas. His letter fell on the desk of Dr. Seaman A. Knapp, special agent for the USDA in Lake Charles, Louisiana. Dr. Knapp, who was in charge of the model farm program, replied that no funds were available for such a project in Texas. Not to be discouraged, Mr. McKay replied that the Farmer's Institute of Terrell, Texas, was interested in pursuing the idea and would be willing to underwrite any costs incurred. Dr. Knapp agreed to travel by train from Louisiana to meet with these Texas producers. The year was 1903 (Hutchison, 1982; Anonymous, n.d.).

Dr. Seaman Knapp had been a minister, president of a seminary, a farmer, and a publisher before being elected president of the Iowa State Agricultural College in Ames in 1884. However, two years later Knapp resigned his presidency and joined a private concern engaged in the agricultural development of the southwestern part of Louisiana. Here he was successful in encouraging hundreds of families from the Midwest to move to Louisiana to begin growing rice and other crops. One of his most effective methods used demonstration farms to show prospective growers how they could profit by farming in Louisiana. The USDA, seeing the success Dr. Knapp had achieved in the private development enterprise in Louisiana, asked Knapp to organize government-funded demonstration farms to disseminate new agricultural practices. Although Knapp was nearly seventy at the time, he accepted. Thus, as director of the USDA model farm program, Dr. Knapp traveled to Terrell, Texas, in early 1903 (Galloway, 1911; Hutchison, 1982).

Meeting at the Odd Fellows Hall in Terrell, Dr. Knapp told the assembled farmers and businessmen that he had no funds to support a model farm program in Texas, but "if you want to help yourselves, I can work with you, but remember the best help always is self-help." Dr. Knapp proposed to demonstrate the best methods of cultivation, crop rotation, and diversification. However, unlike the government model farms, a cooperating farmer in his own fields would carry out this work and the community would set aside funds to protect the cooperating farmer from any loss. The citizens of Terrell embraced the idea and within twenty-four hours raised $900 and selected the W. C. Porter farm as the site to demonstrate and evaluate these new practices. A block of thirty-seven acres of cotton was divided into nine test plots. Here they tested performance of different varieties, and the use of fertilizer and cultivation practices. There is no reference as to the number of replications or the experimental design used in these trials, but the results were obviously convincing. When all of the harvest was in, Mr. Porter announced that his net profit was $700 more as a result of the improved methods. He relinquished all claims to the money the community has raised and declared his intentions to cultivate his full 800-acre farm following the new methods. The success received wide publicity across east Texas and attracted the attention of railroad executives and politicians, including the Secretary of Agriculture who visited the Porter Farm in the fall of 1903 as part of a tour to study the effects of the boll weevil on Texas cotton production (Anonymous, n.d.).

At this same time, the boll weevil was sweeping across central and east Texas. A contemporary wrote: "when the weevil appeared and destroyed the crop, merchants and bankers refused to make advances, and the farmer found himself without credit, without food and without money." The result was financial and agricultural panic. "Laborers left the country, farms were abandoned, stores closed and disaster was apparent everywhere" (B. Knapp, 1911). Facing economic ruin, farmers and businessmen called for federal help. Hunter and Hinds, two USDA entomologists, published a bulletin in 1904 showing how boll weevil could be managed by cultural practices, including early harvest and most importantly, through early crop-residue destruction in the fall (Hunter and Hinds, 1904). The challenge facing USDA was how to get this information to cotton producers before it was too late.

In January, 1904, the department of the Farmers Cooperative Demonstration Work was organized in the USDA Bureau of Plant Industry. Funds for this new department were provided by an appropriation made by Congress to combat the boll weevil (Anonymous, n.d.). Knapp was appointed head of this effort and was given the charge to show farmers how to profitably produce cotton in the presence of the boll weevil (Galloway, 1911). No doubt recalling the success of the Porter Farm work and working from his headquarters in the Old Masonic Temple in Houston, Knapp employed "special agents" to educate and work with willing farmers to practice these new methods in their own fields. Neighbors visited these demonstration farms to see for themselves and word spread quickly about how to stay in business while living with the weevil. Thus, unlike the early government operated "demonstration" farms, this program involved farmers in implementing the practices on their own farms. In 1904, there were twenty-four agents employed in three states working with 7,000 farmer-demonstrators. As the boll weevil marched across the south, Dr. Knapp's army of special agents grew, showing cotton producers how to combat this pest through cultural methods and how crop diversification could replace income lost to weevil. Work expanded to other southern states where agents instructed farm families in crop diversification, livestock production, and home food production and preservation. The programs in the 1920–30s helped stabilize rural economies and "stemmed the devastation wrought by the cotton boll weevil, that often left one-crop cotton farmers completely poverty-stricken" (Lincoln and Blair, 1977).

The success of the Farmers Cooperative Demonstration Program lead to the Smith-Lever Act in 1914, which established the Cooperative Extension Service as part of the land grant university system of teaching and research. The Act states that Extension is to "aid in the diffusing among the people of the United States useful and practical information on subjects relating to Agriculture and Home Economics and to encourage the application of the same." The following year, Texas legislators established the Extension division at Texas A&M. The U.S. agricultural extension model has been adopted by over 100 countries and has been described as "one of the most widely recognized systems for the diffusion of innovations in the world." Probably no other government or private agency can claim to be more successful in transferring technology (Gustafson, 1991).

Dr. Knapp wrote that the "farmer is a natural doubter." Although the USDA published research results and operated model farms at government expense, these methods were often not effective. He concluded farmers could be best convinced of the value of change through demonstrations carried out by farmers themselves on their own farms. Knapp's concept of experiential learning is evident in his statement that, "what a man hears, he may doubt, what he sees, he may possibly doubt, but what he does, he cannot doubt" (Hutchison, 1982).

The power of demonstrations to introduce new farming practices is as valid today as it was almost 100 years ago. Working with growers to demonstrate, evaluate, and implement new technology and practices on their own farms is still a key method of education in Extension programs, including those of Extension Entomology. It is clear the one-on-one teaching, repeated throughout the season, is the most effective way for growers to become knowledgeable and confident regarding a new production practice. The application of the farm demonstration method has been the foundation for the success of the Integrated Pest Management (IPM) programs initiated in the early 1970s in Texas. Given the principles of field scouting, economic thresholds, and the need for integrating cultural control with biological and chemical control methods, extension entomologists introduced and implemented IPM programs for all major field and forage crops, vegetable and horticultural crops, and livestock. In the past several years, Extension IPM programs have been developed for pest control in schools and for control of fire ants (*Solenopsis invicta* Buren).

In spite of these successes, we must constantly ask ourselves, is extension relevant today? After all in 1900, over 50 percent of U.S. citizens were engaged in farming, and there was a real concern about food security. Today, the proportion of those engaged in farming in the U.S. has shrunk to only 3 percent. Dr. Ed Hiler, Texas A&M Vice Chancellor and Dean of Agricultural and Life Sciences, and Director of the Texas Agricultural Experiment Station and Extension Service, recently confirmed the Agricultural Program's pledge to the land grant philosophy stating, "we are committed to providing access to education, to educational outreach, and to sharing the results of research and discovery in ways that help people solve problems and improve their lives or livelihoods."

Changing Role of the Extension Entomologist in Educating the Public

Change is the nature of extension, especially extension entomology, due to the unpredictable nature of insects. Extension entomologists operate across broad scales from an individual's pest problem to problems of local significance and to those of regional and national importance. Programs must be flexible to meet unexpected challenges and opportunities. Some ways in which the role of the extension entomologist is changing in Texas are described here.

The Agricultural Audience

Texas is an agricultural state. It leads the nation in value of livestock production and ranks fifth in crop sales. There are 195,000 farms and ranches in Texas, and about 40 percent of farm operators work full time at this activity. Almost two-thirds of Texas farms and ranches have gross sales of less than $10,000 per year (TASS, 1998). As farm newspapers and magazines discuss weekly, the economic crisis in farming is only getting worse. Commodity prices for many of the major row crops are falling below the cost of production as farmers compete on a world market. In desperation, some farmers are considering crop insurance costs and returns, rather than market prices or yields, in deciding which crops to plant. Under these circumstances farmers have little incentive in pest management and little profit margin to invest in new and costly technology such as transgenic crop varieties or selective insecticides. Extension entomologists are challenged to develop IPM programs, which minimize input costs.

As a result of the Food Quality Protection Act, the U.S. Environmental Protection Agency (EPA) may soon ban organophosphate and carbamate insecticides, yet these represent some of the least expensive insecticides. One option proposed to provide access to these materials while minimizing their health and environmental risks is through prescription application. Under prescription application regulations, a producer would be required to obtain a written recommendation from a licensed pest control advisor before purchasing and applying a "prescription only" pesticide. Extension entomologists could provide leadership in establishing the necessary guidelines and in training the large number of crop advisors that would be required.

Sales of organically produced products continue to increase, and federal laws defining and regulating organic production have been published in the Federal Register. Extension entomologists will be challenged to provide IPM programs compatible with organic production of crop and livestock products. Similarly, independent groups are certifying farms in some states as IPM farms. IPM certification is a marketing tool, which ensures customers that crops have been produced in a responsible manner using IPM practices. Should IPM certification become a reality in Texas, extension entomologists are well positioned to help define IPM certification standards.

New technologies like remote sensing and fertilizer and pesticide application technologies are bringing precision farming into agriculture. However, the value of these technologies to Texas producers is still unknown. Precision farming is bringing remote sensing and precision application of fertilizers and pesticides into farming, but its value to Texas producers is also yet unknown. If precision agriculture proves valuable, extension entomologists will undoubtedly be involved in modifying IPM programs to take advantage of these technologies. At the opposite extreme of precision agriculture, where the measurement unit is a few square feet within a field, is the concept of area-wide pest management programs operating over multiple farms and counties. The best-known example of an area-wide program is the boll weevil eradication program now underway in many parts of Texas. As with this program, extension entomologists will serve an important role in planning future area-wide programs and educating growers about these programs.

Austin Cargill, the vice president of Cargill, recently stated, "genetically engineered crops will make the industrial revolution pale by comparison." Crops with genetically engineered resistance to insects and expressing engineered insecticides will capture much of the attention of extension entomologists in the future. From the commercialization of transgenic *Bt* cotton (expressing *Bacillus thuringiensis* toxins), we have learned that factors such as plant development, growing conditions, and others, can influence the levels of gene expression and pest control achieved. Extension entomologists will play a key role in evaluating transgenic crops under field conditions and integrating them into effective IPM programs.

The proliferation of entomology consultants working independently and through agrochemical companies and suppliers has raised concerns about the future of extension entomologists. Consulting opportunities for entomologists will continue to grow as agriculture becomes more competitive and complex, and growers seek a level of service beyond that which public extension programs can provide. However, extension entomologists will be needed to support and complement consultants by coordinating insect monitoring networks, organizing community-wide and area-wide pest management programs, conducting on-farm research and educational workshops, and evaluating and introducing new technology. These extension activities are not likely to be assumed by consultants and do not compete with consultant services to individual growers. Rather, extension activities build partnerships between growers, consultants, and extension entomologists.

Biological and cultural control methods, unlike pesticides and transgenic plants, often do not have a commercial value yet are the foundation of many IPM programs. Because these practices lack an immediate economic return, neither industry nor consulting entomologists will invest in their development. Extension entomologists have the unique responsibility of educating the public about biological and cultural control methods (Knutson and Smith, 1997).

The Urban Audience

Despite the economic importance of agriculture to the state, 80 percent of the population of Texas lives in urban areas. Bugs are big business in Texas' cities where urban insect pests such as termites, fire ants, fleas, and many others thrive. To deal with these pests, there are over 13,000 licensed pesticide applicators working for more that 3,000 pest control businesses in the state. As the urban population grows, and insect pest control becomes more sophisticated, the pest control industry will look more and more to extension entomologists to provide training for its employees and for educating the public.

Texans, especially urban Texans, love to garden and care for their landscapes. On any weekend in the Dallas–Ft. Worth Metroplex, for example, there are three lawn and garden radio programs on the air providing advice on plant selection and care, including insect control. In 1991, nearly one-half million listeners tuned in to at least one of these three shows on any given weekend (Tomaso, 1991). This enthusiasm is evident in the success of the statewide Master Gardening Program in which extension entomologists provide training to participants who in return volunteer time in support of county extension programs. As the population continues to age and the baby boomer generation begins to retire, this interest in gardening and landscapes will likely increase. Extension entomologists will be called upon to provide nonchemical or least toxic solutions to garden and landscape pest problems. In addition, clientele are interested not in just pests but also in enjoying and encouraging insects, such as butterflies, fireflies, and dragonflies into their backyards.

Extension entomologists also face the task of mass marketing educational information in the highly competitive and demanding urban setting. Expanded efforts to network with city governments and mass media outlets and organized extension marketing programs will be needed to deliver extension information.

Disseminating Information: The World Wide Web

The Internet is a remarkable avenue for the delivery of extension information and is changing how extension entomologists educate the public. Extension has long relied on written bulletins and other types of conventional publications. These are usually expensive and difficult to keep current due to the time involved in revising, editing, and printing, and the cost of distribution. Electronic documents published using the Internet are less expensive to maintain, can be readily updated, are instantly available, and can reach multiple audiences. The National Agricultural Statistics Service reports that in 1997, 41 percent of Texas farms had computers and 18 percent had Internet access. Only California had broader Internet use with 23 percent of the growers having Internet access (Pocknee et al., 1998). However, since farmers in Texas can just as easily access Web sites at the University of California or University of Florida for entomological information, the Internet also increases the potential for competition. To maintain clientele, extension entomology publications delivered through the Internet must provide high quality information to best fit the needs of Texas' citizens.

Extension Entomologists as Researchers

The old model of integration between research and extension functions of land grant universities noted a clear demarcation of responsibilities. Extension appeared to have to wait for Experiment Station researchers to address critical problems, and until researchers or industry provided solutions, extension could only lament the need for results. However, the situation has dramatically changed since then. For example, Steffey and Gray (1992), stated, "in today's land-grant university system, extension specialists who do not conduct research and publish, jeopardize their job security." Involvement of extension scientists in research programs is a major shift in the traditional roles of extension. This is particularly true in Texas where extension entomologists have only recently been able to receive joint appointments in research and extension.

Why is there a need for extension entomology to be involved in research? The answer is simple, for both practical and professional reasons. First, there are simply not enough research entomologists to address all of the research needs that extension entomologists are asked to address. New technology must be evaluated, modified, and adapted to fit local situations. As an example, transgenic corn varieties were developed in the Midwest for control of the European corn borer (*Ostrinia nubilalis* Huebner). In Texas, the primary stalk borer attacking corn is a different species, the southwestern corn borer (*Diatraea grandiosella* Dyar). Corn growers in Texas needed to know how effective these commercial varieties were against southwestern corn borer. Extension and research entomologists evaluated these varieties in grower fields to answer this question. New insecticides must be evaluated to determine their optimum use (rate, frequency, impact on secondary pests and beneficial arthropods) under local cropping conditions. And, as mentioned earlier, biological and cultural control practices and the use of pheromones for monitoring or mating disruption are not likely to be developed by private enterprise, yet may be key tactics in pest management. The collaboration between extension and research scientists in the design of applied research programs will play an increasingly important role in the development and implementation of these practices.

The second reason for extension entomology to be involved in research is professional. From a practical point of view, involvement of extension entomologists in research and publication is expected in those seeking promotion within their professional ranks. This requirement recognizes that research activities maintain scientists current, stimulates creativity, and improves the abilities of extension agents to deliver information as involvement leads to understanding of and trust in research results. Participation of the extension entomologist in research planning and execution is also important as it brings a greater appreciation of the strengths and weaknesses of the research and its application to problem solving.

Although desirable from a professional and program standpoint, activity in applied research can be challenging. Most extension positions in Texas do not provide laboratory or greenhouse space, access to experiment station field plots, or the necessary funds to support field technicians or research associates. Significant time and effort must be devoted to writing grants and maintaining grant funding, training and supervising research technicians and graduate students, and writing publications. This time commitment greatly competes with the extension entomologist's ability to provide individual service to clientele and county extension agents. Extension entomologists involved in research are thus faced with the challenge of balancing an applied research program with the demands and needs of agents and clientele.

Extension Entomologists as a Source of Unbiased, Science-Based Information

A unique characteristic of the Cooperative Extension Service is that as an agency it represents a source of unbiased, science-based information. This characteristic has helped ensure the survival of the agency during controversial times. Extension entomologists have often found themselves in the proverbial situation "between a rock and a hard place." For instance, the recent and continuing controversy over the boll weevil eradication program in Texas, and claims by some that IPM is nothing more than a change from "spray and count" to "count and spray." Harry S. Smith, a pioneer in biological control at the University of California, wrote, "even entomologists are not proof against the dictates of fashion" (Smith and Flanders, 1931). His reference at that time was to the fervor among entomologists for the parasitoid wasp *Trichogramma* as a biological control agent. The "fashion" today may be transgenic crop varieties, novel insecticide chemistry, or compost tea. Interest in organic pest control is a current example relevant to urban areas where the public is eager for nonchemical pest control, regardless of their efficacy. While extension can be sympathetic to this desire and to the goals of organic pest control, we cannot embrace its methods unless proven by scientific investigations.

Our responsibility to providing science-based information is even more critical today than in the past. Some authors have observed a trend in society toward a general distrust of science and a loss of faith in science and technology as means for the creation of a better world. May Berenbaum has stated that the general public has an alternate reality of what entomology and science are. She observed that movies and television programs shape the public's image of science and reflect society's concerns. Extension entomologists have a unique opportunity to demonstrate the power of the scientific method in finding practical solutions to problems. The increasing need to provide unbiased, science-based entomological information will continue to be the hallmark of extension entomologists in educating the public about one of our most formidable and interesting competitors on earth, the insects.

REFERENCES CITED

Anonymous. n.d. Extension's Root in Texas. Texas Agricultural Extension Service Miscellaneous Publication.

Galloway, B. T. 1911. Seaman Asahel Knapp. In USDA Yearbook of Agriculture, 151–54. Washington D.C.: U.S. Government Printing Office.

Gustafson, D. J. 1991. The Challenge of Connecting Priorities to Performance: States' Response to the Forces for Change in U.S. Extension. In *Agricultural Extension: Worldwide Institutional Evolution and Forces for Change,* ed. W. M. Rivera and D. J. Gustafson. New York: Elsevier.

Hunter W. D., and W. E. Hinds. 1904. The Mexican Boll Weevil. USDA Division of Entomology Bulletin 45. Washington D.C.: U.S. Government Printing Office.

Hutchison, J. E. 1982. The Texas Agricultural Extension Service: A Historical Overview. In *Southwestern Agriculture: Pre-Columbian to Modern,* ed. H. C. Dethloff and R. M. May Jr., 124–35. College Station: Texas A&M University Press.

Knapp, B. 1911. Some Results of the Farmer's Cooperative Demonstration World. In USDA Yearbook of Agriculture, 285–96. Washington D.C.: U.S. Government Printing Office.

Knutson, A., and J. W. Smith. 1998. Biological Control Program, 1996–97. Report. Texas Agricultural Extension Service Miscellaneous Publication.

Lincoln, C., and B. D. Blair. 1977. Extension Entomology: A Critique. *Annual Review of Entomology* 22: 139–55.

Pocknee, S. A., M. Wrona, M. Lange, K. Brinkley, and D. Vivien. 1998. The Internet as a Farming Tool. *Cotton Physiology Today* 9.

Smith, H. S., and S. E. Flanders. 1931. Is *Trichogramma* Becoming a Fad? *Journal of Economic Entomology* 24: 666–72.

Steffey, K. L., and M. E. Gray. 1992. Extension-Research Synergism: Enhancing the Continuum from Discovery to Delivery. *American Entomologist* 38: 204–205.

Tomaso, B. 1991. The Garden Gurus. *Dallas Morning News* (Dallas Life Magazine), May 26, 1991.

TASS (Texas Agricultural Statistics Service) 1998. Texas Agricultural Statistics 1997. Austin: United States Department of Agriculture and Texas Department of Agriculture.

Chapter 9

Technology Transfer

The Art and Science

THOMAS W. FUCHS
Texas Cooperative Extension
San Angelo

The land grant university concept, developed in the nineteenth century and formally established in 1862, was one of the greatest and most valuable concepts ever conceived. I believe that it is as relevant today as it was then. Dr. Ed Hiler, Texas A&M University Vice Chancellor for Agriculture and Life Sciences, in his address to the Texas A&M University System Agricultural Program Conference in January 1999, stated that "our commitment to the land grant philosophy means that we are committed to providing access to education, to educational outreach and to sharing results of research and discovery in ways that help people solve problems and improve their lives or livelihoods." An integral and critical component of this system is transferring technology from where it is developed to where it is put to use to improve the lives or livelihoods of clientele who support the land grant system. The scope of this discussion will be limited to the interface of technology developers and technology extenders and the extension of technology to end-users. I will use technology transfer in row crop agriculture as the model since that is where most of my knowledge and experience lie.

THEORETICAL BASIS FOR TECHNOLOGY TRANSFER

A strong theoretical connection exists between the transfer of technology and behavioral sociology. Several behavioral models are used to make the connection. One such model is Bennett's KASA model, which states that people adopt new practices or technology after progressing through a systematic hierarchy of changes in knowledge, attitudes, skills, and aspirations (Bennett, 1977). The KASA model recognizes that mere development and availability of innovative technology in no way guarantees that it will be used, nor is technical training that adds to a client's knowledge and skills by itself sufficient to bring about adoption. Theories of behavioral motivation, models for adult learning, and principles of communication are all germane to this topic but beyond the scope of this discussion.

Much of the literature regarding innovation and technology transfer is based on diffusion theory. Diffusion in this context was defined by Rogers (1995) as the process by which an innovation is communicated through channels over time among members of a social system. Rogers (1983) identified five attributes that most affect the adoption of a new idea. These include *relative advantage,* or why the innovation is better than the current way of doing things in terms of profitability, prestige, convenience and satisfaction; *compatibility,* how the innovation matches the experience, values, and needs of intended users; *complexity,* the level of difficulty required to understand and use the innovation; *trialability,* the degree to which an innovation can be tried on an experimental or trial basis; and *observability,* or how visible the results of the innovation are to others. Copp et al. (1958) identified the stages in the adoption process as follows: (1) *awareness,* where the client hears of the innovation; (2) *interest,* where the farmer, for example, feels the innovation is a workable solution to a farm problem; (3) *acceptance,* where the farmer feels the innovation would be useful on his or her farm; (4) *trial,* where the farmer tries the innovation on his or her farm; and (5) *adoption,* when the farmer continues to use the innovation.

The distribution of farmers among five categories of innovation adoption approaches a normal population distribution (Fliegel, 1993; Rogers, 1995). Farmers were classified into groups based on the time of adoption of innovations (Figure 9.1). They described farmers more than two standard deviations earlier in time of adoption than the average farmer as *innovators* and those between one and two standard deviations earlier than average as *early adopters.* Farmers within one standard deviation earlier or later than average were labeled as *early* and *late majority,* respectively. Finally, those more than one standard deviation later than average in adoption were labeled *late adopters or laggards.* One problem with this classification is that it ignores those who never adopt the innovation. Rarely is 100 percent adoption of technology achieved.

Each of the adopter categories exhibits certain characteristics or values.

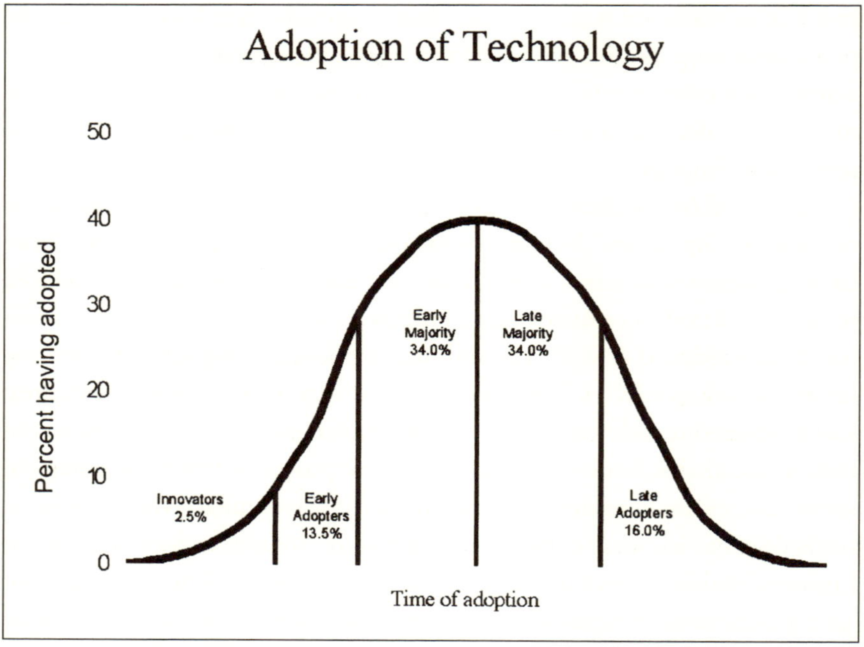

Figure 9.1. Distribution of farmers among five categories according to time of adoption. Adapted from North Central Rural Sociology Committee, Subcommittee for the Study of Diffusion of Farm Practices. 1961. North Central Regional Extension Publication, no.13. East Lansing: Michigan State University Cooperative Extension Service.

Innovators, for example, are often called venturesome, daring, or risk takers. Innovators may not be highly respected by members of the local system but they play an important part in the diffusion of innovations by acting as a gatekeeper in the flow of new ideas into a production region. Innovators will try almost any new technology and strive to be the first to do so. Early adopters tend to be highly respected opinion leaders in an area. They tend to have large farms, be well educated, and have more financial resources than later adopters. Potential adopters often look to early adopters for advice and information about the innovation because they will generally make the innovation work. Early majority adopters tend to be very deliberate but usually adopt innovations just before average members of the system. Members of this group tend to interact frequently with peers but are seldom the real opinion leaders. Late majority adopters adopt new ideas just after the average member of the system. They tend to be skeptical and may adopt only when adoption becomes an economic necessity for them or after considerable peer pressure to adopt. Finally, late adopters or laggards are the last to adopt. This group usually looks back at the way things have been done in the past and are very reluctant to change. They tend to be suspicious of both new innovations and the change agents attempting to implement innovations. They must be virtually certain the new idea will not fail before they are willing to adopt. They possess almost no opinion leadership.

The success of a change agent in bringing about adoption of innovations by clients is positively related to the extent that he or she works with opinion leaders or early adopters of technology (Rogers, 1995). The Cooperative Extension Service currently and historically spends a great deal of time and effort in identifying opinion leaders and early adopters and has been very successful in this effort. Rogers (1995) states, "the Agricultural Extension Service of the United States is reported to be the world's most successful change agency. Certainly it is the most admired and widely copied." The challenge for Extension and other change agencies is to maintain this reputation in this era of rapid change and innovation in agriculture.

Because the adoption of an innovation follows a normal, bell-shaped curve when plotted over time on a frequency basis, when the accumulative numbers of adopters is plotted over time, the result is an S-shaped (or sigmoid) curve as seen in figure 9.2 (Rogers, 1995; Dent, 1998). This curve is also very familiar to biological scientists as is the normal bell-shaped curve. In the

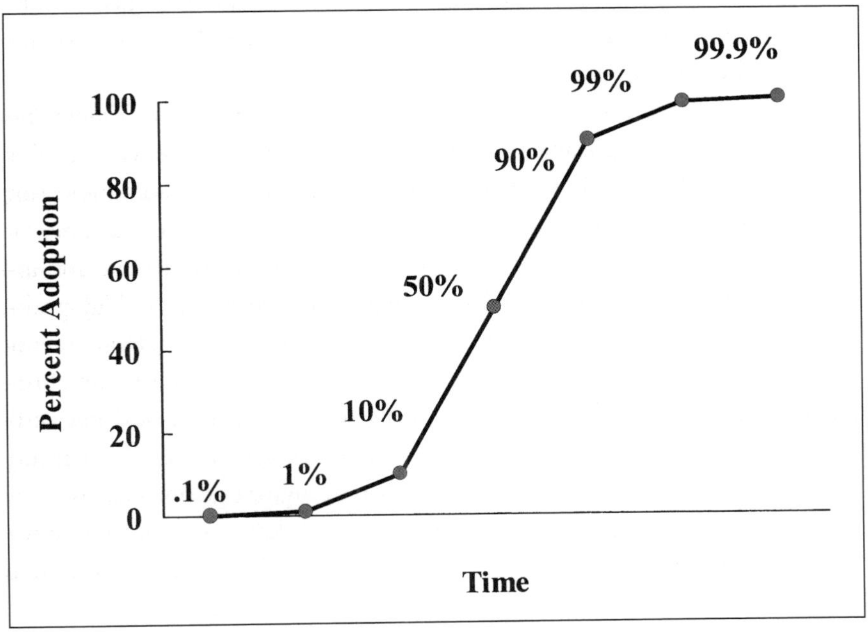

Figure 9.2. Accumulative percent adoption of innovations through time. Adapted from Dent, H. S., Jr. 1999. *The Great Boom Ahead.* New York: Simon and Schuster.

biological literature, the S-shaped curve is frequently called the logistic growth curve to describe increases in populations over time until the carrying capacity of the resource is reached. The S-shaped adopter distribution curve indicates that early adoption tends to proceed at a relatively slow rate. After approximately 10 to 20 percent have adopted, the rate accelerates until 80 to 90 percent of the individuals who will adopt have done so. Then adoption increases at a gradually slower rate as fewer and fewer remaining individuals adopt the innovation. Conventional wisdom indicates that in addition to the theoretical consideration there are numerous practical constraints to adoption of innovations. These include lack of resources, limited access to information, aversion to risks, inadequate farm size to justify equipment needed, insufficient human capital, and inappropriate infrastructure (Feder et al., 1985).

IMPORTANCE OF THE RESEARCH/EXTENSION/CONSULTANT INTERFACE

Using the land grant model in which research develops technology and Extension extends the technology to end-users, I would like to address the importance of the interface of these groups in technology development and transfer. In addition, we must consider independent consultants since this group is increasingly important in transferring agricultural technology and generally has good working relationships with research and Extension faculty. Although commercial consultants, those associated with product sales, are important to technology transfer, much of their impact is with technology surrounding proprietary products rather than those developed directly through land grant institutions, and their contributions will not be considered here.

During my years working in a land grant institution, I have often heard it said from research faculty that innovations they developed have not been properly extended to end-users and thus have not had the impact they justify. Conversely, I have heard from those extending the information that the new technology worked only under a limited set of conditions, was not practical, or was not nearly ready to extend to end-users. Who is correct? I think that generally both groups are correct. Too often researchers seem to believe that once the journal article is accepted for publication, their job is finished. Rarely is the technology ready for implementation at that time. This philosophy is particularly true in today's research environment where research is largely supported by one- or two-year grants and often conducted only in the laboratory, greenhouse, or in small field plots. Rarely has technology been tested under the wide diversity of environmental conditions or production systems under which growers might attempt to implement the technology. I believe that there is a plausible solution to this dilemma. First, the expectation for researchers should be that their job is not complete until the technology is well tested at the end-user level. Extension specialists should be expected to test technology at the regional and/or local levels and be provided resources to do so. The researcher and extender should work together not only during the development phase but also for one to two production seasons in varied production systems to test the innovation under a wider array of conditions. This would allow time for the extender to fully understand the intricacies of the technology and the researcher to more fully understand the environmental conditions and production systems under which the technology will be required to operate. There are currently significant constraints to this happening.

Two changes to the traditional land grant university model, however, are currently facilitating the process. One is the increasing role of extension specialists and specialty agents such as IPM agents in conducting adaptive research in cooperation with research scientists. The second is the increasing number of joint research/extension appointments in many universities including Texas A&M University. Both of these changes are facilitating technology transfer to end-users.

ATTRIBUTES OF INNOVATIONS FOR SUCCESSFUL IMPLEMENTATION

In addition to theoretical considerations of the attributes of innovations discussed previously, there are a number of practical considerations relative to successful implementation. For a technology to be readily adopted by clientele it must be something that is genuinely needed by the clients. This is the reason that Extension spends so much time and energy with end-user groups in listening to concerns and prioritizing clientele needs. The recent Texas Communities Futures Forum held in each Texas county by the Texas Agricultural Extension Service is one example of this type of activity where clientele groups, in facilitated workshops, were asked to identify and prioritize issues most important to them. Extension does this continually. Each of the extension IPM units in Texas, for example, has a steering committee made up of growers, agribusiness representatives, and often, private consultants who help identify and prioritize critical issues and goals for the program. Priorities include topics to be addressed at educational meetings, seminars, and workshops and priorities for applied research and demonstration efforts. The land grant university system should have a system in place that allows for this input to be brought to the attention of researchers who are developing technology.

While such a system has been attempted in the entomology department at Texas A&M University through activities of the Entomology Science Conference, efforts have been marginally successful. Lack of base funding for research projects that has eroded through the past twenty years in favor of grant- and grantor-driven funding has been the primary cause of its ineffectiveness. Often research programs out of necessity respond more to what is fundable than to that which clientele identify as priority problems. This is a weakness in the current land grant university system.

Research is assumed to add to the stock of scientific and technical knowledge that presumably is always appropriate to the needs of potential end-users (Kalaitzandonakes and Bullock, 1998). This is not always the case. Compuware Corporation in the 1999 Annual Report to stockholders stated, "Our strategy is to provide practical solutions that deliver immediate and measurable returns to our clients" (Anonymous, 1999). Perhaps that would be an appropriate strategy for the land grant university system also.

The technology to be transferred and implemented must provide benefit to the grower over current practices. Benefits may be in terms of increased yields, reduced inputs, increased net profits, reduced risks, more convenience, reduced environmental impact, or a number of other advantages, but there must be some measurable advantage over what is currently done. This forms the incentive to implement.

In order to be successfully adopted, the innovation must be compatible with the current production system, that is, it must fit into the current production system or add enough value to economically justify the necessary changes in the production system. Changes in the production system are usually not easily made and may necessitate different crop rotations, added equipment, or increased labor requirements.

The efficacy and cost effectiveness of the technology must be evaluated over time and space with early adopters and under local conditions. This is one of the strengths of Cooperative Extension Services that maintain faculty at the local level who can screen and demonstrate appropriate new technology under localized conditions. One must be able to present data to growers indicating that the technology is effective and that it will make them money. Data generated locally plus knowledge acquired through educational programs often result in the information growers require to make meaningful changes.

States that have opted to reduce county-level positions due to budgetary constraints have largely suffered in grower, commodity group, and industry support as a result.

Growers are currently very concerned about production risks and are very reluctant to adopt technology that they believe will increase risks. New technology is often viewed by producers as increasing risks unless local data indicate otherwise. Those helping to implement technology must be willing to work closely with growers through the entire implementation process to reassure him or her that things are going as planned or be willing and ready to make modifications necessary to insure acceptable production.

ATTRIBUTES AND SKILLS OF SUCCESSFUL CHANGE AGENTS

The transfer of technology from developers to end-users is both an art and a science. As with any other occupation some change agents are more successful than others. What characteristic do successful change agents possess and what makes them successful? Having both participated and observed others in implementing technology, I have noted most successful change agents have certain common characteristics.

Let us first define a change agent. A change agent is an individual who influences client's innovation decisions in a way deemed desirable by the change agency (Rogers, 1995). A change agent generally seeks to secure adoption of certain new ideas or practices. Change agents may include researchers, extension specialists, county extension agents, consultants, and others.

The first skill that most successful change agents have is that of being a good communicator. Communication in this context is the process in which participants create and share information with one another in order to reach mutual understanding. Communication includes both listening and exchanging information. Good change agents are not always good public speakers but almost always they are good communicators. A good change agent can communicate technical information in the language and in ways that clients understand. He or she has the skill to take complex information and make it easily understood.

Good change agents know and understand clientele needs. They must recognize the need for change and be able to point out alternatives to existing problems. Equally important is that clientele must respect and have full confidence in the change agent. They must believe that the person understands their needs, production system, local soil and weather conditions, and limitations, and have confidence that the change agent has their best interest in mind. A good change agent, then, views the production system from the producer perspective keeping in mind the producer's yield goals, crop mix, crop insurance status, farm program considerations, landlord relationships, financial status, and other important considerations. One of the considerations that make extension agents–IPM so effective in implementing change is that they live and work in the same communities as clientele, and they become trusted members of the community. Clientele know that they will be there to help the producer work through any problems or shortcomings of the technology.

Good change agents have strong technical backgrounds and a thorough understanding of the technology they are attempting to transfer. They are able to answer theoretical and practical questions about the technology. They must be able to use both deductive and inductive reasoning. Change agents are often asked to extend the data beyond published research, that is, to use their expertise and training to provide expert opinion beyond areas where data currently exist.

Successful change agents are aware that timing is critically important. Not only is recognition of the need to change important but also there are times when producers are more attentive to suggested changes than others. A good change agent will be aware of the proper time to suggest the change and spend time and energy preparing clientele to be receptive to the change.

METHODOLOGY IN TRANSFERRING TECHNOLOGY

The proportion of knowledge that is effectively transferred is dependent on the conditions and methods of transfer. A variety of modes and mechanisms can be used to transfer knowledge with varying degrees of effectiveness (Kalaitzandonakes and Bullock, 1998). These include passive methods such as technical reports, newsletters, newspaper articles, and the Internet, and active transfer such as presentations at meetings, seminars and workshops, and collaborative efforts such as field meetings, applied research plots, or demonstrations. Research shows that the more participatory the mode of knowledge exchange is, the higher the proportion of knowledge that is transferred (Kalaitzandonakes and Bullock, 1998). This is a powerful model for technology transfer and perhaps has application to our teaching methodology in the classroom also.

Passive methods are effective in creating interest, awareness, and perhaps acceptance that the technology might be useful on a client's farm. However, in order to get the technology adopted, more active methods are usually necessary. Personal consultations with growers where the change agent extends his or her knowledge and discusses data from the fields in question to assist the grower in making management decisions is a powerful exchange of information.

I believe that applied research and demonstrations are the most powerful and effective tools available to transfer technology. The technology a producer witnesses and participates in on his or her own farm makes an immediate and lasting impact. A skeptical producer after participating in a successful demonstration often becomes a highly effective promoter of the same technology. An added advantage to applied research and demonstration activities on producer farms is a high likelihood that the producer will suggest improvements either in the technology itself or in the way it is applied to the farm.

CURRENT TRENDS THAT IMPACT TRANSFER OF TECHNOLOGY

Several current trends will have significant impact on technology transfer in the future. We need to recognize these trends and develop strategies that will allow public institutions to be successful as changes occur in agriculture. Over the past several decades, agricultural production has become increasingly science-based and complex. As this trend continues, and as agriculture becomes further vertically integrated, the demand for technology "packages" is likely to increase. Producers will demand more systems-based research and technology transfer. Since research on technology packages is more complex, requires more interaction among disciplines and with clientele, and perhaps is likely to result in fewer journal publications, research scientists may be more likely than ever to turn toward more basic research and disciplinary science that they understand better and over which they have more control. This is likely to be perceived negatively by producer clientele and may reduce the willingness of farm organizations and commodity groups to fund research projects. Additionally, more quality science is currently being generated in private industries than in the past. Private sector investment in agricultural research is reported to be 1.67 times that of the public sector (Huffman and Evenson, 1993). Much of this science leads to proprietary products, knowledge, and information.

Providers of key farm inputs such as pharmaceuticals, chemicals, and genetics have become critical suppliers of information along with these inputs, leaving the traditional Extension Service and land grant systems at a significant disadvantage in terms of providing the latest technology and information. Similarly, private consultants and farm cooperative personnel have become increasingly important in providing information to producers (Ortman et al., 1993).

The combination of two trends, increased private and less public involvement in farming systems, represents privatization of production information services (Wolf, 1995). A trend toward privatization of Extension now being observed in parts of the United States and is more advanced in certain countries such as England and New Zealand. While this transition has not been very successful in terms of continued public support, the trend appears to be continuing. A somewhat similar trend is being predicted for research in public agricultural institutions as was pointed out by Steven C. Blank in his book *The End of Agriculture in the American Portfolio* (1998).

Because of the dominant role of the public sector in providing information to producers in the past, information was common property and data not proprietary or confidential (Boehlje, 1998). However, as suppliers, consultants, and service firms collect an increasing amount of data in production agriculture, and the private sector continues to play a larger role in providing both data and information, private property rights replace common property concepts. Private property rights enable individuals who have those rights to capture value, that is, dollars from those who use that property. There is also an increasing trend for input suppliers to hire some of the best scientists employed by land grant universities including those who transfer technology to the public. This brings instant recognition and credibility to the supplier while allowing the supplier to limit access to proprietary information to paying customers.

Significant structural changes are also occurring in agriculture. Not only is farm size and ownership changing but also the level of linkages and coordination of farm production with input suppliers and businesses that purchase farm products is also changing rapidly. More business is being done through personally negotiated contracts or ownership arrangements rather than in the open market. Although numerous forces and drivers are contributing, information and knowledge are playing an important role. Many of these large vertically integrated farms are not willing to share information with public-sector specialists who may allow public access to the information. As in other industries characterized by contractual or ownership linkages, those individuals with unique and accurate information and knowledge have increasing power and control in the production system. Since much of the technology and information transferred by extension specialists has historically come from state-of-the-art production systems, this is a significant change.

Another important implication of the increased privatization of information for public sector change agents will be more questions regarding open access to information, especially information based on projects funded by private industry. As more information is privatized, there will also be increased questions regarding the appropriate role of public-sector institutions. The private sector has incentive to provide information only when it can extract value or dollars for doing so, yet society may judge it inappropriate to discriminate in terms of access to information. This was, in fact, part of the original justification for the development of the land grant system.

WHAT WILL BE THE ROLE OF TECHNOLOGY TRANSFER SPECIALISTS IN THE FUTURE?

I believe that two roles will continue to be critical, the role of providing unbiased, science-based information and the role of localized adaptive research. Unbiased, science-based information will continue to be valuable to producers and to the general public including policy-makers. Private companies will provide information only where they have vested interests or where they can get financial returns. Even in these areas, growers and the public will continue to look toward public institutions as the unbiased source of information. Also, areas that have to do with public policy, the public good, and common access resources such as the environment, wildlife, clean air and water, and public recreational resources will continue to depend on public institutions.

The role of conducting localized research will be critical to growers, especially in areas not specifically targeted by vertically integrated conglomerates. One example is that of companies producing transgenic varieties. Texas, for example, was not targeted as a primary target area for *Bt* transgenic cotton and, as a result, most transgenic varieties were not well adapted for Texas conditions. There are also questions relative to gene expression levels under some of the physiological stresses to which Texas cotton is exposed. Only through localized adaptive research were well-adapted varieties identified and cost effectiveness of the technology determined.

REFERENCES CITED

Anonymous. 1999. Compuware Corporation 1999 Annual Report. Farmington Hills, Mich.: Compuware Corporation.

Bennett, C. F. 1977. Analyzing Impacts of Extension Programs. Publication ESC-575, Extension Service. Washington, D.C.: United States Department of Agriculture.

Blank, S. C. 1998. *The End of Agriculture in the American Portfolio.* Westport, Conn.: Quorum Books.

Boehlje, M. 1998. Information and Technology Transfer in Agriculture: The Role of the Public and Private Sectors. In *Privatization of Information and Agricultural Industrialization,* ed. S. A. Wolf. Boca Raton, Fla.: CRC Press.

Copp, J. H., M. L. Sill, and E. J. Brown. 1958. The Function of Information Sources in the Farm Practice Adoption Process. *Rural Sociology* 23: 146–57.

Dent, H. S., Jr. 1998. *The Roaring 2000s Investor: Strategies for the Life You Want.* New York: Simon and Schuster.

Feder, G., R. Just, and D. Zilberman. 1985. Adoption of Agricultural Innovations in Developing Countries: A Survey. *Economic Development and Cultural Change* 33: 255–98.

Fliegel, F. C. 1993. *Diffusion Research in Rural Sociology: The Record and Prospects for the Future.* Westport, Conn.: Greenwood Press.

Huffman, W., and R. Evenson. 1993. *Science for Agriculture.* Ames: Iowa State University Press.

Kalaitzandonakes, N., and J. B. Bullock. 1998. Technology and Information Transfer in U.S. Agriculture. In *Privatization of Information and Agricultural Industrialization,* ed., S. A. Wolf. Boca Raton, Fla.: CRC Press.

Ortman, G., G. F. Patrick, W. N. Musser, and D. H. Doster. 1993. Use of Private Consultants and Other Sources of Information by Large Cornbelt Farmers. *Agribusiness: An International Journal* 9: 391–402.

Rogers, E. M. 1983. *Diffusion of Innovations,* 3d ed. New York: Free Press.

———. 1995. *Diffusion of Innovations,* 4th ed. New York: Free Press.

Wolf, S. 1995. Cropping Systems and Conservation Policy: The Roles of Agrichemical Dealers and Independent Crop Consultants. *Journal of Soil and Water Conservation* 50: 263–69.

Chapter 10

Adoption of IPM by End-Users

Methods and Motivation

JAMES P. TETTE, CURTIS H. PETZOLDT, CARRIE KOPLINKA-LOEHR, MARGARET HAINING COWLES, J. KEITH WALDRON, JOSEPH KOVACH, AND KAREN ENGLISH
New York State IPM Program
Cornell University

What motivates an end-user to adopt Integrated Pest Management (IPM) principles and practices? We believe that there are many forces leading to adoption besides the knowledge and methods developed in the science-based programs of land grant universities. Is the adoption rate affected significantly by the form of delivery of science-based information from these universities to end-users? To some extent, yes, but two factors have significantly changed the face of information delivery to the detriment of critical educational methods: the downsizing of Cooperative Extension and the desire for federal funds to be shared across states and regions. Are the major motivators regulations, or higher profitability, or something else? These questions, as well as approaches to insure grower adoption of IPM principles and practices, have been explored over the past thirty years as part of an ambitious IPM Program in New York, at Cornell University, the state's land grant institution. We hope to share our experiences with farmers' adoption of IPM, specifically, motivating factors and effective ways of delivering information.

BUILD CONFIDENCE AND A PROCESS FIRST

When land grant universities accepted the challenge of educating farmers in IPM practices, the Cooperative Extension arm was given the charge of implementing IPM. What did that mean? For us, it meant that Extension needed to engage in at least four steps of the implementation process: both *evaluation* and *validation* of new IPM research knowledge on commercial farms; *demonstration* of the new knowledge to commercial farmers; and *engagement* in the process of helping the farmers adopt the new knowledge. These steps, which were central to building confidence among Extension and farmer groups, were often undertaken as separate steps and sometimes they were addressed simultaneously. The confidence building among Extension educators and farmers clearly outweighed the importance of providing new management options from research.

AMERICAN CULTURE INFLUENCES ADOPTION

Part of our task in New York has been to alter the way Americans approach solutions to problems. In agriculture, just as in the field of human health, our culture continues to seek the silver bullet to cure its ills. So, to some extent, a land grant university can't seem to make an impact with new methods unless they are very simple to employ. Scientists can create all types of pest management options only to find that forces outside their control can cause huge shifts in adoption. We saw this in potato pest management, where, after ten years of intense research efforts to develop an array of options for managing the Colorado potato beetle, farmers have chosen instead to use the newest insecticide on the market.

DELIVER INFORMATION IN OLD AND NEW WAYS

Thoughtful use of both old and new delivery methods has spurred a higher level of IPM adoption. On the one hand, for example, we have found that traditional one-on-one interactions have greatly increased growers' confidence and experience with IPM methods. Those working in educational programs such as nutrient management have found similar results through one-on-one interactions (Shephard, 1999). On the other hand, in our experience, the use of electronic technology can supplement and sometimes replace face-to-face education. Daily fax messages, e-mail messages to farmers, or Web sites that can automatically retrieve and process data are means of relaying information that have interested farmers in IPM practices more than traditional means have.

INVOLVE STAKEHOLDERS

Another vital motivating force has been to involve farmers and other stakeholders in setting IPM research and outreach priorities for the years ahead. Stakeholder or end-user involvement has been critical to success in New York's IPM programs and in other sustainable efforts as well (Shephard, 1999). This process is yearly and ongoing. Similarly, the involvement of these same people in IPM implementation projects and in continuous educational efforts has fostered greater adoption. Cornell University recently included farmers and others in the process of defining IPM through a set of IPM elements or guidelines. The elements for each of sixteen crops in New York have been derived from our science base and proven in our Extension on-farm evaluations. Each year farmers join with research and extension scientists to discuss and negotiate a new set of IPM elements, or to revise an existing set.

A typical set of IPM elements, those used for strawberries in New York, is shown in appendix 10.1. In this example one can see the contributions of horticultural scientists, weed scientists, plant pathologists, and entomologists. What can't be seen in any current list of elements are the new options that are in some phase of implementation. Many of these new options are considered in yearly discussions and some of them are eventually added to a current set of elements. These elements allow New Yorkers to clearly define IPM for a given crop, and using these definitions, document more precisely the extent of IPM adoption.

FOUR FORCES MOTIVATE GROWERS

Some of us in New York feel that despite efforts to engage farmers in defining IPM elements and implementing on-farm research and educational projects, many farmers will adopt IPM methods only at a very slow pace. Nonetheless, we have observed that if the end-user is motivated, knowledge will be adopted. Thus, it may be that success or failure of an IPM program is related to motivation of the particular end-user. In our experience we have observed at least four forces that have caused farmers to adopt IPM practices in rapid fashion: government regulations, profitability, consumers' avoidance, and marketplace interests.

Government Regulations

Impending government regulations are an agent for change in farmers' pest management practices. There is no doubt that the implementation of the Food Quality Protection Act (FQPA), for example, has profoundly influenced farmer pesticide use, leading to greater adoption of IPM practices. In New York, we are seeing regulations that require prior notification of pesticide applications, pesticide phase-out regulations, and outright efforts to ban pesticide use. Most of these regulations have come about because of individuals' concern over pesticide residues in water and exposure to pesticide residues. Similar concerns challenge agriculture in many parts of the nation (Zalom et al., 1999).

Profitability

The potential for higher profitability has been an obvious motivating force from the very first days of IPM. Any time farmers can see that an IPM method may reduce costs or time, or improve product quality, they are very likely to adopt that method.

Consumers' Avoidance

Another interesting motivator is what has been called "consumer avoidance" (Zalom et al., 1999). This concept is best illustrated by the ongoing efforts of Gerber Baby Foods to assure customers that their products are free of certain pesticide residues. Gerber lost a significant share of the market when consumers, reacting to environmental reports of the dangers of pesticide residues affecting children, ceased purchasing Gerber products (Hether, 1999). These factors, whether emotional or based in fact, have led both Gerber and its growers to rapidly adopt IPM practices. Gerber has been willing to share the cost of IPM adoption with the farmers.

Marketplace Interests

Our recent experience with still another motivating force, one of an expressed preference by the marketplace for IPM products, has probably been the strongest motivator to date. This is a very different pressure from the Gerber scenario in that it is a proactive movement from outside the farming community rather than a defensive response on the part of farmers or food processors. In New York we currently have supermarkets and food processors guaranteeing farmers a market share if they will grow according to sets of IPM elements. These same groups and some individual farmers are also using their own resources to ensure that they are meeting the requirements associated with the adoption of IPM practices, primarily through record keeping. In addition, a supermarket chain and a food processor in New York are using their corporate advertising dollars to tell the public about IPM. Their efforts are greatly aiding the educational endeavors of New York's land grant university.

Measuring IPM Adoption

In the past few years, New York's IPM Program teamed up with the state-level USDA Agricultural Statistics Service to determine what motivates farmers to adopt IPM and how long the knowledge shared in Cornell Cooperative Extension programs is retained. Questions regarding adoption of IPM were based on a set of IPM elements. Results were depicted on a continuum that runs from chemically intensive pest management to more sustainable pest management systems (Benbrook, 1996). The further along one is on the continuum, the greater is one's adoption of IPM. In the first example, figure 10.1, a survey of field crop growers in 1997 revealed that adoption increased, as did the extent of education of farmers, with consistent involvement with our programs (Waldron, 1998). Survey respondents who had participated in Cornell Cooperative Extension education in IPM and in an intensive Extension education program called "Tactical Agriculture Teams" ("TAg" Teams) were the most advanced in IPM adoption.

Similarly, a 1998 survey with strawberry growers showed significant adoption of IPM methods (Figure 10.2) and even greater adoption of IPM methods when some of those farmers decided to identify their product in the marketplace as IPM-grown (J. Kovach, personal communication). The interplay of marketplace forces and intensive educational efforts through Cooperative Extension is evident in this figure. The New York State Berry Growers Asso-

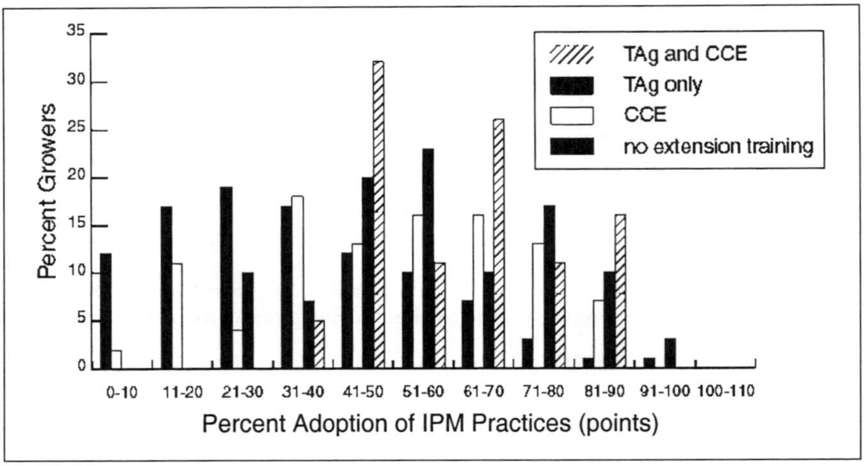

Figure 10.1. Data from statewide field crop survey show that farmers' adoption of IPM practices correlates positively with education through the IPM-based TAg ("tactical agriculture") program and contact with Cornell Cooperative Extension.

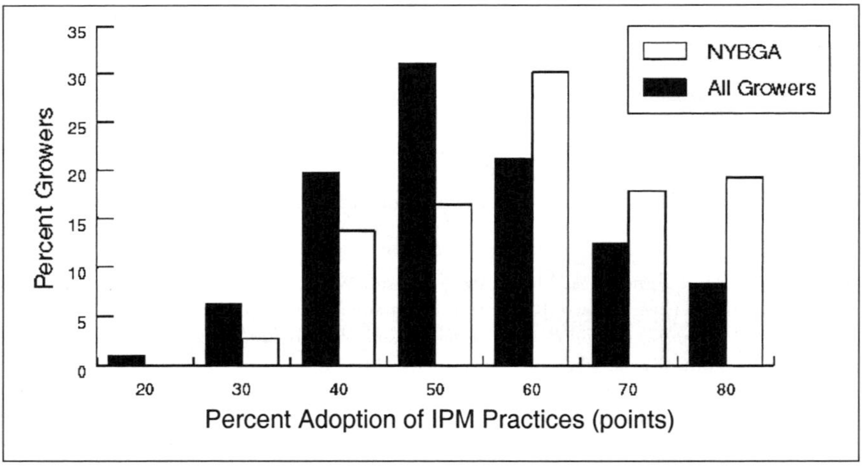

Figure 10.2. Percent adoption of IPM methods among New York strawberry growers in 1998.

ciation offered its members the opportunity to provide an IPM label on their products. Many members chose to participate in this program, and the association sponsored special educational sessions, conducted by Cornell Cooperative Extension, to help its members understand the application of a set of elements. The association also helped these farmers meet the requirements for third-party verification of adherence to the elements. The data clearly show that IPM adoption is higher when both marketplace factors and more intensive education are present.

A third example: fresh-market sweet corn farmers were surveyed in 1996, answering questions based on a set of IPM elements (Petzoldt, personal communication). The progress in adoption was judged to be significant, given that extension educational efforts directed at these farmers were only two years old and did not reach into all areas of the state (Figure 10.3). Concurrently, the supermarket that encouraged its growers to provide sweet corn grown under a set of IPM elements determined that all participating farmers adopted IPM almost completely (> 80 percent on the continuum) in just one year. Today this supermarket continues to cosponsor IPM educational workshops with Cooperative Extension to teach its farmer-suppliers about IPM. Similar cooperative educational efforts with food processors are underway.

IPM ADOPTION ENHANCES ENVIRONMENT

Our measurements of adoption of IPM have been very interesting but really lack value unless something can be said about what impact the adoption has had on the environment. Since the inception of the IPM Program in New York, we have been measuring pesticide use as one form for stating the impact of the Program. In the 1990s we determined that pesticide use measurements were not the only way to express impact. One of our colleagues developed an Environmental Impact Quotient (EIQ) to assist in measuring the envi-

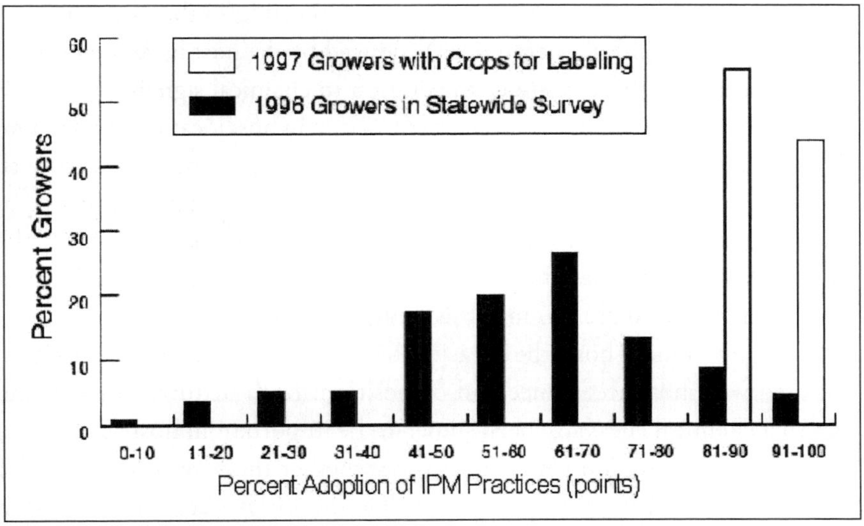

Figure 10.3. Percent adoption of IPM methods among fresh-market sweet corn growers in 1996 and 1997.

ronmental impacts of our cropping systems (Kovach et al., 1999). This measurement was the first of its kind in the United States and it provided a relatively easy way to educate farmers about the potential impacts their choices of inputs, especially pesticides, might be having on the environment. Measuring the EIQ of various cropping systems points us to additional data that help us to speak to impact. Examples of these impacts can be seen in table 10.1, which illustrates the impact of the IPM Program on strawberry production. Since 1987 the use of pesticides has decreased in all categories of pesticides. Applications are down, as are pounds per acre. Similarly, results can be seen in several other crops. In addition, we have been able to measure EIQ values for several cropping systems over two years, and find that most of these values are decreasing (Table 10.2).

CONCLUSIONS

Our experience and data show that IPM adoption can be advanced more completely by:

1. Providing end-users with a setting for continuous education.
2. Keeping end-users connected to the land grant university's research and extension base.
3. Providing end-users with experience in the use of IPM methods.
4. Designing IPM programs that will meet anticipated regulatory actions.
5. Forging IPM Programs that will meet marketplace interests.

Table 10.1. **Comparison of Pesticide Use in New York Strawberry Fields from the Previous Decade to the Present Shows the Impact of the New York State IPM Program on Pesticide Reduction among Strawberry Producers.**

	Pounds of Formulated Pesticide Per Acre Per Year	
Pesticide	*1987–1989 (average use over 3 years)*	1998
Fungicide	5.9	2.8
Insecticide	4.5	1.1
Herbicide	10.6	5.4
Totals	21.0	9.3

Table 10.2. **Environmental Impact Quotient (EIQ) Values, Which Reflect Pesticide Choices, Are Dropping for Most of the Crops Shown Here, Indicating Good News for Water Quality and Human Health.**

Crop	*EIQ Values 1996*	*EIQ Values 1997*
Beets	72	66
Carrots	258	173
Kraut Cabbage	45	74
Peas	23	27
Snap Beans	114	110
Sweet Corn	136	119

Even though land grant universities have advanced the adoption of IPM among the farmers and other end-users in agriculture, the next frontier is to bring IPM principles and practices to community settings: homes, schools, parks, buildings, recreational areas, landscapes, and other nonagricultural areas where pests need to be managed. The land grant universities need to recognize their options in pest management. They can either serve what *is* or *shape what is to be.* In New York, we have enjoyed shaping the future of Integrated Pest Management.

REFERENCES CITED

Benbrook, C. M. 1996. *Pest Management at the Crossroads.* Yonkers, N.Y.: Consumers Union.

Hether, N. 1999. The Gerber Slide Presentation, online at http://www.ecologic-ipm.com/Gerber/index.htm. Presented at the EPA-USDA Workshop on Transition: Accelerating Adoption of Safer and Diversified Pest Management Systems and Reducing Reliance on High Risk Pesticides, July.

Kovach, J., C. Petzoldt, J. Degni, and J. Tette. 1992. A Method to Measure the Environmental Impacts of Pesticides. *New York's Food and Life Sciences Bulletin,* Number 139.

Shephard, R. 1999. Making Our Nonpoint Source Pollution Education Program Effective. *Journal of Extension* 37.

Waldron, J. K. 1998. Personal communication.

Zalom, F. G., M. N. Oliver, and D. E. Hinton, 1999. Alternatives to chlorpyrifos and diazinon dormant sprays. Final report to the California State Water Resources Control Board, September.

Appendix 10.1. Set of defining IPM Elements for strawberry growers, developed by growers and specialists from the New York State Integrated Pest Management Program.

1999 NEW YORK STATE INTEGRATED PEST MANAGEMENT ELEMENTS FOR STRAWBERRIES

Soil and Nutrient Management and Cultural Practices

1. Land was out of strawberries for at least one year prior to planting and was without a crop in the following families: Rosaceae (strawberries, raspberries, etc.), Solenaceae (potato, tomato, peppers, eggplant).

2. A water use plan that minimizes disease development, optimizes water-use efficiency, and minimizes erosion and runoff is used.

3. Fertilizer recommendations are based on soil and leaf analysis when appropriate.

4. Strawberries are mulched with weed-free rye hay, or other biodegradable mulch (e.g., straw).

5. To optimize air drainage and circulation, planting patterns use a narrow matted or ribbon row. Harvest canopy width does not exceed two feet.

6. In a healthy planting, beds are renovated and leaves are mowed or incorporated to reduce disease inoculum.

7. The bulk of nitrogen is applied after renovation to reduce *Botrytis* problems.

Arthropod Management (Insect and Mites)

Arthropod monitoring methods and thresholds should conform to state IPM program guidelines. Records should be kept of all monitoring information collected and threshold used.

1. Tarnished plant bug nymph monitoring, using flower truss counts, begins at bloom and continues at least weekly until harvest.

2. Applications of insecticides for tarnished plant bug are applied only if thresholds are exceeded.

3. Strawberry bud weevil (clipper) monitoring begins when flower buds emerge and temperature is above 65°F.

4. Border rows are sampled. Sprays are applied only if thresholds are exceeded.

5. Two spotted spider mite populations are monitored. Miticides are applied only if thresholds are exceeded.

6. Identification and monitoring of other troublesome pests (spittlebugs, root weevils, leafhopper, etc.) are conducted using Extension recommendations and appropriate controls are applied.

7. No insecticides are applied when bees are active.

8. Sampling dates and thresholds used are recorded.

Disease Management

1. Where a history of red stele, *Verticillium* wilt, or leaf diseases warrants, cultivars that have tolerance or resistance to the appropriate disease are used.

2. In appropriate soils and sites, strawberry plantings are made on beds that are raised enough to provide water drainage.

3. Applications of fungicides are based on weather and disease pressure.

Weed Management

1. A weed survey is conducted and appropriate strategies are implemented. Herbicide rate and selection are based on these results.

2. Cultivation and hand weeding is used as a substitute for herbicides.

3. Plant growth within thirty feet of field border is controlled (i.e., mowing) to reduce weed seeds, improve air drainage, and reduce insect infestation.

Pesticides and Pesticide Records

Only pesticides registered in the state and approved for the target pest and crop will be used. Records of pesticide applications including date, field identification, targeted pest, pesticide name and EPA number, formulation, and rate and number of acres treated are maintained.

1. Insecticide/fungicide sprayer is calibrated and dated at least once a year.

2. Herbicide sprayer is calibrated and dated at least once a year.

3. Spray records are maintained and organized.

4. Pesticide choice is based on preserving natural enemies as well as efficacy.

5. No sprays are applied to fruit after it is formed.

Part IV

Entomology: A Model for Pest Management Research and for the Advancement of Science

Chapter 11

Ecological Theory and Biological Control of Insect Pests

WILLIAM W. MURDOCH
Department of Ecology, Evolution, and Marine Biology
University of California–Santa Barbara

The Department of Entomology at Texas A&M University has played an important role in the development of many areas of entomology that have benefited society, including the area of pest control, which will be my main topic. The area of entomology I will discuss is population ecology, and especially population dynamics, which studies the abundance of organisms. Within that broad discipline, ecology-in-entomology has served human society most directly through helping to control pests in agriculture, especially via biological control of pests, which will be the focus of my talk.

Before discussing biological control, I want to point out that entomology has made a much broader contribution to society. First, entomologists are probably the main group of scientists to have developed and expanded the theory of population dynamics, and such theory underlies successful resource management. For example, the successful management of fisheries relies on concepts such as density-dependence in whose development entomologists played a key role. Perhaps less obvious is that the modern theory of biological conservation has its origins in entomological studies. This is not broadly appreciated, since we usually think of conservation as applying to cute and charismatic organisms like cheetahs and the spotted owl. However, the central problem in conservation today is how to conserve species and biodiversity in an environment that is becoming progressively fragmented by agriculture, development, and the destruction of habitat.

The theory behind conservation biology began with an entomological problem. The origin of the theory is modeled in 1969 by Levins, who was interested in the problem of how best to control a pest population that existed in a patchy environment, such as a set of agricultural fields. He was interested in what happened if you managed to drive the pest extinct in some fields that were later re-colonized from other areas. These are models for what is now called metapopulation dynamics. Extensions of this theory are the basis of modern approaches to the design of habitat reserves for the entire range of conservation targets from the spotted owl in the United States to native forests and mammals in Australia.

Furthermore, spatial studies of insects have been the major source of observational and experimental data that have tested metapopulation theory. Insects thus serve as the main model for conservation biology. A famous example is provided by a checkerspot butterfly (*Euphydryas editha* [Boisduval]) in Finland. In Finland this species is found only in dry meadows distributed across a small set of islands in the Baltic Sea. Studies over the past twenty years have shown that the butterfly is present at any one time in only about 65 percent of the 2,300 meadows. Each year, the butterfly goes extinct on 10 to 20 percent of these meadows. But the entire population persists because each year the butterfly also colonizes about the same number of new sites. The regional population thus persists in spite of local extinction. This study provides evidence that we might be able to preserve species on fragments of habitats, provided they are sufficiently interconnected.

Let's turn now to my main topic, the link between population ecology and biological control of insect pests. For those of you unfamiliar with biological control, it is broadly the use of natural enemies, such as predatory and parasitic insects, to control pests, mainly insects. In classical biological control, which I will focus on, the problem is that an insect that is not a pest in its area of origin, China for example, is accidentally introduced into a new area, such as the United States. It becomes a pest, and entomologists reason that it must be suppressed to low densities in its area of origin by its natural enemies—insect predators and parasites—and has escaped such control in the new area. They therefore go to China, find natural enemies, and import and release them to attack the pest. Success occurs when the pest is reduced to a low density, and remains there over long periods.

Classical biological control has been successful for more than 100 years. More than 100 species (36 percent of those targeted) have been brought under control. And of course a single success actually represents many successes. For example, the cottony-cushion scale (*Icerya purchasi* Maskell), a pest of citrus, has been controlled by the famous *Vedalia* beetle (*Rhodolia cardinalis* [Mulsant]) in many areas in each of fifty countries for up to 110 years. Furthermore, biological control is enormously effective economically. Costs are typically small, and once success is achieved, you can literally walk away and let the natural enemies get on with the job. The benefit to cost ratio is enormous (32 to 1), and there is no doubt that billions of dollars are saved each year through biological control.

Yet there have also been many failures in classical biological control—indeed, more failures than successes. What is astonishing to the nonentomologist is that the process has been almost purely trial and error. A pest arises; entomologists look for natural enemies and release one or more species. If these are successful, they move on to the next project. If they fail, they try some others, and so on. There have been almost no attempts made to find out why some attempts fail and others succeed. The California red scale (*Aonidiella aurantii* [Maskell]), a pest of citrus, was an extreme example. Fifty-one species of natural enemies were released over a period of almost seventy years, until the parasitoid *Aphytis melinus* DeBach succeeded in controlling red scale in the late 1950s. That success has been repeated all over the world. But we do not know why this enemy succeeded while the others failed. This is not a criticism of entomologists; they have done a wonderful job. But they are almost never given the opportunity to investigate failures and successes. Yet surely there is something to be learned from this process, which amounts to ecological experiments on a grand scale. It is not likely that entomologists will be allowed to go on releasing natural enemies at will. A backlash is developing against this process because of potential effects on native, nontarget species. It behooves us to try to develop a scientific basis for classical biological control.

The failures are of two broad types. Some natural enemies are introduced but fail to become established—perhaps because the climate is inappropriate. These we will not be concerned with here. The second class of failures is the one that interests us—the natural enemy becomes established, but for some reason fails to control the pest. Here is where ecological theory might help us find successful agents more frequently.

My major point is that the time is ripe for population ecology to provide scientific guidance for biological control. In particular, mathematical theory is now realistic enough to be useful. It can do two things. First, it can help, together with experiments, to uncover the processes and mechanisms that cause some natural enemies to be successful while others fail. I will illustrate this with my work on red scale and its controlling parasitoid *A. melinus.* Second, once we have developed reliable theory, we can use it to guide future projects in biological control, for example as a guide in selecting a natural enemy.

I will motivate and illustrate the theory with the system I work on in lemon and grapefruit groves near Santa Barbara, California. In these groves, California red scale is under perfect control by its parasitoid, *A. melinus.* Adult female red scale produces crawlers that generally move only a few inches, insert their mouthparts, produce a protective cover, and never move again unless they are adult males. They pass through three instars, increasing in size as they age, and have two to three overlapping generations per year. Red scale was introduced from China to California over 100 years ago. They can reach enormous densities on a tree and kill it; they almost destroyed the citrus industry in California in the mid-twentieth century.

In the late 1950s, entomologists introduced the parasitoid *A. melinus* from Iran. This is a small parasitic wasp about 1 mm long. The female drills a hole in the scale cover. She either oviposits an egg on the host body under the cover, or feeds on the host's hemolymph. This is called host feeding, and the female uses the nutrients to mature additional eggs. *Aphytis melinus* is a spectacular success and the scale has been under control for forty years in Southern California and in many other parts of the world. The scale is suppressed to about 1/1000th of its potential density, and in the area where I work, populations are very stable in the sense that numbers have changed little over several decades. Indeed, we have shown that it is one of the least variable insect populations ever recorded. Both the low density and the dynamic stability are crucial.

The first question we need to ask is: what is the spatial scale at which the control mechanisms operate? Are metapopulation dynamics important, as in the checkerspot butterfly, meaning that control is a regional process? Or do the mechanisms operate at the scale of a grove or on a single tree? In red scale, we answered this question with a couple of simple experiments. We enclosed trees in cages, isolating the local red scale/*A. melinus* interaction from the rest of the grove, and asked if this local isolated system was less stable than the population in the grove as a whole. The answer is that the degree of control and stability was the same in an isolated cage as in trees connected to the rest of the grove over four red scale and twelve *A. melinus* generations. So we have local stable pest control. This is a very important result—it tells us that whatever mechanisms cause suppression and stability, these mechanisms operate very locally.

In a pilot experiment (which was cut short for logistical reasons), we then showed that the relevant spatial-scale was probably only a single branch of the tree. We caused an outbreak of scale on such a branch, inducing scale densities 1,000-fold higher than in the grove. Scale density had come back to within thirty-fold higher than ambient within a few months, when we had to stop the experiment. *A. melinus* responded very rapidly, and increased about 300-fold in response to the outbreak. It caused up to 80 percent parasitism, whereas that in control trees was less than 10 percent. So we suspect this density-dependent parasitism by *A. melinus* is the force that was bringing the scale back to ambient densities. These experiments tell us that the mechanisms are very local, and that we need to focus on the details of the interactions, that is, on behavior, not on large-scale spatial processes. We need to look at the interaction between individual parasitoids and hosts.

To understand how *A. melinus* and the host interact, we look at the details of scale life history. Adult female scale are invulnerable to attack (a very common pattern in parasitoid-host interactions) and produce immatures that settle and grow through various stages, virtually all of which are attacked by *A. melinus.* Each stage increases in size as it ages, and it becomes a better and better host for the parasitoid. Thus, smaller stages are host-fed—they are eaten and the nutrients later turned into new eggs. Somewhat older (and larger) stages are parasitized but produce only male parasitoids. Finally, the oldest and largest stages produce female parasitoids, and if the scale is very large, may produce two or more females. Only female parasitoids kill scales, so attacks on smaller stages (producing males) do not directly increase the future population of attacking females. We see that the gain to the future female population increases with the age of the host attacked. Such increasing gain with host age is also a widespread feature of parasitoid-host interactions. The results of this study have quite general implications. Our models of this system thus have quite general application.

The female's response to her own state is also important. Parasitoid decisions depend not only on the size or age of the host encountered but also on the state of the parasitoid, and in particular how many mature eggs she has. For example, evolutionary theory tells us she should host feed when she has few eggs, and parasitize when she has many. We have done experiments and confirmed that this is indeed the case. We can also include this behavior in our models.

We developed mathematical models of the *A. melinus*/red scale interaction, and we enter any amount of life history detail that we think is important. Most recently, for example, we have developed a model (based on experimental data) that includes more than ten classes of scale and different parasitoid reactions to their encounter. The model also includes state-dependent parasitoid decision-making. In addition to this complex model, we have also developed much simpler versions. The important point here is that new mathematical techniques allow us to base theoretical simulations directly on natural history, behavior, and ecology of the insects. This is a substantial advance in population theory in the last two decades.

Simulations suggest models may be reasonable descriptions of the real system as these models generate dynamics that look similar to those in the field. For example, the model predicts damped cycles, and indeed we see very small cycles with a period close to that predicted. It thus appears that the stage-structured interactions we have in the model may well be the key to explaining this system. We are now doing a large field experiment in which we have created outbreaks of scale at different densities within caged trees. We are measuring scale densities and are also monitoring *A. melinus* responses and behavior to test the predictions and assumptions of the model.

The second use for such simulation models is in making decisions. For example, we can use the theory to ask which of two parasitoid species, considered to be released as biological control agents, is better. Because the models contain detailed biology and life histories, we can and must ask a whole set of new questions not covered by older theory. Suppose we want to know if a parasitoid that attacks the egg stage of the pest (an egg parasitoid) will be better for control than a parasitoid that attacks a larval stage of the pest (a larval parasitoid). Our model must specify the pest's life history: How long does the egg stage last? What is its background death rate? How long does the larval stage last, what is its background death rate? It must also specify the properties of the two parasitoids. What are the attack rates? the maturation rates? and the juvenile and adult death rates of each parasitoid? And so on. We immediately obtain new kinds of results. For example, the parasitoid that wins in competition may not be the better control agent because it may minimize the density of the wrong pest stage, so yielding worse control. Thus we need to match the pest and the enemy. Again, it turns out we can trade lower attack rate by adult parasitoids for higher survival of the parasitoid juveniles or adults.

When we analyze a model of this sort we find the following answers when we specify that we need to minimize the adult pest stage. First, it is almost always better to release the egg parasitoid, rather than the larval parasitoid. In general, it is better to release a parasitoid that attacks an earlier stage of the pest. When the pest's egg and larval stages have the same duration the larval parasitoid is never better then the egg parasitoid. This is even more the case when the egg stage is longer than the larval stage. Only when the egg stage of the pest is briefer than the larval stage is it possible under some circumstances to get better control by using the larval parasitoid. Notice, however, that in many pests the larval stage is longer than that of the egg. Second, when the pest's larval stage lasts longer it is best to release the egg parasitoid when there is not much other mortality on the pest eggs and larvae. Thus we are beginning to see rules of thumb emerge, relating parasitoid type and pest's life history and demography.

This is all fine in theory, but does it work in practice? There is some modest cause for optimism. In fact, two species of *Aphytis* were introduced into citrus groves in Southern California. *Aphytis lingnanensis* Compere produced some control in the Los Angeles area, but it was not enough. So in 1959 *A. melinus* was introduced as well. Three things happened: it replaced the existing unsatisfactory species (*A. lingnanensis*) almost everywhere; it did it with great speed (about one scale generation per grove); and it improved biological control. A remarkable feature of the displacement process is that these two parasitoid species are virtually identical and have very similar properties when studied in the laboratory. How, then, could *A. melinus* be so much better? Luck and Podoler detected one difference: *A. melinus* is able to get female offspring from somewhat smaller (younger) scales than can *A. lingnanensis;* that is, *A. melinus* has a greater gain in this small age class of hosts. The difference is very small, but they thought it might give *A. melinus* an advantage and explain the results. We therefore developed a standard, stage-structured model that has the two parasitoid species competing and differing in the feature found by Luck and Podoler. The model is able to account for all three stages of the process observed in California. First, we let it go to equilibrium with *A. lingnanensis* only. Then we introduce a small number of *A. melinus.* This species replaces *A. lingnanensis* and it does so very quickly—about one scale generation. Finally, red scale density is reduced about threefold. In fact, this stage-structured approach has now been applied to three examples and the model has correctly identified the parasitoid species that appears to have been the better control agent. The other two examples are two important pests in Africa. The key is that either one parasitoid attacks an earlier stage than the other, or again that it gets female offspring from an earlier (smaller) stage.

In summary, I have discussed an area of pest control, biological control, where entomology has made an enormous contribution to human society. I have tried to show that recent developments in ecological theory provide an opportunity to make biological control a more scientific approach to pest management, and have provided some evidence that this can work in real systems. More remains to be done, of course.

Finally, entomology has contributed to human society less directly in other ways, for example in stimulating the modern theory of conservation biology. This is happening again, in yet another important field, ecotoxicology. Toxicology is very good at predicting the effects of various toxic substances on individual organisms. But if we want to know the effects on ecological populations and communities we need to be able to go from descriptions of effects on individuals to predicting effects on populations of organisms. This is precisely what the stage-structured theory and its relatives are designed to do, and my colleagues and I are now working on models and experiments to use the theory for just that purpose.

CHAPTER 12

Metapopulations, Fragmentation, and Connectivity

Implications for Insect Populations

DANIEL SIMBERLOFF
Department of Ecology and Evolutionary Biology
University of Tennessee–Knoxville

The role of space in molding patterns and processes has lately moved to the forefront of ecology (Hanski and Simberloff, 1997). It is seen as a crucial aspect of more powerful theories at the gene, individual, population, and community levels, and fieldwork on spatial matters is seen as a major challenge (Kareiva, 1990). Although there were other early considerations of the role of space in structuring nature, the current interest springs largely from the equilibrium theory of island biogeography of MacArthur and Wilson (1967) and the widely noted increasing fragmentation of all habitats (Meffe and Carroll, 1994). It has produced two competing paradigms, both heavily criticized: the metapopulation concept (Hanski and Simberloff, 1997) and landscape ecology (Wiens, 1997a). Most of the interest in the spatial component of ecology, under the equilibrium theory and the two new paradigms, arises from conservation concerns, and rather little attention has been focused on their implications for pest management. I will review the status and accomplishments of all three paradigms and discuss their potential role in pest management.

THE EQUILIBRIUM THEORY

The equilibrium theory of island biogeography generated enormous interest among ecologists and biogeographers because it depicted nature as complicated and dynamic, yet susceptible to modeling using an easily gathered statistic (number of species) and simple mathematics (Simberloff, 1974). The reason a complex entity could be treated so easily is that the theory featured spatial heterogeneity. Nature was seen as divided into more or less discrete microcosms, separated from one another by inhospitable habitat. By focusing on the separate microcosms, one could hope to grasp the essential elements of structure and dynamism. The key propositions were (1) local extinction would occur in each insular microcosm, (2) it would be redressed by occasional immigration into the microcosm, (3) the extinction and immigration would balance each other (so there would be an equilibrium number of species), (4) extinction rate would be determined by area of the site, and (5) immigration rate would rest on its isolation from potential source areas.

Very quickly the theory was expanded in two directions. First, even the original conception recognized "island" as a metaphor for any insular habitat, such as a lake, a field surrounded by forests (or vice versa), or any sort of patchily distributed habitat. Second, the theory was seen as enormously relevant to conservation. Wilson and Willis (1975) proposed a set of refuge design principles, ostensibly based on the theory, to maximize species richness; these became the basis for many important management plans (e.g., IUCN, 1980). More fundamentally, the equilibrium theory, by positing the widespread occurrence of local extinction, ultimately led conservation biologists to consider the significance and causes of such extinction and what might prevent it (Haila and Järvinen, 1982).

The general enthusiasm for the theory was such that it was inevitably cited as potentially applicable to agricultural pest control (references in Levins and Wilson, 1980), but it was apparently never applied. Probably the reason for the difference from the response in conservation is that, in agriculture, the economic effect of a particular management program is immediate and apparent, discouraging major innovation that does not seem to have a high probability of being effective. In conservation, the results of management are not often so apparent and may be greatly delayed (Simberloff, 1986b). Some proposed applications of the equilibrium theory for pest control would give anyone pause. For example, for soybean insects, Price (1976) argued that a group of small fields might be a better strategy than a single large one because they would equilibrate faster. Stenseth (1981) suggested that, if it is possible to reduce immigration of a pest into suitable habitat greatly, all resources should be devoted to that strategy.

In the face of substantial criticism of the theory of island biogeography (e.g., Gilbert, 1980; Williamson, 1989), it eventually receded into the history of ecology and conservation, and the rules of refuge design based on it were rejected as not generally founded in the theory and empirically unreliable (Simberloff, 1986a; Soulé and Simberloff, 1986). Part of the problem was that conservationists do not usually aim to maximize the number of species in each refuge; rather, they want to maintain particular species of conservation concern somewhere in the entire system of refuges. An even greater problem was that the theory implied that small reserves were of little use because their populations were doomed to rather quick extinction, but conservationists were loath to abandon small reserves. In any event, probably the main view of the equilibrium theory now among both ecologists and conservationists is that it was heuristically useful as it focused attention on spatial heterogeneity and the causes of local extinction (Haila and Järvinen, 1982), but that its direct applications to nature are limited. In agriculture, it never seems to have gotten off the ground and has left no legacy.

A PARADIGM SHIFT

In a remarkably rapid and clear-cut paradigm shift, metapopulation thinking has replaced the equilibrium theory of island biogeography as the governing paradigm in conservation biology (Hanski and Simberloff, 1997). The two key notions of the metapopulation paradigm are that populations are spatially structured into assemblages of local breeding subpopulations, and that movement among local subpopulations affects but does not dominate local dynamics, including possibly reestablishing a subpopulation after local extinction. Of course, a crucial concern is exactly how much movement makes an ensemble of individuals one big population as opposed to a metapopulation of subpopulations; I will return to this point.

The two key early antecedents to metapopulation thinking were by Wright (1940) in evolution and Andrewartha and Birch (1954) in ecology. The similarity of Wright's evolutionary model to the metapopulation paradigm is just now being recognized (Hanski and Simberloff, 1997). Andrewartha and Birch, studying insect population fluctuations, observed wild fluctuations, apparent local extinctions, and also re-colonization of vacated sites. Thus they perceived a spatial component for population dynamics and espoused a view of nature very similar to that now labeled metapopulation dynamics. However, this aspect of their ecological thinking had little impact, probably because they almost totally rejected the idea that density-dependence could play a role in governing insect populations (Hanski and Simberloff, 1997).

Although they anticipated the metapopulation paradigm, they did not inspire it.

The formal beginning of the current burst of interest in metapopulation thinking came from the work of Levins (1969), who conceived of space as divided between patches of (identically) suitable habitat on the one hand, and an unsuitable matrix on the other. His quantitative theory rested on three other assumptions: habitat patches have equal size and isolation, subpopulations have uncorrelated dynamics, and the movement of individuals among patches is so rare that it has minimal impact on local dynamics. The definition of metapopulation has since been somewhat expanded. Current versions of the classical Levins model have four conditions (Hanski, 1997): (1) suitable habitat is found in discrete patches; (2) all populations, even those in the largest patch, have a substantial risk of extinction; (3) occasional movement occurs between patches; and (4) population dynamics in the different patches are asynchronous.

Several reasons explain why the metapopulation paradigm so rapidly and completely replaced the equilibrium theory. At first it may seem surprising that this happened, as the two conceptions of nature have much in common. Both see nature as divided into discrete entities located in a spatially homogeneous matrix, with local extinction and movement between patches. In both conceptions, an arbitrary element is just how much movement occurs between patches for the models to remain appropriate. However, different emphases between metapopulation theory and equilibrium theory may well bear on the paradigm shift. The very term equilibrium shows that the emphasis is on the static feature (number of species) rather than on the dynamic underlying changes, and the paradigm shift coincided with a general shift among ecologists toward a conception of a nonequilibrium nature (references in Hanski and Simberloff, 1997). Second, by focusing on individual populations rather than the single statistic of species richness, metapopulation thinking fits better with the growing involvement of genetics in conservation biology (Simberloff, 1988), as well as the growing concern with such population processes as demographic stochasticity. Finally, and perhaps most importantly, metapopulation theory rescued small sites as useful for conservation. After all, if much of nature consists only of small subpopulations (and this is increasingly true as all sorts of habitats are progressively fragmented), any conservation measures will have to focus on small sites. Metapopulation models even suggest that a certain number of unoccupied sites is required for metapopulation persistence (e.g., Lande, 1988; Hanski, 1997). Thus, conservationists no longer had to justify their interest in preserving small sites. And the metapopulation paradigm seemed perfectly suited to a fragmented world.

APPLICATIONS OF METAPOPULATION THEORY

Although there are a variety of types of single-species metapopulation models (Hanski and Gilpin, 1997), most attention has been directed at three distinct sorts of spatial arrangements, and several modeling approaches are applicable to each. First is the original Levins model, in which all patches are conceived as equally small and equally isolated, and movement between all pairs of subpopulations is equally likely. Second is the mainland-island model of Boorman and Levitt (1973) in which there is a large group of patches, but a central one of them is much larger and produces a stream of propagules that flow to the surrounding small sites, each of which houses an ephemeral subpopulation. Each subpopulation eventually disappears, but its patch is ultimately recolonized from the large central one. This model is basically a multi-island, single-species version of the equilibrium theory. Finally, Pulliam (1988) and Pulliam and Danielson (1991) suggest a source-sink model, in which all sites are small, but, because of variation in site quality, some sites are always more productive than others and export propagules to the others, which would not persist at all were it not for this immigration. The source-sink model can be construed as a version of the mainland-island model in which the mainland is replaced by a group of patches with productive, persistent subpopulations.

Harrison (1991; 1994) has suggested the classic Levins metapopulation model does not approximate the structure of many species in nature. Rather, she has found that field data (which are generally weak for movement of individuals) construed as demonstrating classical metapopulations usually show something else entirely: (1) In many instances, there is so much movement and so little division into clear patches that an entire region can be seen as containing one large, interbreeding population. Even when a species occupies a patchy habitat, movement rates may be so high that there is effectively one large population. For example, insects on fallen fruit, rotting logs, dung, or carrion may usually be so mobile that, even though they have been cited as classic metapopulations (e.g., Hanski, 1997), they may not be (Harrison and Taylor, 1997). Similarly, host-specific insects on a patchily distributed plant within a field need not be metapopulations, though such species may display metapopulation dynamics at a larger scale, such as different fields or forests

(Harrison and Taylor, 1997). (2) Many situations conform approximately to the mainland-island model, with a large, permanent source population and ephemeral sink subpopulations whose frequent disappearance is irrelevant to the persistence of the species. Checkerspot butterflies (*Euphydryas editha* [Boisduval]) on patches of serpentine soil conform to this model (Harrison et al., 1988). (3) In some situations, a progressive destruction and fragmentation of habitat has left subpopulations more or less isolated on small patches, and there is movement among at least some patches. But the entire system is in decline; more subpopulations are disappearing than are being established by re-immigration to empty sites, and the appearance of a persistent metapopulation is simply an artifact of the fragmentation of formerly continuous habitat. Approximately ten of the ninety-four resident butterfly species in Finland conform to this nonequilibrium extinction model because of recent habitat destruction (Hanski and Kuussaari, 1995).

It is important to emphasize this last point. Just the fact that habitat fragmentation has left a formerly continuously distributed species occupying a series of discrete patches does not mean that it constitutes a metapopulation or that metapopulation dynamics are crucial to the survival of the species. Fragmentation is such a pervasive phenomenon and has aroused so much conservation interest that it has become a very confused concept, with many conceptions of what constitutes a fragment and how fragmentation affects species (Bunnell, 1999). But perhaps the most widely held view, epitomized by many maps showing progressive fragmentation (e.g., Curtis, 1956; Moore, 1962) is that fragments are patches of a habitat in a uniformly inhospitable sea (just as in the metapopulation model) and that populations can survive in these patches so long as movement occurs among them. Even documented movement between patches does not guarantee that we are dealing with a classic metapopulation situation; we would have to know that this movement is important. The apparent match between a fragmented landscape and metapopulation models is a key attraction of the latter for conservationists. It inspires optimism—perhaps a species will persist in spite of the habitat loss, through metapopulation dynamics. And this might well be the case for particular species. For many others, the fact that they look like metapopulations may be a cruel delusion, as they may be doomed. As habitat destruction increasingly fragments any habitat, and while the process is still recent enough that the habitats remain largely inhabited and are not so isolated as to stem all movement, almost any species is bound to look as if it is a classic persistent metapopulation, when it may in fact be in a transient state and ultimately doomed to extinction.

Of course, a species may conform to different models in different regions, depending on environmental circumstances. For example, the silver-studded blue butterfly (*Plebejus argus* Linnaeus) in Wales shows a continuum from classical to mainland-island metapopulations (Thomas and Harrison, 1992), while the checkerspot butterfly constitutes a mainland-island metapopulation in coastal California but more of a classical metapopulation structure in montane areas (Harrison et al., 1988; Singer and Thomas, 1996). The Glanville fritillary (*Melitaea cinxia* Linnaeus), discussed below, conforms to the classic metapopulation model in the Aland Islands but has been extinguished in many parts of Europe as the density of suitable patches declined from grazing of its habitat (Hanski, 1997).

The species whose population dynamics most convincingly conform to those of the Levins metapopulation tend to occupy habitat patches that inevitably change because of succession or other processes. A number of British invertebrates occupy such habitats (Thomas and Morris, 1995), and many butterfly species may adhere to this model (Hanski and Thomas, 1994; Hanski and Kuussaari, 1995; Thomas and Hanski, 1997). In such species, understanding metapopulation dynamics is crucial to conservation, so information is required on movement and local extinction rates and how these are affected by patch size and isolation (Hanski et al., 1995). A key overriding fact is that restriction of conservation just to currently occupied sites would likely be fatal, as these sites will eventually become unsuitable and there must be other sites that will become suitable.

Not all convincingly depicted classical metapopulations use ephemeral habitat. For example, in the Aland Islands the Glanville fritillary uses dry meadows, which occur as discrete, small patches (Hanski, 1997, and references therein). In this archipelago, these patches are currently not being destroyed, and they do not undergo succession on a time scale relevant to this discussion. Even the largest subpopulations occasionally disappear, so there is no "mainland" population. Patches that lose their subpopulations are recolonized by individuals from other patches, and the population dynamics of different patches are asynchronous. Thus, this metapopulation approximates many features of the ideal Levins model. Even in cases like these, when deterministic processes such as succession do not change sites, modeling suggests that a certain number of currently unoccupied sites are necessary for persistence of classic metapopulations (Lande, 1988; Hanski, 1997). A current research activity, conducted primarily theoretically, is an attempt to determine the minimum number of sites and/or subpopulations that will allow a classic metapopulation to persist (Hanski, 1997). At some point, the network of subpopulations becomes sufficiently sparse and isolated that there is insufficient re-colonization of vacant sites, and the entire metapopulation collapses. This point may be a sharp inflection; as number of sites drops gradually, the total metapopulation size may drop more or less apace, until suddenly total extinction occurs. A recent expansion of metapopulation modeling is spatially realistic modeling of particular populations in which a simulation is used, with parameters derived from field data, to predict the fate of a particular population, or to compare the likely fates of a species in several possible refuge systems (Hanski and Simberloff, 1997).

With respect to applications, almost the entire metapopulation literature, both theoretical and empirical, is devoted to conservation—saving populations. Yet the principles seem germane to pest control—getting rid of populations. For example, much attention has been paid to whether metapopulation dynamics contribute to the persistence of two-species assemblages (generally parasite-host or predator-prey). In a sense, the ideal operation of classical biological control conforms to the classic Levins model, as in the Australian interaction between prickly pear cactus (*Opuntia* spp.) and the moth *Cactoblastis cactorum* (Berg). In this instance, each species exhibits metapopulation dynamics because of the spatial structure and the interaction.

Harrison and Taylor (1997) point out that one species may have a stable population, at least in part of its range, independently of spatial structure of a second, interacting species, while the second species may be a metapopulation. Thus the winter moth (*Opheroptera brumata* Linnaeus) persists indefinitely as a classic metapopulation in apple orchards with its parasitoids, and it disperses into forests where local extinctions are common—this is a form of mainland-island metapopulation. Conversely, a parasitoid can have a stable population even if its host populations go extinct (in metapopulation fashion) so long as it has alternative hosts, as in the oak gall wasp *Xanthoteras politum* (Basse H) and its parasitoids (Washburn and Cornell, 1981). For very few multispecies assemblages have adequate data been gathered to determine which species are acting as metapopulations. Perhaps the best-studied example is that of ragwort (*Senecio jacobaea* Linnaeus) with the cinnabar moth (*Tyria jacobaeae* Linnaeus) and its braconid parasitoid *Cotesia popularis* in the Netherlands (van der Meijden and van der Meijden-van Wijk, 1997). Here the entire species complex is maintained regionally, in spite of frequent local extinction, because of spatial effects related to the impact of habitat heterogeneity on ragwort (which has some persistent refuge sites) combined with limited dispersal of the parasitoid.

The examples so far have been well studied primarily for academic reasons—to understand persistence or elimination—rather than for applied purposes. Yet one could imagine potential application to agricultural or silvicultural pest control. For example, one widely used metapopulation model (Nee et al., 1997) predicts that habitat destruction and barriers (like roads) make it less likely that parasitoid-host metapopulation dynamics will allow stable persistence of both species. Or one could imagine spatially realistic modeling, of the sort suggested for refuge design (Hanski and Simberloff, 1997), applied to the layout of agricultural fields. Modeling of dispersal dynamics in a metapopulation context might suggest which biological control agents would be more effective in controlling their target pests. Of course, parameterizing such models would be the key difficulty for agriculture as for conservation.

A key difference between the predictions of the equilibrium theory and those of classical metapopulation theory is the role of connectivity. Applications of the equilibrium theory to conservation see connectivity as an unalloyed good. The rules of refuge design invoking proximity of clusters of reserves and corridors connecting them derive the idea that increased immigration into sites can only reduce extinction rates, thus increase the equilibrium number of species present in the community as well as lengthen the time it takes for any component population to disappear. Although the reasons why local populations go extinct were not accentuated in the original presentation of the theory by MacArthur and Wilson (1967), such factors as demographic and environmental stochasticity as well as genetic drift and inbreeding depression have been frequently invoked (Simberloff, 1988). This emphasis on the fundamental importance of connectivity in order to stem the action of all these forces has led to a prominent movement in conservation to construct or maintain corridors between refuges, on scales from the meso to macro. The fact that the empirical evidence of their utility is very scant (Hobbs, 1992; Simberloff et al., 1992) has not led to a waning of this movement (Beier and Noss, 1998).

Classical metapopulation theory sees connectivity very differently. Some movement between sites is needed to redress or even prevent local extinction, but too much movement between sites may lead the entire metapopulation to crash because it would tend to synchronize the population dynamics of all subpopulations (Hanski, 1997), and the asynchrony is crucial to the persistence of the ensemble. Thus classical metapopulation theory demands a range of movement rates—not too little, but not too much. It seems that if agriculturally important species are indeed structured as metapopulations, this feature could be important. On the one hand, for persistence of a biological control agent, for example, perhaps host patches (say, of a crop plant) ought to be located so that separation distances fall in a narrow range. On the other hand, possibly a pest, such as a weed or a phytophagous insect, could be driven to extinction not only by being forced into too sparse a metapopulation but also too dense a metapopulation. Specific predictions would require much empirical information on basic population biology, including movement rates.

LANDSCAPE ECOLOGY

Landscape ecology is the other paradigm recently advanced to introduce space into ecological structure and dynamics (e.g., Naveh and Lieberman, 1994; Forman, 1995). Although both metapopulation dynamics and landscape ecology have attracted enormous attention, and both deal with the ecological role of space, there has been remarkably little interaction between the practitioners of the two fields. Although Wiens (1997a) locates a key difference between the two paradigms in the scales at which they are applicable (metapopulations at the micro- or mesoscale, landscapes at the macroscale), many published applications of both approaches are at the mesoscale. The key difference is rather that metapopulation modelers typically conceive of the matrix between patches as a homogeneous, inhospitable, featureless mass, while landscape ecologists view this matrix as a mosaic of various kinds of habitats of varying degrees of suitability for each species (Wiens, 1997b).

The main concerns of landscape ecologists have been (1) patch quality, (2) how the surrounding mosaic of patches conditions phenomena within any patch, (3) flows of individuals and processes across boundaries between patches, and (4) the degree and nature of connectivity between patches (Wiens, 1997a). There has been much less formal mathematical modeling of landscape dynamics than of metapopulation dynamics, and this fact probably accounts for the relative dearth of attempts to parameterize models using empirical data, for example, on rates of movement. The main theoretical prediction derives from percolation theory (Gardner et al., 1987; 1989), in which the entire landscape is reduced to two sorts of habitat patches, suitable and unsuitable. The prediction is that, above some critical threshold, cells of the suitable habitat are likely to form a continuous mass that spans the landscape, and an individual in this "percolating cluster" can "percolate" across the landscape. Below it, if such connectivity is important, the population disappears. It is obvious that this result is very similar to that in classical metapopulation dynamics in which the density of suitable patches in an otherwise homogeneous, inhospitable matrix must remain above some threshold or the entire metapopulation may collapse. That percolation theory also rests on just two habitat types may help to explain the similarity in the results. It is difficult to see exactly what data could be gathered to distinguish which, if either, of these two models represents some field situation. Wiens and colleagues (references in Wiens, 1997b) have used tenebrionid beetles to investigate the general problem of how mosaic structure affects movement through a landscape. The main results are that spatial heterogeneity affects beetle movement differently at small scales (radius of ca. 42 cm or less) than at large ones, and that different species respond differently to various mosaic patterns.

With respect to conservation planning, it is not obvious when landscape models would be more useful than metapopulation models and give different results. In generating an overall stance toward conservation matters, metapopulation thinking automatically leads one to consider refuges as the dominant response to conservation issues, the network of refuges corresponding to the sites potentially inhabited by the metapopulation. However, many workers (e.g., Franklin, 1989) have argued that refuges are insufficient, and that in some way the entire landscape must be managed simultaneously for a range of purposes, including conservation. Certainly landscape ecology conduces to this sort of thinking, with its variety of habitat types and varying degrees of connectivity among them. However, for now, aside from simply saying that we ought to manage the entire landscape, there are few specific suggestions for doing so, few detailed guides to what exactly a land manager should do in any particular case.

As with the metapopulation paradigm, the bulk of the landscape ecology literature is concerned with its relevance to conservation, and how it might affect agricultural pest control is barely touched. By its very acceptance of the fact that nature is a mosaic of habitats, and a general sense that humans affect almost the entire landscape, landscape ecology deals with agriculture (e.g., Bunce et al., 1993). However, the concern is how to make agriculture support conservation better, rather than how to raise production (but see Ryszkowski et al., 1993). Of course, there is an old tradition of interplanting crop species with other plants that may harbor predators or parasitoids (e.g., Flaherty, 1969) or attract pests from a target crop (e.g., Stern, 1969). Similarly, hedgerows can harbor natural enemies that suppress agricultural pests such as *Brevicoryne brassicae* Linnaeus, the cabbage aphid (Southwood, 1971). However, the advent of landscape ecology does not seem to have informed this approach to pest management or suggested new approaches to designing such variegated landscapes.

CONCLUSIONS

The contributions of the dynamic equilibrium theory of island biogeography to conservation were inspirational. It cemented in the public mind the metaphor of refuges as islands of habitat in a sea of developed landscape, and it forced biologists to consider the frequency and causes of local extinction. It did not suggest valid specific management procedures. Several speculative academic suggestions about possible applications to agricultural pest control were never tested and quickly forgotten.

Now, in conservation, the metapopulation paradigm has replaced the equilibrium theory paradigm. This change was probably greatly spurred by the increasing fragmentation of many habitat types and the apparent match between metapopulation thinking and a fragmented landscape. The classical metapopulation model is quite similar to the equilibrium model, except there is no mainland, only an archipelago of islands. Metapopulation thinking is popular among academic ecologists as well as conservation biologists, but there is little evidence that much of nature is structured as a classical metapopulation. A large fraction of those species that appear to be structured in this way may be insects. Metapopulation models suggest some fairly specific conservation imperatives, such as the need to preserve unoccupied sites, and even, for some models, the minimum number of unoccupied sites to save (e.g., Nee et al., 1997). One of its predictions—that too much connectivity is bad for conservation—directly contradicts equilibrium theory. Although there is much discussion in the conservation literature and even lay press about metapopulations, metapopulation modeling has not yet been applied to conservation planning frequently. One reason is surely the dearth of empirical information (particularly on movement rates) necessary to make specific recommendations. As such information becomes available, spatially realistic metapopulation models may be expected to be employed to give specific recommendations. The possibility of their application to pest control has barely been considered, though, if pest populations are really structured in this way, some metapopulation models suggest specific approaches to controlling them.

Landscape ecology became popular about the same time as metapopulation thinking did, and it has some points in common with metapopulation models, particularly the conception of nature as a mass of habitat patches. The key difference is that the matrix in metapopulation models is homogeneous, whereas in landscape models it is heterogeneous—the landscape is a mosaic of many kinds of patches. Unlike metapopulation dynamics, landscape ecology has produced few quantitative models, and its application to both conservation and agriculture to date is largely as a way of thinking rather than a set of specific prescriptions.

REFERENCES CITED

Andrewartha, H. G., and L. C. Birch, eds. 1954. *The Distribution and Abundance of Animals.* Chicago: University of Chicago Press.

Beier, P., and R. F. Noss. 1998. Do Habitat Corridors Provide Connectivity? *Conservation Biology* 6: 1241–52.

Boorman, S. A., and P. R. Levitt. 1973. Group Selection on the Boundary of a Stable Population. *Theoretical Population Biology* 4: 85–128.

Bunce, R. G. H., L. Ryszkowski, and M. G. Paoletti, eds. 1993. *Landscape Ecology and Agroecosystems.* Boca Raton, Fla.: Lewis Publishers.

Bunnell, J. L. 1999. Let's Kill a Panchreston: Giving Fragmentation Meaning. In *Forest Fragmentation: Wildlife and Management Implications,* ed. J. A. Rochelle, L. A. Lehmann, and J. Wisniewski. Leiden: Brill.

Curtis, J. T. 1956. The Modification of Mid-Latitude Grasslands and Forests by Man. In *Man's Role in Changing the Face of the Earth,* ed. W. L. Thomas, 721–36. Chicago: University of Chicago Press.

Flaherty, D. L. 1969. Ecosystem Trophic Complexity and Densities of the Willamette Mite, *Eotetranynchus willamettei* Ewing (Acarina: Tetranychidae). *Ecology* 50 (5): 911–16.

Forman, R. T. T. 1995. *Land Mosaics: The Ecology of Landscapes and Regions.* New York: Cambridge University Press.

Franklin, J. F. 1989. Towards a New Forestry. *American Forests* (November–December): 1–8.

Gardner, R. H., B. T. Milne, M. G. Turner, and R. V. O'Neill. 1987. Neutral Models for the Analysis of Broad-Scale Landscape Pattern. *Landscape Ecology* 1: 19–28.

Gardner, R. H., R. V. O'Neill, M. G. Turner, and V. H. Dale. 1989. Quantifying Scale-Dependent Effects of Animal Movement with Simple Percolation Models. *Landscape Ecology* 3: 217–27.

Gilbert, F. S. 1980. The Equilibrium Theory of Island Biogeography: Fact or Fiction? *Journal of Biogeography* 7: 209–35.

Haila, Y., and O. Järvinen. 1982. The Role of Theoretical Concepts in Understanding the Ecological Theatre: A Case Study on Island Biogeography. In *Conceptual Issues in Ecology,* ed. E. Saarinen, 261–78. Dordrecht, Netherlands: Reidel.

Hanski, I. 1997. Metapopulation Dynamics: From Concepts and Observations to Predictive Models. In *Metapopulation Biology: Ecology, Genetics, and Evolution,* ed. I. A. Hanski and M. E. Gilpin, 69–91. San Diego: Academic Press.

Hanski, I. A., and M. E. Gilpin, eds. 1997. *Metapopulation Biology: Ecology, Genetics, and Evolution.* San Diego: Academic Press.

Hanski, I., and M. Kuussaari. 1995. Butterfly Metapopulation Dynamics. In *Population Dynamics: New Approaches and Synthesis,* ed. N. Cappuccino and P. W. Price, 149–72. San Diego: Academic Press.

Hanski, I., J. Pöyry, T. Pakkala, and M. Kuussaari. 1995. Multiple Equilibria in Metapopulation Dynamics. *Nature* 377: 618–21.
Hanski, I., and D. Simberloff. 1997. The Metapopulation Approach: Its History, Conceptual Domain, and Application to Conservation. In *Metapopulation Biology: Ecology, Genetics, and Evolution,* ed. I. A. Hanski and M. E. Gilpin, 5–26. San Diego: Academic Press.
Hanski, I., and C. D. Thomas. 1994. Metapopulation Dynamics and Conservation: A Spatially Explicit Model Applied to Butterflies. *Biological Conservation* 68: 167–80.
Harrison, S. 1991. Local Extinction in a Metapopulation Context: An Empirical Evaluation. In *Metapopulation Dynamics: Empirical and Theoretical Investigations,* ed. M. E. Gilpin and I. Hanski, 73–88. London: Academic Press.
———. 1994. Metapopulations and Conservation. In *Large-Scale Ecology and Conservation Biology,* ed. P. J. Edwards, N. R. Webb, and R. M. May, 111–28. Oxford, United Kingdom: Blackwell.
Harrison, S., D. D. Murphy, and P. R. Ehrlich. 1988. Distribution of the Bay Checkerspot Butterfly, *Euphydryas editha bayensis:* Evidence for a Metapopulation Model. *American Naturalist* 132: 360–82.
Harrison, S., and A. D. Taylor. 1997. Empirical Evidence for Metapopulation Dynamics. In *Metapopulation Biology: Ecology, Genetics, and Evolution,* ed. I. A. Hanski and M. E. Gilpin, 27–42. San Diego: Academic Press.
Hobbs, R. J. 1992. The Role of Corridors in Conservation: Solution or Bandwagon? *Trends in Ecology and Evolution* 7: 389–92.
IUCN. 1980. *World Conservation Strategy.* Morges, Switzerland: International Union for the Conservation of Nature / United Nations Environmental Program / World Wildlife Fund.
Kareiva, P. 1990. Population Dynamics in Spatially Complex Environments: Theory and Data. In *Population Regulation and Dynamics,* ed. M. P. Hassell and R. M. May, 53–68. London: Royal Society.
Lande, R. 1988. Demographic Models of the Northern Spotted Owl (*Strix occidentalis caurina*). *Oecologia* 75: 601–607.
Levins, R. 1969. Some Demographic and Genetic Consequences of Environmental Heterogeneity for Biological Control. *Bulletin of the Entomological Society of America* 15: 237–40.
Levins, R., and M. Wilson. 1980. Ecological Theory and Pest Management. *Annual Review of Entomology* 25: 287–308.
MacArthur, R. H., and E. O. Wilson. 1967. *The Theory of Island Biogeography.* Princeton, N.J.: Princeton University Press.
Meffe, G. K., and C. R. Carroll. 1994. *Principles of Conservation Biology.* Sunderland, Mass.: Sinauer.
Moore, N. W. 1962. The Heaths of Dorset and Their Conservation. *Journal of Ecology* 50: 369–91.
Naveh, Z., and A. S. Lieberman. 1994. *Landscape Ecology: Theory and Application.* New York: Springer-Verlag.
Nee, S., R. M. May, and M. P. Hassell. 1997. Two-Species Metapopulation Models. In *Metapopulation Biology: Ecology, Genetics, and Evolution,* ed. I. A. Hanski and M. E. Gilpin, 123–47. San Diego: Academic Press.
Price, P. W. 1976 . Colonization of Crops by Arthropods: Non-Equilibrium Communities in Soybean Fields. *Environmental Entomology* 5: 605–11.
Pulliam, H. R. 1988. Sources, Sinks and Population Regulation. *American Naturalist* 132: 652–61.
Pulliam, H. R., and B. J. Danielson. 1991. Sources, Sinks, and Habitat Selection: A Landscape Perspective on Population Dynamics. *American Naturalist* 137: 50–66.
Ryszkowski, L., J. Karg, G. Margarit, M. Paoletti, and R. Zlotin. 1993. Above-Ground Insect Biomass in Agricultural Landscapes of Europe. In *Landscape Ecology and Agroecosystems,* ed. R. G. H. Bunce, L. Ryszkowski, and M. G. Paoletti, 71–82. Boca Raton, Fla.: Lewis Publishers.

Simberloff, D. 1974. Equilibrium Theory of Island Biogeography and Ecology. *Annual Review of Ecology and Systematics* 5: 161–82.

———. 1986a. Design of Nature Reserves. In *Wildlife Conservation Evaluation,* ed. M. B. Usher, 315–37. London: Chapman and Hall.

———. 1986b. Island Biogeographic Theory and Integrated Pest Management. In *Ecological Theory and Integrated Pest Management Practice,* ed. M. Kogan, 19–35. New York: Wiley.

———. 1988. The Contribution of Population and Community Biology to Conservation Science. *Annual Review of Ecology and Systematics* 19: 473–511.

Simberloff, D., J. A. Farr, J. Cox, and D. W. Mehlman. 1992. Movement Corridors: Conservation Bargains or Poor Investments? *Conservation Biology* 6: 493–504.

Singer, M. C., and C. D. Thomas. 1996. Evolutionary Responses of a Butterfly Metapopulation to Human- and Climate-Caused Environmental Variation. *American Naturalist* 148: S9–S39.

Soulé, M. E., and D. Simberloff. 1986. What Do Genetics and Ecology Tell Us about the Design of Nature Refuges? *Biological Conservation* 35: 19–40.

Southwood, T. R. E. 1971. Farm Management in Britain and Its Effect on Animal Populations. In Proceedings of the Third Tall Timbers Conference on Ecological Animal Control by Habitat Management, 29–51. Tallahassee, Fla.: Tall Timbers Research Station.

Stenseth, N. C. 1981. How to Control Pest Species: Applications of Models from the Theory of Island Biogeography in Formulating Pest Control Strategies. *Journal of Applied Ecology* 18: 773–94.

Stern, V. M. 1969. Interplanting Alfalfa in Cotton to Control Lygus Bugs and Other Pests. In Proceedings of the First Tall Timbers Conference on Ecological Animal Control by Habitat Management, 55–69. Tallahassee, Fla.: Tall Timbers Research Station.

Thomas, C. D., and S. Harrison. 1992. Spatial Dynamics of a Patchily Distributed Butterfly Species. *Journal of Animal Ecology* 61: 437–46.

Thomas, J. A., and R. G. Morris. 1995. Rates and Patterns of Extinction among British Invertebrates. In *Extinction Rates,* ed. J. H. Lawton and R. M. May, 111–30. Oxford: Oxford University Press.

Thomas, C. and I. Hanski. 1997. Butterfly Metapopulations. In *Metapopulation Biology: Ecology, Genetics, and Evolution,* ed. I. A. Hanski and M. E. Gilpin, 359–86. San Diego: Academic Press.

van der Meijden, E., and C. A. M. van der Veen-van Wijk. 1997. Tritrophic Metapopulation Dynamics: A Case Study of Ragwort, the Cinnabar Moth, and the Parasitoid *Cotesia popularis.* In *Metapopulation Biology: Ecology, Genetics, and Evolution,* ed. I. A. Hanski and M. E. Gilpin, 387–405. San Diego: Academic Press.

Washburn, J., and H. V. Cornell. 1981. Parasitoids, Patches, and Phenology: Their Possible Roles in the Local Extinction of a Cynipid Gall Wasp Population. *Ecology* 62: 1597–1607.

Wiens, J. A. 1997a. The Emerging Role of Patchiness in Conservation Biology. In *The Ecological Basis of Conservation,* ed. S. T. A. Pickett, R. S. Ostfeld, M. Shachak, and G. E. Likens, 93–107. New York: Chapman and Hall.

———. 1997b. Metapopulation Dynamics and Landscape Ecology. In *Metapopulation Biology: Ecology, Genetics, and Evolution,* ed. I. A. Hanski and M. E. Gilpin, 43–62. San Diego: Academic Press.

Williamson, M. H. 1989. The MacArthur-Wilson Theory Today: True but Trivial? *Journal of Biogeography* 16: 3–4.

Wilson, E. O., and E. O. Willis. 1975. Applied Biogeography. In *Ecology and Evolution of Communities,* ed. M. L. Cody and J. M. Diamond, 523–34. Cambridge, Mass.: Harvard University Press.

Wright, S. 1940. Breeding Structure of Populations in Relation to Speciation. *American Naturalist* 74: 232–48.

Chapter 13

Coevolution of Resistance and Virulence in Host-Parasitoid Interactions

H. C. J. GODFRAY, M. D. E. FELLOWES, A. R. KRAAIJEVELD
Centre for Population Biology
Department of Biology
Imperial College at Silwood Park
Ascot, Berkshire,
United Kingdom

The study of the coevolutionary interaction between organisms and their natural enemies, particularly their parasites and pathogens, is an extremely active area of contemporary evolutionary biology. Because hosts and prey are often under strong natural selection to avoid predation and disease, and because their natural enemies in turn are under strong selection to counter any such adaptation, the resulting interaction is likely to be dynamic at the genetic level, and also to impact upon the population dynamics of the two species (Godfray, 1994). These considerations have led to intense research on how natural selection will mold optimum levels of resistance and virulence (Frank, 1993), how the population dynamic interaction between a species and its natural enemies may be influenced by reciprocal adaptation (Thompson, 1998), and serious discussion about whether parasites and pathogens may be responsible for sexual reproduction (Ryan, 1998) and spectacular adaptations such as extreme secondary sexual characters (Hamilton and Zuk, 1982; Andersson, 1994). From being a backwater of evolutionary research twenty years ago, it is now very much in the mainstream.

Progress in evolutionary research, as in many other branches of science, comes from a combination of research that utilizes the huge diversity of animals, plants, and microbes available for study on the planet, and studies of chosen model systems that are subject to concerted effort by numerous laboratories. Until recently, evolutionary biologists have used model systems rather sparingly, at least compared to those working in genetics and development. However, the ever-increasing information that is available for the classic model systems of genetics and developmental biology makes them more and more attractive for evolutionary study (Powell, 1997).

Of course, if one's area of research is predator-prey or host-pathogen interactions, then choosing a model organism as a subject only provides half the solution to many problems. Nevertheless, there are a growing number of studies that use *Escherichia coli* and its pathogenic bacteriophages, *Arabidopsis* and its fungal pathogens, and mice and their protozoan and helminth parasites to address evolutionary questions.

In this chapter we want to explore the value of *Drosophila* and its parasitoids as a model system for investigating general questions in coevolution, as well as specific issues concerning host-parasitoid coevolution. We begin with a brief description of how hosts and parasitoids interact behaviorally and physiologically, and in the context of the present volume it is a pleasure to acknowledge the major contributions of the Entomology Department at Texas A&M to this field. In the second section, we review some of the main strands of research that have utilized *Drosophila* and its parasitoids, and then go on to discuss in a little more detail recent work from our laboratory. We finish by speculating on further questions that studies on *Drosophila* and its parasitoids may help resolve.

HOST-PARASITOID INTERACTIONS

Parasitoids are insects that lay their eggs in or on the bodies of other insects, their hosts; the eggs hatch and the parasitoid larvae develop on their hosts, eventually killing it (Godfray, 1994; Quicke, 1997). Each larva obtains all the resources it needs to become an adult from a single host, like most parasites, but invariably kills the host, like a predator. The majority of parasitoids are wasps in the order Hymenoptera but a substantial minority are flies (Diptera) with smaller numbers found amongst the beetles (Coleoptera) and other orders. Parasitoids are ubiquitous in nearly all terrestrial ecosystems, and are thought to have a significant effect on the population dynamics of many of their hosts, often being the chief regulatory factor (Hassell, 2000). Estimates of the number of species of parasitoid on earth vary widely, with most workers going for figures in the order of 0.5 to 2 million. Much research on parasitoids is driven by their importance in pest management, and they have been used successfully on many occasions as classical or inundative biological control agents (Holt and Hochberg, 1997).

Different species of parasitoid attack the host egg, larval (nymphal) or pupal stages (or combinations of these), with adult hosts being subject to parasitism by a relatively small fraction of species. One major division in parasitoid life histories is between species that feed as ectoparasitoids from outside the host and those that feed internally as endoparasitoids (Godfray, 1994). A second major division concerns those species that attack growing stages of the host: those that kill or permanently paralyze the host at the time of oviposition (idiobionts), and those that may temporarily paralyze the host during oviposition (koinobionts), but it recovers and continues to feed and grow in size while the parasitoid persists in a state of suspended development. The significance of the koinobiont life history is that it allows the parasitoid to attack hosts that are still too small to provide all the resources that the developing larva requires, though it also exposes the parasitoid to the host's immune system for an extended period of time.

Parasitoids are small organisms that have to search a very complex environment for hosts, which themselves are small organisms. One of the triumphs of the parasitoid biology over the last half-century has been the discovery of many of the adaptations that parasitoids use for host location, for example. Many of these involve the detection of chemical cues emitted from the microhabitat or food plant used by the host, or from secretions or waste products derived from the host. In this area of chemical espionage the host will be under selection to provide as few cues as possible for the parasitoid, but some essential chemical pheromones produced by the host have been shown to be hijacked by the parasitoid. In addition to chemical signals, parasitoids frequently use visual and vibrational information in host location, while a few, for example the dipteran parasitoids of stridulating grasshoppers and crickets, make use of sounds produced by the host. It is also now apparent that parasitoids are not simple programmed automata responding to a fixed hierarchy of cues. Many species have been shown to associate new stimuli with the presence of hosts and to use this in subsequent host location.

Having found a host, the parasitoid must make a series of decisions associated with host acceptance and oviposition. Both textural, surface, and internal chemical cues have been shown to be important in host acceptance, while the size of the host may determine whether the parasitoid uses it for oviposition, or for food (host feeding). One specific aspect of host discrimination is the question of whether to lay further eggs on a previously parasitized host (superparasitism). Almost always fitness is reduced when superparasitizing, and at one time superparasitism was always assumed to be an error, but if the opportunity costs are low enough this behavior may be favored by natural selection. Many parasitoids lay more than one egg per host and have to decide the size of clutch they lay on a particular host. Experiments have shown that both the size and quality of the host, and the opportunities for locating other hosts in the environment, both interact to influence the female's clutch size behavior (Godfray, 1994). Finally, hymenopterous parasitoids have to decide what sex eggs to lay (Hamilton, 1967; Charnov et al., 1981; King, 1987; Hardy, 1994; Godfray, 1994). Hymenoptera are haplodiploid, that is fertilized eggs give rise to female progeny and unfertilized eggs to male. The female thus has proximate control over offspring sex ratio and this means that natural selection can produce locally adaptive sex ratios in a way that it is not possible for organisms locked into a chromosomal sex determination mechanism. The two main factors influencing sex ratio in hymenopterous parasitoids are host size or quality, and sibling mating. Parasitoids tend to lay male eggs on poor quality hosts that give rise to small adults and females on good quality hosts that give rise to large adults. The reason for this is that males gain less from being large than females because male reproductive success is normally less contingent on size than female fecundity (Charnov et al., 1981). As Fisher showed in the 1930s, we expect equal sex ratios to evolve in an outbred population because the fact that all individuals have a mother and a father means that investment in the rarer sex is always the most efficient way of getting genes into future generations, and hence the evolutionary stable strategy is equality of investment. This argument breaks down when a significant proportion of mating takes place amongst siblings (local mate competition) as now sons are competing with each other for reproductive opportunities, and resources "wasted" on sons could be used to produce more daughters (and hence further mates for the remaining male offspring). This argument accounts for the strong correlation between sibling mating and extremely female-biased sex ratios in parasitoid wasps (Hamilton, 1967).

The theory of local mate competition developed by Hamilton in the 1960s was a watershed in evolutionary ecology as it introduced game theoretic thinking into the subject for the first time. Hamilton, an expert entomologist, supported his argument by a comparative survey of largely parasitoid insects, and much of the subsequent experimental tests of his ideas have used hymenopterous parasitoids. Indeed, the study of the adaptive significance of variant sex ratios has been one of the greatest success stories of modern evolutionary analysis, and a substantial proportion of the experimental work in this area has involved parasitoids. There has also been a very fruitful interplay between experimental work with parasitoids and the development of theoretical ideas in a number of areas of evolutionary behavioral ecology, for example in the study of host acceptance and patch use, and in the study of optimal clutch size (Godfray, 1994).

An idiobiont parasitoid can develop safely on a dead or permanently disabled host, but a koinobiont parasitoid during its period of suspended development has to prevent itself from being destroyed by the host's immune system. The chief defense of insects against parasitoids is a cellular immune response called encapsulation (Salt, 1970; Strand and Pech, 1995). The molecular details of this process are not yet understood but parasitoids or other foreign bodies are recognized as nonself, and this causes the aggregation of hemocytes (blood cells) to form a capsule. The walls of the capsule cells break down, and the whole structure melanizes and hardens, resulting in the death of the parasitoid egg or larva, either by suffocation (Salt, 1970) or through chemicals produced by the capsule (Nappi et al., 1995; Nappi and Vass, 1998). Humoral defenses appear also to be important in some species, though cellular encapsulation is the major resistance mechanism.

Parasitoids overcome encapsulation by both passive and active means. Two main passive means of counterattack are the deposition of eggs in tissue such as the brain where they are not encountered by circulating hemocytes, and camouflage in which the parasitoid is not recognized as foreign, perhaps through molecular mimicry of the host's basement membrane or other structures. More active means of counterattack frequently involve the injection of toxins at oviposition that damage or disable the immune system. A more subtle strategy is the injection of viruses or viruslike particles. The best-studied examples of these are the polydnaviruses of ichneumonid and braconid parasitoid wasps (Edson et al., 1981). The polydnavirus genome is incorporated in the wasp genome but in the cells of the female reproductive system is highly amplified and encapsidated within a viral particle (Fleming, 1992; Shelby and Webb, 1999). The virus is injected into the host where its genes are expressed, their products damaging the host immune system. However, the virus does not replicate in the host. Viruses may also be involved in camouflaging the parasitoid, something that is especially likely to be true of the viruslike particles found in some species that contain no DNA (Schmidt and Schuchmann-Feddersen, 1989; Rizki and Rizki, 1990; Rizki et al., 1990). The juvenile parasitoid itself may contribute to its own defense. Some species are known to secrete substances directly into the hemocoel, while in other species the cells of the embryonic membranes dissociate and float free in the hemolymph where they become enormously enlarged (hence their name teratocytes) with a well-developed endoplasmic reticulum and a presumably secretory function (Dahlman and Vinson, 1993).

DROSOPHILA AND ITS PARASITOIDS

Drosophila melanogaster Meigen needs little introduction to a biological audience, though its congeners and its parasitoids may be less familiar. The majority of work on *Drosophila* parasitoids has involved host species that chiefly feed on fermenting substrates such as *D. melanogaster* itself and its close relatives such as *D. simulans* Sturtevant and *D. yakuba* Burla, but also other sap and fruit feeding species in the *obscura* group such *D. subobscura* (Meigen) and *D. pseudoobscura* Frolova. The parasitoids of mushroom-feeding species, and of species that feed in decaying vegetation have also been studied, especially in work looking at microhabitat specialization by their parasitoids.

The two chief types of *Drosophila* parasitoids are koinobiont larval parasitoids that delay development until the host has pupated, and idiobiont pupal parasitoids that attack, kill, and emerge from the pupa. The larval parasitoids belong to two taxonomic groups of parasitoid wasps, the braconid subfamily Alysiinae and the cynipoid family (or subfamily) Eucoilidae. Quite a range of genera from both taxa can be reared from drosophilids, but the majority of laboratory studies have involved just two genera, the alysiine *Asobara* of which *A. tabida* (Nees) is the most important species, and the eucoilid *Leptopilina* of which *L. boulardi* (Barbotin et al.) and *L. heterotoma* (Thompson) are the most important. *A. tabida* and especially *L. heterotoma* have a reasonably broad range of hosts within the genus, while *L. boulardi* is more of a specialist of *D. melanogaster* and its close relatives (Carton et al., 1986). The pupal parasitoids are taxonomically distinct, belonging to the chalcidoid family Pteromalidae, and the proctotrupoid family Diapriidae. The only parasitoid to be studied extensively so far is the pteromalid *Pachycrepoides vindemniae* (Rondani) which has a very broad host range, attacking the pupae of many fly species, and which can sometimes be a pest of geneticists' *Drosophila* cultures.

Until recently, research into the ecology and evolution of *Drosophila* parasitoids has largely been concentrated in continental Europe, particularly the Netherlands and France, though with important physiological studies in the United States. A group at Leiden in the Netherlands including Bakker, van Lenteren, Vet, and especially van Alphen have studied extensively host location, host acceptance, and sex ratio in a variety of *Drosophila*-parasitoid interactions. This group has been one of the pioneers of the application of behavioral ecology to parasitoid studies, in particular using optimal foraging theory to understand parasitoid patch use and oviposition behavior (e.g., van Alphen and Drijver, 1982; van Alphen and Janssen, 1982; van Alphen and Galis, 1983; van Alphen and Vet, 1986; van Alphen and Visser, 1990; Driessen and Hemerik, 1992).

Many parasitoids search for patches of hosts that they then deplete as they discover and parasitize fly larvae. This has parallels with predators searching for prey in a patchy environment. Patch use theory was developed to predict when individuals should leave a patch and begin searching for a second patch, assuming that natural selection maximizes the rate of energy intake. A simple substitution of rate of offspring production for rate of energy intake led to its successful application to parasitoids searching for hosts. Another strand of optimal foraging theory are the diet models that predict when a predator should include relatively poor prey items with the optimal diet set. Again the Leiden group applied these ideas to parasitoids searching for hosts of different size or species (Janssen, 1989). However, for both patch use and host acceptance, there are aspects of parasitoid biology that required new ideas and theoretical development. The presence of previously parasitized hosts in a patch or in the environment in general has no direct counterpart in simple foraging situations. Parasitized hosts may simply represent a new class of less valuable host, but their presence also provides evidence about the quality of the environment and the activities of other searching wasps. As mentioned in the previous section, superparasitism was once thought only to occur accidentally when host discrimination failed, but application of ideas from behavioral ecology led to a reassessment of this behavior and the realization that parasitoids will often be selected to superparasitize (van Alphen and Visser, 1990). A further complication is that a parasitoid encountering a previously parasitized host may not "know" whether it contains her own or another female's eggs. Some species seem to be able to detect self-superparasitism (Visser, 1993), others not. A consequence of this is that some wasps may be unwilling to superparasitize when searching alone in a patch, but avoid superparasitizers if they are aware of other searching conspecifics (Visser et al., 1990).

Another area in which the Dutch group has profitably exploited *Drosophila*-parasitoid interactions is in studying comparative host location strategies. The two main taxa of larval parasitoids use rather different host location strategies (Vet and van Alphen, 1985). *Asobara* females stand motionless and detect the presence of larvae by listening for their vibration, while females of most *Leptopilina* species search more actively, stabbing the medium to locate hosts. Longer-distance host location chiefly involves volatile chemical emanating from the microhabitat. Parasitoids are finely attuned to detecting cues from the microhabitats of their favored host, and indeed studies of host location in *A. tabida* led to the discovery of a sibling species, *A. rufescens* Foerster, that specialized on drosophilids feeding in decaying plant material (Vet et al., 1984). However, *Drosophila* parasitoids are also adept at learning new stimuli and if exposed to hosts in a new microhabitat will subsequently orientate to cues associated with this new source of hosts (Vet et al., 1993).

Drosophila defend themselves against parasitoids by cellular encapsulation. Exactly how their parasitoids counteract host defenses is not clearly understood but in *Asobara* it appears at least partially to involve adhesion of the egg to fat bodies and other host tissues where they achieve protection from circulating hemocytes (Kraaijeveld and van der Wel, 1994; Eslin et al., 1996). Early work showed that *L. boulardi* injected a substance into the host that helped disable the immune system, and more recently this has been shown to be a viruslike particle, though its precise mode of action is not known (Rizki and Rizki, 1990; Rizki et al., 1990; Dupas et al., 1996). In northern and western Europe, the chief host of *A. tabida* is *D. subobscura,* and this fly appears incapable of encapsulating *A. tabida* (or any other *Drosophila* parasitoid so far tested). In the same areas, *D. melanogaster* is also subject to *A. tabida* attack, and its ability to encapsulate the parasitoid egg is relatively low. However, there appears to be a geographical pattern in resistance of *D. melanogaster* to *A. tabida* across Europe, with higher levels found in central southern Europe; this pattern can partly be explained by *D. subobscura* being less common in southern Europe leading to higher rates of parasitism by *A. tabida* on *D. melanogaster* (Kraaijeveld and van Alphen, 1995). Genetic analysis of this trait suggests that the majority of the geographical variation is due to a single locus (Orr and Irving, 1997). There also appears to be a cline in *A. tabida* virulence, with strong strains found in the Mediterranean and weaker strains in the more northern parts of Europe. Again, this cline can largely be understood by the availability of a nonencapsulating host species (*D. subobscura*) to ovipositing parasitoid females in more northern parts of Europe (Kraaijeveld and van Alphen, 1994; Kraaijeveld and Godfray, 1999).

The second major group in continental Europe that has exploited *Drosophila*-parasitoid interactions is the French group of Boulétreau and especially Carton. Building on pioneering work by Walker in the 1950s, they have shown that it is possible to select for changes in the level of resistance in laboratory populations of *D. melanogaster,* and using isofemale lines that there is considerable genetic variation for both resistance and virulence within and across geographical populations of fly and wasp (Carton and Kitano, 1981; Carton and Boulétreau, 1985; Carton and Nappi, 1991; Carton et al., 1992; Carton and Nappi, 1997; Carton and Dupas, 1999). Like the Dutch group, they too have found interesting geographical patterns in host resistance and have also concluded that the structure of the local *Drosophila*-parasitoid community may be critical in explaining these patterns. More recently, and in collaboration with Nappi's laboratory in the United States, they have used isofemale lines to isolate genes involved in the defensive reaction against both *A. tabida* and *L. boulardi* and are mapping their positions in the *Drosophila* genome (Carton and Nappi, 1997; Benassi et al., 1998; Dupas et al., 1998). We return to this work at the end of the next section after we have discussed our work using *Drosophila* and its parasitoids.

THE COSTS AND BENEFITS OF RESISTANCE

Why are hosts not better adapted to destroying their parasitoids? One explanation is that there is a physical or biochemical constraint limiting adaptation: the parasitoids have a trick that enables them to overcome host defenses that the host is just not able to overcome. In population genetic terms no additive variation in resistance exists or can exist upon which natural selection can act. However, the facility with which insects develop resistance should make us suspicious of this *deus ex machina,* while concrete evidence for additive genetic variation in resistance contradicts this hypothesis.

A second explanation is that the host would develop better resistance but has not yet had time. But parasitoid-imposed selection pressures are often very strong, and host-parasitoid interactions are very old, and so this explanation only really makes sense if the target keeps changing: if the parasitoid is always counteradapting to the latest host adaptation so that the interaction is dynamically changing over evolutionary time. Such an explanation raises a whole suite of further questions: what limits the rates of adaptation in the host and the parasitoid and hence what determines the evolutionary lag between the two species?

Yet a third explanation is that the host could evolve better resistance but that greater protection against parasitoids comes at a cost to other components of the host's fitness. In this view the levels of resistance observed in the field are the result of a balance of the costs and benefits of greater and lesser resistance. Additive genetic variation in resistance might exist, but stabilizing selection keeps the optimum at one level. Of course, over time the optimum might shift, and this and the last explanation can be seen as opposite ends of a continuum.

Our laboratory has been trying to understand the different selection pressures acting on levels of host resistance in *Drosophila.* The main cost of host resistance is investment in building and maintaining the encapsulation machinery. Costs of resistance are assumed in virtually all arguments and models of the evolution of host defenses against pathogens, parasites, and parasitoids but have been extremely difficult to measure (Gemmill and Read, 1998). *Drosophila* has proved very valuable for estimating related costs in other branches of life history theory, and in the first subsection we describe how we have sought to demonstrate costs of resistance against parasitoids. Given that you are attacked, defense is essential if you are to survive, but if successful defense results in a severely impaired adult host, then the value of investing in resistance is reduced. Estimating the benefits of successful defense is comparatively easy, and we describe our work in this area in the second subsection. Finally, the value of resistance is influenced by the probability of parasitoid attack, a figure that is set by the population dynamics of the species involved. In the third subsection we describe how this probability may be measured or estimated, and we finish with a brief account of some ongoing experiments.

The Costs of the Machinery

Ideally one would like to manipulate directly using genetic engineering the extent of investment in resistance and then observe the consequences for other components of fitness. Such direct manipulation is not yet possible in *Drosophila* though it is being used to explore the genetically much simpler situation of fungal resistance in *Arabidopsis* (Stahl et al., 1999a; Stahl et al., 1999b). Instead we manipulate resistance levels using artificial selection. The potential drawback of this approach is that one may cause other changes to the selection lines beyond the loci that affect resistance. One minimizes this by rigorously treating selection and control lines in exactly the same way (apart, of course, for the artificial selection), and in maintaining as large as possible populations in the experimental lines to reduce the amount of linkage disequilibrium between resistance and other loci.

In separate experiments we have selected for increased resistance in *D. melanogaster* against *A. tabida* and *L. boulardi* (Kraaijeveld and Godfray, 1997; Fellowes et al., 1998a). In both cases we had four replicate populations of selected flies, each paired with a control population. All lines were maintained at population sizes of at least 100 to reduce inbreeding, and were treated identically in pairs except that the selection lines were subject to parasitism. Selection line flies were exposed to parasitoids in circumstances where approximately 90 percent of larvae were encapsulated (any higher would have led to superparasitism). Pupae of flies that had successfully encapsulated a parasitoid egg could be identified because the melanized capsule is clearly visible through the pupal wall. Only flies that had survived parasitism were used as parents of the next generation. The frequency of encapsulation was assayed each generation in the selection lines, though for logistic reasons slightly less often in the control lines. The parasitoids used in the experiments were from laboratory strains of the two species and were treated in a way to prevent adaptive change in virulence over the course of the experiments. After selection we looked for differences in the fitness of flies kept under stressed and nonstressed conditions.

In both experiments there was a clear response to selection (Figure 13.1). The base population had been established from wild-caught flies in the Netherlands a couple of generations before the start of the experiment, and survived parasitism by *A. tabida* and *L. boulardi* at rates of 5 percent and 0.5 percent respectively. After five generations, the rate of survival had risen to about 60 percent and 45 percent respectively, where it leveled off. There is thus considerable additive genetic variation for resistance in this *D. melanogaster* population (Kraaijeveld and Godfray, 1997; Fellowes et al., 1998a). Calculating the narrow sense heritability is not straightforward for a threshold trait such as this, but a fairly naïve application of the standard formula gives a figure in the range of 20 to 25 percent.

To look for correlated responses to selection we first measured standard demographic parameters such as survival, development time, size, fecundity, and longevity under conditions of unlimited food and little or no larval competition. No differences were found in either experiment between control and selected flies. This was not unexpected as other trade-offs in *Drosophila* have only been detected in situations where the flies were stressed. We thus adapted a protocol of Lewontin (1955) to examine larval survival under conditions of increasing competition for food: fifteen experimental larvae (*either* from control or selection lines) were placed on a measured amount of food with fifteen larvae of a genetically marked tester strain. The relative survival of experimental flies compared to tester flies allows the construction of a competitive ability index. In both cases, no differences were found between control and selection lines when plenty of food was present, but as resources became more

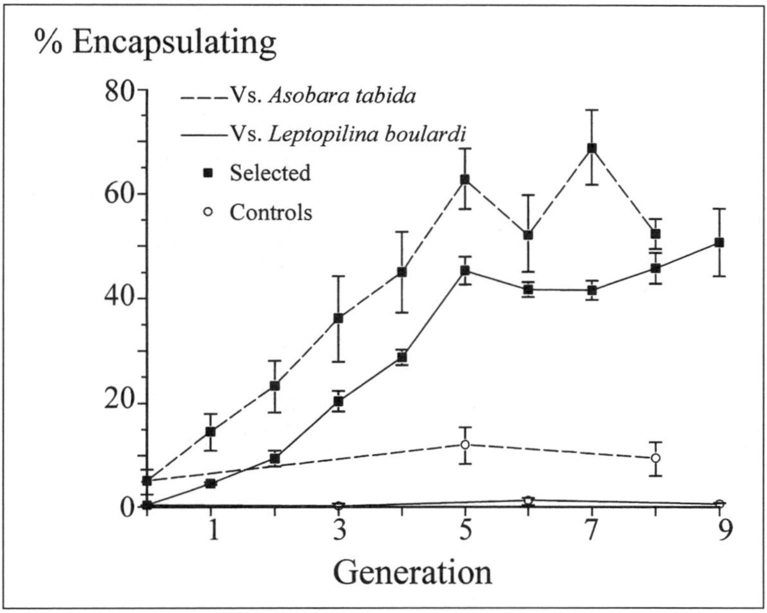

Figure 13.1. Changes in encapsulation ability (vs. *Asobara tabida:* broken lines; vs. *Leptopilina boulardi:* continuous lines) of lines (selected: filled boxes; controls: open circles) over nine generations of selection (mean ± S.E.).

limiting the rate of survival of selection line flies declined (Figure 13.2). Thus there appears to be a cost of resistance: higher survival after parasitoid attack is associated with lower survival under conditions of larval competition for food (Kraaijeveld and Godfray, 1997; Fellowes et al., 1998a). Moreover, the same cost was found in lines selected against two different (unrelated) parasitoid species. It should be stressed here that *Drosophila* larvae in the wild are frequently found at densities where competition is important. The presence of a trade-off is also supported by the observation that when selection was relaxed, levels of resistance declined (Kraaijeveld et al., 1998).

How might higher levels of resistance impact on competitive ability? One possibility is that the genes favored in the selection lines have direct negative pleiotropic effects on other systems in the fly larva. Alternatively, we may be selecting for a shift in the resource budget of the fly larvae, with reduced investment in trophic function and increased investment in defense. Previous studies of *Drosophila,* selected for increased competitive ability, found that a common response was an increase in the rate of feeding measured by the rate of retraction of the cephalo-pharyngeal skeleton (Joshi and Mueller, 1988; Joshi and Mueller, 1996). This led us to measure feeding rates in our sets of lines (Figure 13.3) and we found that in both experiments increased resistance

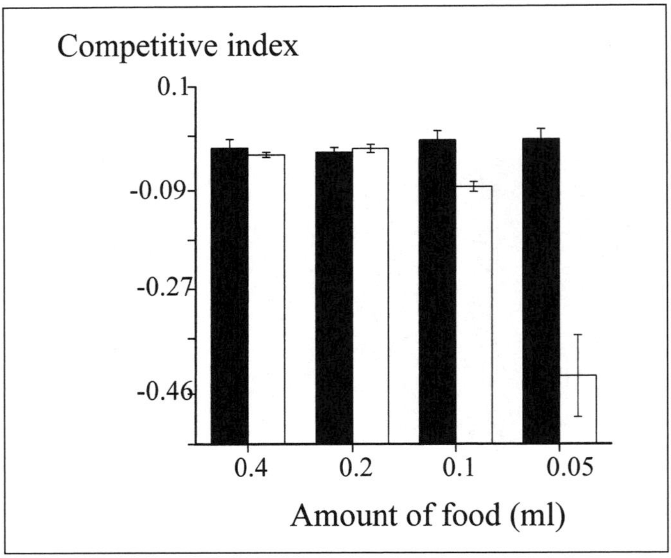

Figure 13.2. Relative competitive ability of lines selected for resistance to *Leptopilina boulardi* (open bars) and their controls (filled bars; mean ± S.E.). Lines selected for resistance to *Asobara tabida* and their controls showed similar patterns.

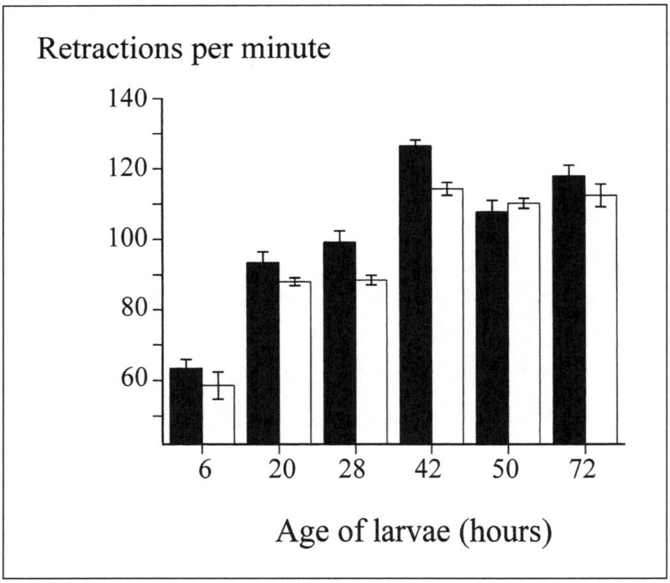

Figure 13.3. Feeding rate (mean sclerite retractions per minute^{-1} ± S.E.) of lines selected for resistance to *Leptopilina boulardi* (open bars) and their controls (filled bars). Lines selected for resistance to *Asobara tabida* and their controls showed similar patterns.

was associated with reduced rates of feeding (Fellowes et al., 1999a). This is consistent with, though does not prove, the hypothesis that we have selected for a rebudgeting of resources from trophic to defensive functions.

An obvious further question to ask is does increased resistance to *A. tabida* impart higher survival after *L. boulardi* attack and vice versa. We extended the experiment to include a third parasitoid, *L. heterotoma* (we also looked at temperature reaction norms in resistance against the three species though we do not discuss this here). The result was surprising: lines selected against *L. boulardi* showed enhanced resistance to *A. tabida* and *L. heterotoma,* while lines selected against *A. tabida* were more resistant to *L. heterotoma* but not *L. boulardi* (Fellowes et al., 1999b). A possible explanation for this is as follows: Recall *A. tabida* and *L. heterotoma* are relatively polyphagous parasitoids of *Drosophila,* while *L. boulardi* is much more of a specialist. Perhaps host resistance has two components, one that is operative against parasitoids in general and a second component that is required, in addition, for successful defense against the specialist *L. boulardi.* Carton and Nappi (1997) and Dupas et al. (1998) have found separate genes that are implicated in defense against *A. tabida* and *L. boulardi,* and it will be interesting to see whether these loci are involved in the response to selection in our experiments.

Consequences of Successful Defense

The first people to examine the consequences of successful defense against parasitoids for adult performance were Carton and David (1983) who studied the fitness of *D. melanogaster* that had successfully encapsulated *L. boulardi.* Female flies with capsules tended to be smaller and to produce fewer eggs. Fellowes et al. (1999c) also looked at *D. melanogaster,* but this time after attack by *A. tabida.* Again female flies containing capsules were smaller and produced significantly fewer eggs than controls. Male flies had smaller thoraxes, though not smaller wing lengths. Capsule-bearing males that were allowed repeated copulations with females did not show a reduction in their number of offspring, but those allowed to copulate only once did. Finally, there was also a reduction in the thickness of the puparial wall (Fellowes et al., 1998b).

What causes the reduced fitness of flies with capsules? We do not yet know but it is likely to be either a direct negative effect of parasitism, or a consequence of the redirection of resources away from growth toward capsule formation. In the second case, the limiting resource might be energy, or it might be the chemicals required for encapsulation. Successful capsule formation requires the deposition of melanin, and it is possible that as a consequence less melanin is laid down in the puparial wall. Interestingly, in isofemale line studies, parasitoid survival increases under conditions of larval crowding (Wajnberg et al., 1985), perhaps because the host has fewer resources to devote to defense.

Because pupae with capsules are smaller, Fellowes et al. (1998b) predicted that they might be relatively less preferred by the pupal parasitoid *P. vindemniae* in comparison with unparasitized pupae. This species is known to be able to discriminate between hosts of different quality (van Alphen and Thunnissen, 1983). In fact the parasitoid did not distinguish between old pupae with and without capsules, while it actually showed a preference for young pupae with a capsule. The reason for this preference is not yet understood, though one possibility is that *P. vindemniae* favor parasitized pupae as it takes less time to drill through their thinner puparial walls (Fellowes et al., 1998b).

A final and counterintuitive consequence of encapsulation was suggested by Carton et al. (1986; referring to unpublished experimental work by Gouyon et al.). They had preliminary data that females preferred to mate with males bearing a capsule, visible through the abdominal wall. Perhaps females choose males that were able to demonstrate their ability to survive parasitism and that could thus pass on these "good" genes to their offspring (though they also demonstrate the fact that they were discovered by searching parasitoids which may be due to "bad" genes). This intriguing idea chimed with modern ideas about the role of parasites in sexual selection (Hamilton and Zuk, 1982). Unfortunately, more recent attempts to demonstrate a preference for males carrying capsules have failed to show this effect (Kraaijeveld et al., 1997).

Probability of Parasitism

The advantages of being able to survive parasitoid attack clearly depend on the probability that this happens in the field. Essentially, this part of the equation is a function of the system's population dynamics, though obviously the nature of the population dynamic interaction between the host and the parasitoid will itself be influenced by how natural selection influences resistance and virulence. One straightforward way of estimating the probability of attack is simply to go into the field and measure it. However, it is here that *Drosophila,* which is such a good model organism in so many other ways, is less than ideal. We know surprisingly little about the ecology of *Drosophila* feeding on fermenting substrates in the wild (Powell, 1997). It is very easy to collect flies and parasitoids by putting out artificial bait, but it is much harder to locate and study natural populations of flies feeding in sap fluxes and such, and very difficult to estimate typical levels of parasitoid attack and whether parasitoids have an important influence on different species population dynamics.

A poor substitute for facts is to investigate the linkage between population dynamics and population genetics using a modeling approach. One of us has recently reviewed models of the evolution of parasitoid resistance and virulence in a population dynamic framework (Godfray, 2000) and here we merely discuss one recent model that has been inspired by our experimental work on *Drosophila* and its parasitoids but is still very much an abstract model rather than an attempt to describe this particular system (Sasaki and Godfray, 1999).

Consider a host-parasitoid interaction where the parasitoid is the major factor regulating the host's population density. We will model the population dynamics using a standard discrete-generation formulation due to Nicholson and Bailey (1935) and May (1978). Essentially the probability of the host avoiding attack by a parasitoid (which may or may not lead to encapsulation) is the zero term of the negative binomial distribution. What this means biologically is that the risk of parasitoid attack increases with parasitoid density, but less steeply than if the process is governed by a random (Poisson) distribution. Specifically, as long as the controlling parameter of the negative binomial distribution is beyond a threshold value, sufficient hosts escape attack by high-density populations of parasitoid that the interaction is stable and not subject to the diverging oscillations that lead to instability of the original Nicholson-Bailey model with random parasitoid search.

We further assume that each host and parasitoid invests a certain amount of resources in resistance and virulence respectively. Then if a host that invests x in defense is attacked by a parasitoid that invests y in virulence, the probability of the host as opposed to the parasitoid triumphing is a function of the difference, $x - y$. In this view, resistance and virulence are graded traits like running speed in a predator-prey interaction: absolute increases in x or y are always beneficial. Were resistance and virulence to be determined by genotypic matching or a gene-for-gene type mechanism, it would be the absolute *difference* between x and y that would determine the outcome. More work needs to be done on this issue, but we believe that the available evidence points strongly toward a graded rather than a matching interaction. Finally, we assume that higher values of resistance and virulence lead to increasing costs in terms of fecundity or survival (though costs are not density-dependent as we found them to be in *Drosophila*), and model the genetics by assuming fifty competing clones each of hosts and parasitoids with different levels of resistance and virulence.

The results are quite complex but we believe show two relevant biological outcomes. In the first, hosts invest nothing in resistance and parasitoids intermediate amount in virulence; these investment levels do not change over time. This outcome occurs especially if host resistance is expensive. In the second, hosts and parasitoids engage in an arms race with spiraling levels of the two traits until a point is reached when the host population "gives up" and ceases to invest in resistance. There is now no selection on costly high levels of virulence and the parasitoid drastically reduces investment in this trait. The host is now selected to begin investing moderate amounts in resistance and the arms race begins again. The second outcome thus predicts continual cycles of arms races and host surrender.

It is fairly obvious how reciprocal selection could stoke an arms race, but why does the host blink first and surrender? The reason for this is the inherent asymmetry in the host-parasitoid interaction: parasitoids have to be able to develop in a host and if the whole host population is investing heavily in defense then the parasitoid population has little alternative but to invest heavily in virulence. But the host population, speaking anthropomorphically, has a choice: it can gamble on not being found by a parasitoid. Under conditions when the outcome is a parasitoid-host arms race, this gamble is only judged worth taking by the arbiter of natural selection when the arms race has driven the cost of effective resistance to very high levels. When the first outcome applies, for example when host resistance is intrinsically expensive, the gamble is always worth taking.

One strong and a couple of tentative conclusions can be drawn for this work. The strong conclusion is that in modeling these issues it is essential to consider both population and genetic dynamics. The outcome of the model depends critically on the probability of parasitoid attack, and this is determined by the population dynamics of the interaction. The first tentative conclusion is that first outcome of the model may correspond to cases such as *D. subobscura,* which are hosts that appear never to defend themselves against their parasitoids; perhaps it is in their interests to save resources and gamble. The second tentative conclusion is that the substantial additive genetic variation maintained in the continual cycles predicted second outcome of the model may help to explain the substantial genetic variation in resistance that has now been found several times in *Drosophila* and also in pea aphid and its parasitoid (*Acyrthosiphon pisum* [Harris] and *Aphidius ervi* [Haliday]) by Henter (1995) and Henter and Via (1995), the one major non-*Drosophila* study of resistance and virulence. However, these conclusions must firmly be treated as tentative as the models neither incorporate alternative hosts and parasitoids nor density-dependent costs of resistance, both factors known to be important in the *Drosophila* system.

Ongoing Work

The studies described so far have identified a cost to resistance and shown that it is associated with reduced rates of feeding. We are currently trying to explore the mechanisms underlying the increased resistance in the selected lines, and whether it is associated with higher hemocyte densities. In an interesting study, Eslin and Prevost (1996; 1998) found a cross-species correlation between hemocyte densities and the average ability of the *Drosophila* species to survive *Asobara tabida* attack. We are also trying to explore the genetic basis of resistance; whether it is mono- or polygenic, and whether the genes responsible in our experiments are the same as those identified by Orr and Irving (1997), Carton and Nappi (1997), and Dupas et al. (1998). To do this, we are using a dense microsatellite map of the *D. melanogaster* genome and looking to see if certain loci have become homozygous in the selection treatments.

Ongoing experiments are trying to select for increased virulence in *A. tabida* and to see whether this is correlated with any other aspects of the wasp's fitness. Additive genetic variance for virulence is substantially less than for resistance. We are also exploring how a very different parasitoid, *Pachycrepoides vindemniae,* the idiobiont that attacks *Drosophila* pupae, might select for different types of resistance or avoidance mechanisms in the host.

CONCLUSIONS

Parasitoids are fascinating in their own right but are also useful as experimental systems for investigating more general questions in biology. In recent years, parasitoid biologists have made important contributions to both evolutionary behavioral ecology and chemical behavioral ecology and to population ecology. Work with parasitoids has illuminated aspects of insect physiology and development, as well as the burgeoning field of non-Mendelian genetic elements. But forty years ago, parasitoid workers were at the forefront of genetics, with large numbers of mutants of *Bracon hebetor* (Say) and *Naso-*

nia (then *Mormoniella*) *vitripennis* (Walker) known, most of which are now lost. Might parasitoids again be useful model system in evolutionary genetics?

Allen Orr has recently made the interesting point that much of recent population genetics has been concerned with the dynamics of deleterious or near-neutral mutants (Orr, 1998). Such alleles are abundant and relatively easy to treat analytically and of great interest in the context of the neutralist-selectionist debate that has dominated so much of the subject since the mid-1960s. But many interesting questions posed in the 1930s such as the nature of the genetics of adaptations have received relatively short shrift (Orr and Coyne, 1992). Orr argues that one of the reasons for this is the difficulty of finding suitable adaptations amenable to experimentation and that parasitoid resistance in *Drosophila* may be an excellent model system (Orr and Irving, 1997). Other questions such as the maintenance of clinal variations and the role of arms races and community structure in natural enemy interactions may also be addressed with this system.

The imminent completion of the *Drosophila* genome project, and the gene chip and other technologies that will come in its wake, will markedly change the way biologists go about their work, and the questions that they are able to address. This is already clear and happening for geneticists, cell biologists, and developmental biologists, for whom the arguments for working with one of the classic model organisms are becoming ever more insistent. We believe that these technical advances will have nearly as great an impact on evolutionary ecologists and will change both their modus operandi and the questions that they are able to ask. We look forward to radical advances in our understanding of the genetic basis of resource-consumer coevolution, and believe that parasitoid studies will play an important role in this adventure.

We finish with an open question. Do parasitoid workers need a model organism on which to concentrate their research; a species for which a detailed genetic map is available—perhaps even a full genome sequence (those serried ranks of sequencers will need something to keep them occupied once man, mouse, fly, worm, and cress are completed!)? Should that parasitoid be one of the old genetic war horses, *Nasonia* or *Bracon;* or one about which a lot is known physiologically, *Campoletis, Microplitus,* or *Cotesia;* or one of immense economic importance, *Trichogramma* or *Encarsia*? But inevitably one would also want the host, and we do already have (surely by the time this is published) *Drosophila* and so is there not an overwhelming argument that we should plump for a *Drosophila* parasitoid, even if physiologists say it is too small and ecologists say it is too difficult to study in the field? But should it be *Asobara tabida* (Nees) or *Leptopilina boulardi* Barbotin et al.?!

REFERENCES CITED

Andersson, M. 1994. *Sexual Selection.* Princeton, N.J.: Princeton University Press.

Askew, R. R., and M. R. Shaw. 1986. Parasitoid Communities: Their Size, Structure, and Development. In *Insect Parasitoids,* ed. J. K. Waage and D. Greathead. London: Academic Press.

Benassi, V., F. Frey, and Y. Carton. 1998. A New Specific Gene for Wasp Cellular Immune Resistance in *Drosophila. Heredity* 80: 347–52.

Carton, Y., and M. Boulétreau. 1985. Encapsulation Ability of *Drosophila melanogaster:* A Genetic Analysis. *Developmental and Comparative Immunology* 9: 211–19.

Carton, Y., M. Boulétreau, J. J. M. van Alphen, and J. C. van Lenteren. 1986. The *Drosophila* Parasitic Wasps. In *The Genetics and Biology of Drosophila,* vol. 3e, ed. E. Ashburner, H. L. Carson, and J. N. Thompson Jr., 347–94. London: Academic Press.

Carton, Y., and J. R. David. 1983. Reduction of Fitness in *Drosophila* Adults Surviving Parasitism by a Cynipid Wasp. *Experentia* 39: 231–33.

Carton, Y., and S. Dupas. 1999. Genetics of Resistance and Virulence in Host-Parasitoid Systems. In *Techniques in Insect Immunology,* ed. A. Wiesner, G. B. Dunphy, V. J. Marmaras, I. Morishima, M. Sugumaran, and M. Yamakawa, 279–88. Fair Haven, N.J.: SOS Publishing.

Carton, Y., F. Frey, and A. Nappi. 1992. Genetic Determinism of the Cellular Immune Reaction in *Drosophila melanogaster. Heredity* 69: 393–99.

Carton, Y., and H. Kitano. 1981. Evolutionary Relationships to Parasitism by Seven Species of the *Drosophila melanogaster* Subgroup. *Biological Journal of the Linnean Society of London* 16: 227–41.

Carton, Y., and A. Nappi. 1991. The *Drosophila* Immune Reaction and the Parasitoid Capacity to Evade It: Genetic and Coevolutionary Aspects. *Acta Oecologia* 12: 89–104.

Carton, Y., and A. J. Nappi. 1997. *Drosophila* Cellular Immunity against Parasitoids. *Parasitology Today* 13: 218–27.

Charnov, E. L., R. L. Los-den Hartogh, W. T. Jones, and J. van den Assem. 1981. Sex Ratio Evolution in a Variable Environment. *Nature* 289: 27–33.

Dahlman, D. L., and S. B. Vinson. 1993. Teratocytes: Developmental and Biochemical Characteristics. In *Parasites and Pathogens of Insects,* vol. 1, *Parasites,* ed. E. N. Beckage, S. N. Thompson, and B. A. Federici, 145–65. London: Academic Press.

Driessen, G., and L. Hemerik. 1992. The Time and Egg Budget of *Leptopilina clavipes,* a Parasitoid of Larval *Drosophila. Ecological Entomology* 17: 17–27.

Dupas, S., M. Brehelin, F. Frey, and Y. Carton. 1996. Immune Suppressive Virus-Like Particles in a *Drosophila* Parasitoid: Significance of Their Intraspecific Morphological Variations. *Parasitology* 113: 207–12.

Dupas, S., F. Frey, and Y. Carton. 1998. A Single Parasitoid Segregating Factor Controls Immune Suppression in *Drosophila. Journal of Heredity* 89: 306–11.

Edson, K. M., S. B. Vinson, and M. D. Summers. 1981. Virus in a Parasitoid Wasp: Suppression of the Immune System in the Parasitoid's Host. *Science* 211: 582–83.

Eslin, P., P. Giordanengo, Y. Fourdrain, and G. Prevost. 1996. Avoidance of Encapsulation in the Absence of VLP by a Braconid Parasitoid of *Drosophila* Larvae: An Ultrastructural Study. *Canadian Journal of Zoology* 74: 2193–98.

Eslin, P., and G. Prevost. 1996. Variation in *Drosophila* Concentration of Haemocytes Associated with Different Ability to Encapsulate *Asobara tabida* Larval Parasitoid. *Journal of Insect Physiology* 42: 549–55.

———. 1998. Hemocyte Load and Immune Resistance to *Asobara tabida* are Correlated in Species of the *Drosophila melanogaster* Subgroup. *Journal of Insect Physiology* 44: 807–16.

Fellowes, M. D. E., A. R. Kraaijeveld, and H. C. J. Godfray. 1998a. Trade-Off Associated with Selection for Increased Ability to Resist Parasitoid Attack in *Drosophila melanogaster. Proceedings of the Royal Society of London, Series B* 265: 1553–58.

Fellowes, M. D. E., P. Masnatta, A. R. Kraaijeveld, and H. C. J. Godfray. 1998b. Pupal Parasitoid Attack Influences the Relative Fitness of *Drosophila* That Have Encapsulated Larval Parasitoids. *Ecological Entomology* 23: 281–84.

Fellowes, M. D. E., A. R. Kraaijeveld, and H. C. J. Godfray. 1999a. Association between Feeding Rate and Selection for Parasitoid Resistance in *Drosophila melanogaster. Evolution* 53: 1302–05.

———. 1999b. Cross-Resistance Following Artificial Selection for Increased Defense against Parasitoids in *Drosophila melanogaster. Evolution* 53: 966–72.

———. 1999c. The Relative Fitness of *Drosophila melanogaster* (Diptera, Drosophilidae) That Have Successfully Defended Themselves against the Parasitoid *Asobara tabida* (Hymenoptera, Braconidae). *Journal of Evolutionary Biology* 12: 123–28.

Fleming, J. A. G. W. 1992. Polydnaviruses: Mutualists and Pathogens. *Annual Review of Entomology* 37: 401–25.

Frank, S. A. 1993. Coevolutionary Genetics of Plants and Pathogens. *Evolutionary Ecology* 7: 45–75.

Gemmill, A. W., and A. F. Read. 1998. Counting the Costs of Disease Resistance. *Trends in Ecology and Evolution* 13: 8–9.

Godfray, H. C. J. 1994. *Parasitoids: Behavioral and Evolutionary Ecology.* Princeton, N.J.: Princeton University Press.

———. 2000. Host Resistance, Parasitoid Virulence, and Population Dynamics. In *Parasitoid Population Biology,* ed. M. Hochberg and A. R. Ives, 121–38. Princeton, N.J.: Princeton University Press.

Hamilton, W. D. 1967. Extraordinary Sex Ratios. *Science* 156: 477–87.

Hamilton, W. D., and M. Zuk. 1982. Heritable True Fitness and Bright Birds: A Role for Parasites? *Science* 218: 384–87.

Hardy, I. C. W. 1994. Sex Ratio and Mating Structure in the Parasitoid Hymenoptera. *Oikos* 69: 3–20.

Henter, H. J. 1995. The Potential for Coevolution in a Host-Parasitoid System. II. Genetic Variation within a Population of Wasps in the Ability to Parasitize an Aphid Host. *Evolution* 49: 439–45.

Henter, H. J., and S. Via. 1995. The Potential for Coevolution in a Host-Parasitoid System. I. Genetic Variation within an Aphid Population in Susceptibility to a Parasitic Wasp. *Evolution* 49: 427–38.

Holt, R. D., and M. E. Hochberg. 1997. When is Biological Control Evolutionarily Stable (or Is It)? *Ecology* 78: 1673–83.

Janssen, A. R. M. 1989. Optimal Host Selection by *Drosophila* Parasitoids in the Field. *Functional Ecology* 3: 469–79.

Joshi, A., and L. D. Mueller. 1988. Evolution of Higher Feeding Rate in *Drosophila* Due to Density Dependent Natural Selection. *Evolution* 42: 1090–93.

———. 1996. Density-Dependent Natural Selection in *Drosophila:* Trade-Offs between Larval Food Acquisition and Utilization. *Evolutionary Ecology* 10: 463–74.

King, B. H. 1987. Offspring Sex Ratios in Parasitoid Wasps. *Quarterly Review of Biology* 62: 367–96.

Kraaijeveld, A. R., D. A. Emmett, and H. C. J. Godfray. 1997. Absence of Direct Sexual Selection for Parasitoid Resistance in *Drosophila melanogaster. Journal of Evolutionary Biology* 10: 337–42.

Kraaijeveld, A. R., and H. C. J. Godfray. 1997. Trade-Off between Parasitoid Resistance and Larval Competitive Ability in *Drosophila melanogaster. Nature* 389: 278–80.

———. 1999. Geographical Patterns in the Evolution of Resistance and Virulence in *Drosophila* and Its Parasitoids. *American Naturalist* 153: S61–S74.

Kraaijeveld, A. R., and J. J. M. van Alphen. 1994. Geographical Variation in Resistance of the Parasitoid *Asobara tabida* against Encapsulation by *Drosophila melanogaster* Larvae: The Mechanism Explored. *Physiological Entomology* 19: 9–14.

———. 1995. Geographical Variation in Encapsulation Ability of *Drosophila melanogaster* Larvae and Evidence for Parasitoid-Specific Components. *Evolutionary Ecology* 9: 10–17.

Kraaijeveld, A. R., J. J. M. van Alphen, and H. C. J. Godfray. 1998. The Coevolution of Host Resistance and Parasitoid Virulence. *Parasitology* 116: S29–S45.

Kraaijeveld, A. R.. and N. N. van der Wel. 1994. Geographic Variation in Reproductive Success of the Parasitoid *Asobara tabida* in Larvae of Several *Drosophila* Species. *Ecological Entomology* 19: 221–29.

Lewontin, R. C. 1955. The Effects of Population Density and Composition on Viability in *Drosophila melanogaster. Evolution* 9: 27–41.

May, R. M. 1978. Host-Parasitoid Systems in Patchy Environments: A Phenomenological Model. *Journal of Animal Ecology* 47: 833–43.

Nappi, A. J., and E. Vass. 1998. Hydrogen Peroxide Production in Immune-Reactive *Drosophila melanogaster. Journal of Parasitology* 84: 1150–57.

Nappi, A. J., E. Vass, F. Frey, and Y. Carton. 1995. Superoxide Anion Generation in *Drosophila* during Melanotic Encapsulation of Parasites. *European Journal of Cell Biology* 68: 450–56.

Nicholson, A. J., and V. A. Bailey. 1935. The Balance of Animal Populations. Part 1. *Proceedings of the Zoological Society of London* 3: 551–98.

Orr, H. A. 1998. The Population Genetics of Adaptation: The Distribution of Factors Fixed during Adaptive Evolution. *Evolution* 52: 935–49.

Orr, H. A., and J. A. Coyne. 1992. The Genetics of Adaptation: A Reassessment. *American Naturalist* 140: 725–42.

Orr, H. A., and S. Irving. 1997. The Genetics of Adaptation: The Genetic Basis of Resistance to Wasp Parasitism in *Drosophila melanogaster. Evolution* 51: 1877–85.

Powell, J. R. 1997. *Progress and Prospects in Evolutionary Biology: The Drosophila Model.* Oxford: Oxford University Press.

Quicke, D. L. J. 1997. *Parasitic Wasps.* London: Chapman and Hall.

Rizki, R. M., and T. M. Rizki. 1990. Parasitoid Virus-Like Particles Destroy *Drosophila* Cellular Immunity. *Proceedings of the National Academy of Sciences USA* 87: 8388–92.

Rizki, T. M., R. M. Rizki, and Y. Carton. 1990. *Leptopilina heterotoma* and *L. boulardi:* Strategies to Avoid Cellular Defense Responses of *Drosophila melanogaster. Experimental Parasitology* 70: 466–75.

Ryan, M. J. 1998. Sexual Selection, Receiver Biases, and the Evolution of Sex Differences. *Science* 281: 1999–2003.

Salt, G. 1970. The Cellular Defence Reactions of Insects. Cambridge Monographs in Experimental Biology No.16. Cambridge: Cambridge University Press.

Sasaki, A., and H. C. J. Godfray. 1999. A Model for the Coevolution of Resistance and Virulence in Coupled Host-Parasitoid Interactions. *Proceedings of the Royal Society of London, Series B* 266: 455–63.

Schmidt, O., and I. Schuchmann-Feddersen. 1989. Role of Virus-Like Particles in Parasitoid-Host Interactions of Insects. *Subcellular Biochemistry* 15: 91–119.

Shelby, K. S., and B. A. Webb. 1999. Polydnavirus-Mediated Suppression of Insect Immunity. *Journal of Insect Physiology* 45: 507–14.

Stahl, E. A., J. Bergelson, G. Dwyer, R. Mauricio, and M. Kreitman. 1999a. Trench Warfare, Not an Evolutionary Arms Race, Governs Disease Resistance Polymorphism in *Arabidopsis Rpm 1. MS.*

Stahl, E. A., G. Dwyer, R. Mauricio, M. Kreitman, and J. Bergelson. 1999b. Dynamics of Disease Resistance Polymorphism at the *Rpm 1* locus of *Arabidopsis. Nature* 400: 667–71.

Strand, M. R., and L. L. Pech. 1995. Immunological Basis for Compatibility in Parasitoid-Host Relationships. *Annual Review of Entomology* 40: 31–56.

Thompson, J. N. 1998. The Population Biology of Coevolution. *Researches in Population Ecology* 40: 159–66.

van Alphen, J. J. M., and R. A. B. Drijver. 1982. Host Selection by *Asobara tabida* Nees (Braconidae: Alysiinae), a Larval Parasitoid of Fruit Inhabiting *Drosophila* Species. I. Host

Stage Selection with *Drosophila melanogaster* as Host Species. *Netherlands Journal of Zoology* 32: 215–31.

van Alphen, J. J. M., and F. Galis. 1983. Patch Time Allocation and Parasitization Efficiency of *Asobara tabida,* a Larval Parasitoid of *Drosophila. Journal of Animal Ecology* 52: 937–52.

van Alphen, J. J. M., and A. R. M. Janssen. 1982. Host Selection by *Asobara tabida* Nees (Braconidae; Alysiinae), a Larval Parasitoid of Fruit Inhabiting *Drosophila* Species. II. Host Species Selection. *Netherlands Journal of Zoology* 32: 194–214.

van Alphen, J. J. M. and I. Thunnissen. 1983. Host selection and sex allocation by *Pachycrepoideus vindemiae* Rondani (Pteromalidae) as a facultative hyperparasitoid of *Asobara tabida* Nees (Braconidae; Alysiinae) and *Leptopilina heterotoma* (Cynipoidea; Eucoilidae). *Netherlands Journal of Zoology* 33: 497–514.

van Alphen, J. J. M., and L. E. M. Vet. 1986. An Evolutionary Approach to Host Finding and Selection. In *Insect Parasitoids,* ed. J. K. Waage and D. Greathead, 23–61. London: Academic Press.

van Alphen, J. J. M., and M. E. Visser. 1990. Superparasitism as an Adaptive Strategy. *Annual Review of Entomology* 35: 59–79.

Vet, L. E. M., C. J. Janse, C. van Achterberg, and J. J. M. van Alphen. 1984. Microhabitat Location and Niche Segregation in Two Sibling Species of *Drosophila* Parasitoids: *Asobara tabida* Nees and *A. rufescens* Foerster (Braconidae: Alysiinae). *Oecologia* 61: 182–88.

Vet, L. E. M., M. B. Sokolowsi, D. E. MacDonald, and H. Snellen. 1993. Response of a Generalist and a Specialist Parasitoid (Hymenoptera: Eucoilidae) to Drosophilid Larval Kairomones. *Journal of Insect Behavior* 6: 615–24.

Vet, L. E. M., and J. J. M. van Alphen. 1985. A Comparative Approach to the Host Detection Behavior of Parasitic Wasps. 1. A Qualitative Study on Eucoilidae and Alysiinae. *Oikos* 44: 478–86.

Vet, L. E. M., and A. W. Groenewold. 1990. Semiochemicals and Learning in Parasitoids. *Journal of Chemical Ecology* 16: 3119–35.

Vinson, S. B. 1985. The Behavior of Parasitoids. In *Comprehensive Insect Physiology, Biochemistry, and Pharmacology,* ed. G. A. Kerkut and L. I. Gilbert, 417–69. New York: Pergamon Press.

Visser, M. E. 1993. Adaptive Self- and Conspecific Superparasitism in the Solitary Parasitoid *Leptopilina heterotoma* (Hymenoptera: Eucoilidae). *Behavioral Ecology* 4: 22–28.

Visser, M. E., J. J. M. van Alphen, and H. W. Nell. 1990. Adaptive Superparasitism and Patch Time Allocation in Solitary Parasitoids: The Influence of the Number of Parasitoids Depleting a Patch. *Behaviour* 114: 21–36.

Wajnberg, E., G. Prevost, and M. Boulétreau. 1985. Genetic and Epigenetic Variation in *Drosophila* Larvae Suitability to a Hymenopterous Endoparasitoid. *Entomophaga* 30: 187–91.

Chapter 14

A New Look at the Coevolution of the Insect-Plant Relationship and Its Relevance to Agriculture

S. BRADLEIGH VINSON
Department of Entomology
Texas A&M University–College Station

When someone brings up a discussion of the relationship between insects and plants, the first thing that comes to the mind of most people, including entomologists, is the interaction between plants and their insect pests. In fact, departments of entomology were developed in land grant institutions as a discipline to deal with man's major competitors for food and fiber—the insects. As discussed by Vinson (1999) the development of our ideas regarding insect-plant relationships formed from the need to manage pest insects. Both biological control (DeBach and Rosen, 1991) and the use of host plant resistance (van Emden and Way, 1973; Harris, 1980) became the foundation of biological pest management. Later, insecticides were developed and have dominated our management practices ever since. Unfortunately, the use of pesticides in many situations replaced biological control agents (predators and parasitoids) with an externally applied toxin (van den Bosch, 1978). Further, the application of an externally applied toxin to the plant is a pest management approach that is similar to the plant's own management strategy, which is the mobilization of toxic secondary metabolites, but in this case, applied internally. From a management point of view, this results in plant pests that are being managed by toxins applied internally (by the plant) and pesticides applied externally (by man). The two strategies do not significantly differ in their approach or in their mechanisms of action (Brattsten et al., 1977), and as a result they both reinforce the selective pressure on the pest insect. Such reinforced selection pressure could force insects to rapidly develop resistance to the toxicant, as has been shown through the use of pesticides (Plapp, 1977; Georghiou and Saito,

1982). The development of genetic engineering may not do much more than further increase the selective pressure toward the loss of chemical management. As Whitman (1981) questioned, it is a paradox as to how plants with relatively long generation times and low recombination rates compared to herbivores, have survived the evolutionary arms race. The current tri-trophic view of a plant's interaction with herbivores is that the plant has evolved various secondary compounds that suppress the herbivores and that the third trophic level may or may not be compatible with a plant's defensive strategies, but if the third trophic level suppresses herbivores, this action is fortuitous (Bottrell et al., 1998; Turlings and Benrey, 1998). This tri-trophic view has a herbivore-centric perspective (Olff et al., 1999) that is fostered by our view of herbivores as pests. As discussed by Vinson (1999) there is an alternative view. Here I will focus on this alternative view and on host specific herbivores, which predominate (Bernays, 1997a, b), and their habitual parasitoids. I will very briefly describe the plant-herbivore, herbivore-parasitoid, and plant-parasitoid relationships and discuss a number of evolutionary conflicts that occur if one examines the present view of the plant-insect relationship from a tri-trophic perspective. I will examine some of these conflicts and then reexamine them from the point of view of a mutualistic, coevolved system.

DI-TROPHIC INTERACTIONS

Plant-Insect Coevolution

Most of the theories concerning plant-insect relationships have focused on two aspects: the role of insects as pollinators, and how plants defend themselves from herbivores. More recently there has been a discussion of the interaction between plants and the third trophic level, parasitoids and predators. Here I will focus on the interactions between herbivores and their host plants.

The plant-herbivore relationship has basically been described as an adversarial one, referred to as evolutionary arms races (Ehrlich and Raven, 1964; Feeny, 1976). This arms race (Figure 14.1, I) has been described in terms of the herbivore locating the plant and the plant defending itself. The discovery phase involves the plant trying to hide or be inconspicuous (Figure 14.1, IA) while the herbivore tries to locate a plant (Figure 14.1, IB). In fact, the risk of discovery was considered by Feeny (1976) to be a major factor in whether a plant evolves a commitment to plant defense and the type of defense that it evolved. From the ideas discussed by Feeny (1976) and Futuyma and Gould

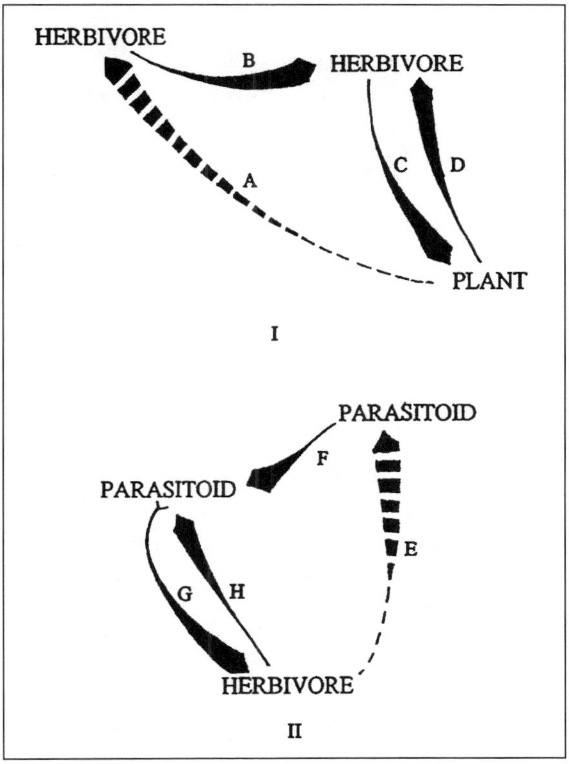

Figure 14.1. The general pre-tri-trophic view of the plant-herbivore and herbivore-parasitoid relationship. I. Although plants attempt to hide (A), their specialist herbivores locate (B) and attack the plant (C), which defends against the herbivore (D). The relationships (A) and (B), and (C) and (D) are adversarial. II. Although herbivores try to hide from parasitoids (F), their habitual parasitoids locate (E) and attack the herbivore (G), which attempts to defend itself. As a result the relationships (F) and (G), and (G) and (H) are adversarial.

(1979), the chemical defenses of plants can be divided into two strategies. One strategy involves relatively short-lived plants (unapparent), which are protected by chemicals that act qualitatively (toxins), generally at low concentrations and produced at a low cost. The other strategy involves plants that are long-lived (apparent) and protected by high concentrations of costly chemicals that act quantitatively (digestibility reducers) in a dose-dependent manner. The plant, once found, evolves to express secondary metabolites or defensive chemicals (Figure 14.1, ID) that reduce herbivore fitness (Cates and Rhoades, 1977) with some herbivores countering (Figure 14.2, IC) by adapting to the plant's defensive compounds (Feeny, 1976; Rosenthal and Janzen, 1979). The result has been the evolution of numerous secondary chemical factors that reduce herbivore activity (Harborne, 1988) and the evolution of resistance to these various factors by the insect (Caprio and Tabashnik, 1992). Further, the discipline of host plant resistance is predicated on an understanding of plant-herbivore defenses and utilizing them to increase plant fitness. Our view of host plant defenses may be true and work for generalist herbivores, but would not appear to be true in the long term for the specialist herbivores.

Insect-Insect Coevolution

There are many types of insect-insect interactions, but here we are interested in the insects as hosts for parasitoids (herbivore-parasitoid interaction). Herbivores are attacked by a number of carnivores. Among these are the parasitoids, which tend to be more specialists than the predators. The specialist parasitoids, like the herbivores, must find their host (Figure 14.1, IIF) and these hosts attempt to hide or be inconspicuous (Figure 14.1, IIE). Once the host is located the host may defend against the parasitoid (Gross, 1993), but if the host (herbivore) is attacked, it can mount an internal defense (Figure 14.1, IIH) (Vinson, 1990a, b; Strand and Pech, 1995). However, most habitual or specialized parasitoids have evolved an effective means to evade or suppress the immune system (Vinson, 1990b) of their herbivore host (Figure 14.1, IIG) and some have even evolved to subvert the immune system to their own benefit (Vinson, 1990b). As a result there is an "arms race" between G and H (Figure 14.1, II). However, there is one important difference between the plant-herbivore interaction and the herbivore-parasitoid interaction, the parasitoid ultimately kills its host while most herbivores function more like true parasites and do not kill the host plant. Thus, parasitoids would appear to place a heavier selection pressure on the herbivore than the herbivore on the plant. Although herbivores try to hide (Figure 14.1, IIE), they must feed, rid themselves of waste products, and interact with other organisms and thus betray their location. However, herbivores have evolved to minimize the release of cues that function over a long range. Thus, the challenge for parasitoids is to find the general location of potential hosts. Once a female parasitoid reaches a potential host patch she switches from orientating to long-range cues to responding to cues that function over a short range that the host can not eliminate (Vinson, 1985), the detectability-reliability paradigm (Vet et al., 1995). Consequently, the parasitoid has evolved to respond to long-range cues that do not come directly from the host but arise from associated organisms, other host stages, or the host's food (Vinson, 1985). This eliminates any direct selective pressure between the parasitoid and its habitual host concerning long-range cues (Vinson, 1998). In the case of parasitoids of herbivores, the challenge of locating herbivores at long range appears to be solved by parasitoids responding to plant volatile compounds (Vinson, 1981).

TRI-TROPHIC INTERACTIONS, CURRENT VIEW

Price et al. (1980) suggested that studies of plant-herbivore interactions should consider the third trophic level. When this is done there are several additional interactions that can occur as diagramed in figure 14.2. One of these is a direct interaction between the plant and the parasitoid (Figure 14.2, J) that can be positive, neutral, or negative depending on the interrelationships or path of the interaction. For example, a parasitoid successfully attacking a herbivore (Figure 14.2, G) benefits the plant (Figure 14.2, K), but plant toxins (Figure 14.2, D) can result in herbivores (Figure 14.2, L) that are toxic to parasitoids (Barbosa, 1988). Vinson (1975, 1981) suggested that plants play a pivotal role

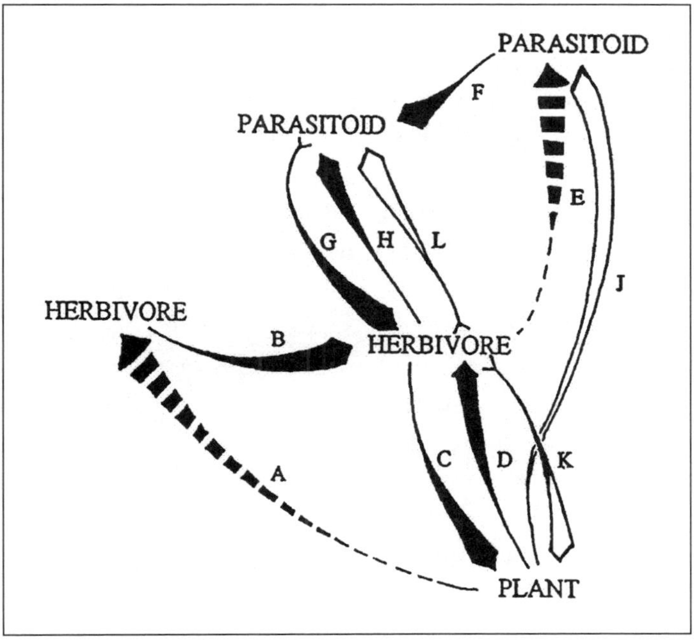

Figure 14.2. The basic tri-trophic view combines I and II from figure 2, which is simplified from the concept from Price et al. (1980). As a result, attention was directed to plant-parasitoid interactions. For example, parasitoids are attracted to plants and as a result attack herbivores (G), which benefits the plant. However, plants produce herbivore toxins (D) that can poison parasitoids (J).

in host selection, thus the carnivore-herbivore interaction should consider the first trophic level. This suggestion was based on many earlier studies that showed that parasitoids in the field would respond to plants (Vinson, 1975), or parasitoids were attracted to plant extracts or volatiles (Altieri et al., 1981). Further, several studies had demonstrated that parasitoids were preferentially attracted to damaged plants (Reed et al., 1970). Since these studies, the role of the plant in biological control has received a lot of attention (McAuslane, 1994; Turlings and Benrey, 1998). Most of this emphasis has focused on the attraction of parasitoids to plants, but the quality of the herbivore has also been examined in a few cases.

Plant Attraction

Plant volatiles consist of blends that vary with the plant's genotype, age, physiological state, condition, and the time of day (Turlings et al., 1998). It is clear that volatiles released by the plant attract parasitoids, and these volatiles play a critical role in the initial host location sequence (Turlings and Tumlinson, 1991; Vinson, 1998), although plant shape and color can also play a role (Vinson, 1985; Waeckers et al., 1998). Innate plant volatile compounds are released by the plants that are attractive to certain parasitoids (Figure 14.3, J). As a result, parasitoids are preferentially attracted to and search among plants that serve as food for their hosts (Udayagiri and Jones, 1992). Parasitoids have even been observed to "patrol" host plants lacking herbivores in the field (Vinson, 1975; Waage, 1983) and greenhouse (Altieri et al., 1982). This is important as it suggests that these parasitoids are part of the plant system regardless of the presence or absence of hosts. However, it is also clear that parasitoids are preferentially attracted to damaged plants (Figure 14.3, N), and there are now many examples demonstrating the significantly enhanced response of female parasitoids to volatiles released by herbivore-damaged plants (McAuslane et al., 1991; Vinson and Williams, 1991; Turlings et al., 1998). As a result of the attraction to damaged plants, the parasitoids are brought to the plant (Figure 14.3, F), herbivores are attacked (Figure 14.3, G) and suppressed resulting in a benefit to the plant (Figure 14.3, K). Further, uninfested plants may benefit from the infested ones (Bruin et al., 1991).

Herbivore Quality

Defensive secondary plant chemicals have been established as important in the evolution of plant-herbivore interactions (Price, 1991; Rosenthal and

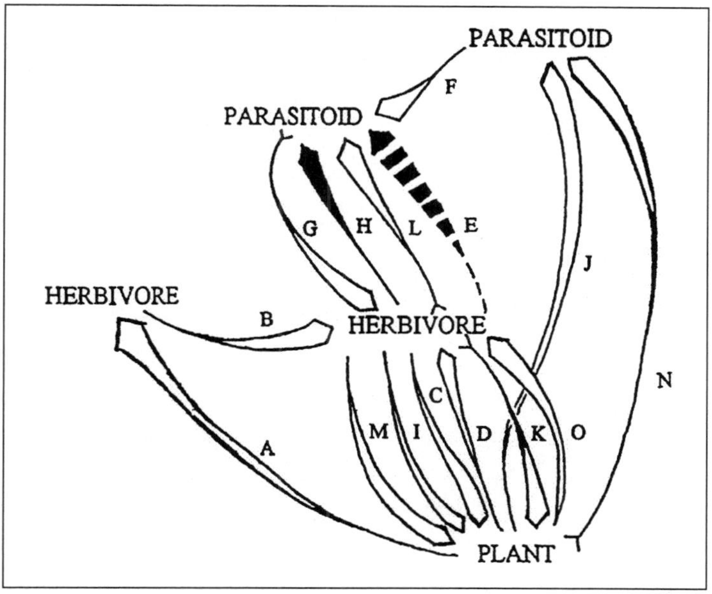

Figure 14.3. Present herbivore-centric tri-trophic view of the insect-plant relationship (modified from Vinson, 1999) includes the view that herbivores that may induce the plant (K) to release volatiles (L) that orient some parasitoids to the plant (E). These parasitoids may attack (G) and if they do, whether their progeny survive depends on the suitability of the herbivore (J) due to the secondary compounds in the plant (D). Whether the plant benefits (I) depends on all the interactions indicated by the open arrows being positive. However, most of the interactions either result in an arms race or a conflict, as indicated: [A. plant hides (unapparent) + B. herbivore searches for host = arms race.] [C. herbivore feeds on plant + D. plant produces secondary chemicals to defend against herbivores = arms race.] [E. parasitoids search for hosts + F. herbivore hides (unapparent) = arms race.] [G. parasitoid attacks and suppresses host immune system + H. herbivores immune system suppresses parasitoid = arms race] [I. parasitoid attack of herbivore benefits plant (fortuitous) + J. herbivore toxic to parasitoid = incompatible]. [K. herbivore induces plant (D) to increase toxins = herbivore conflict]. [K. herbivore induces plant to attract parasitoid (L) which attacks herbivore = herbivore conflict]. [L. plant attracts parasitoid + D. plant produces herbivore toxin = plant conflict].

Berenbaum, 1991; Roitberg and Isman, 1992) and in host plant resistance (Panda and Khush, 1995), and include the toxins, hormone mimics, growth suppressants, antifeeding compounds, or digestibility-reducers (Rhoades and Cates, 1976) that, at the very least, reduce herbivore health (Reese, 1977; Berenbaum, 1983). But these products or the consequence of their action on herbivores can affect the third trophic level in many ways.

Morgan (1910) was the first to suggest that a parasitoid might be harmed by plant's secondary products (Figure 14.3, D–J). He suggested that the reduced parasitism of *Manduca sexta* on tobacco by *Apanteles* spp. was due to nicotine in the plant and subsequently in the host. More recent reports have demonstrated that plant secondary substances that confer plant resistance to herbivores can result in these herbivores being less suitable for parasitoid development (Duffey et al., 1986; Barbosa, 1988). In addition, there is a number of herbivore feeding studies on resistant and susceptible plant varieties that show resistant varieties are often not compatible with biocontrol agents (Orr et al., 1985; Ruberson et al., 1989). There is the possibility that levels or types of secondary plant toxins occur that result in parasitoid death but have little impact on the herbivore. For example, parasitoids of California red scale are encapsulated when their host is feeding on certain plants (Smith, 1957). But how does this incompatibility fit in with the concept of plants attracting the third trophic level?

The discovery that injured plants release volatiles that are attractive to the third trophic level led Whitman (1988) and Vinson (1991) to suggest that plants, in effect "cry out for help." However, such a response would not evolve if the plant is also chemically defended and the attracted parasitoids are unable to reproduce in the host located. Further, there are a number of reports where herbivore damage leads to increased titers of plant secondary substances directed against the herbivore (van Dam and Bhairo-Marhe, 1992), which should lead to increased problems for both the herbivore and parasitoid. There are four major ways around this problem: (1) plants could either be attractive to the third trophic level or they could be chemically well defended, but not both; (2) female parasitoids could detect poor quality or poisoned herbivores and reject them; (3) the developing parasitoid larva could avoid the toxins in the host, which may be possible in herbivores that sequester the toxins in certain tissues; or (4) the developing parasitoid could adapt to the toxin. All of these possibilities may occur, but in most cases we have an inadequate understanding of the interactions between the three trophic levels in regard to these secondary chemicals to determine which of these various scenarios are most prevalent. For example, the ability of females to reject hosts on the basis of size or because they have been previously parasitized is well known; however, their ability to detect the presence of toxic compounds in their hosts (Vinson and Iwantish, 1980) or diseased hosts (Vinson, 1990c) is not as well established.

Herbivore Induction of Plant Processes

There is a body of evidence that demonstrates that insect herbivores induce systemic changes in the physiology of the plants on which they feed (Kogan and Fischer, 1991; Kuc, 1982). Although Zucker (1983) suggested that induced secondary chemicals protect plants from disease but have no effects on herbivores. However, a number of studies demonstrate that the induced secondary chemicals are the same as those involved in the plant's herbivore resistance (van Dam and Bhairo-Marhe, 1992), and these changes also apply to the volatiles released by the plant (Vinson and Williams, 1991; Udayagiri and Jones, 1992; Turlings et al., 1998). It has also been demonstrated that herbivores can "induce" the plant to produce and release novel volatile compounds, and that different herbivores can trigger the release of different blends (Pare and Tumlinson, 1997; DeMoraes et al., 1998). Further, the herbivore-induced changes on one plant may affect other plants (Bruin et al., 1992). Thus, herbivores induce changes in the plant (Figure 14.3, M) that result in the plant releasing volatiles that are attractive to the herbivores habitual parasitoids (Figure 14, 3N). However, these plant insect interactions may often be incompatible with herbivore or plant interests.

Several authors have suggested that the interactions between plants and the third trophic level are either beneficial, detrimental, or neutral, and that these interactions are not predictable (van Emden, 1986; Bottrell et al., 1998). However, the predictability depends on a careful analysis of the interaction and the presence of a conceptual framework that can be used to examine the interaction. If we examine the herbivore, the induction of its host plant to increase the concentration of toxic secondary compounds (Figure 14.3, arrows M, O) is not a herbivore compatible strategy. Further, the induction of the plant to release parasitoid attractant volatiles that results in increased mortality for the herbivore (Figure 14.3, arrows M, N, F, G) is also not a herbivore compatible strategy. The production and release of volatiles may attract parasitoids that in turn attack herbivores that are not suitable for parasitoid development. This scenario is not a plant compatible strategy for either the plant's innate volatile release (Figure 14.3, arrows J, F, G, D) or for the plant's herbivore-induced volatile release (Figure 14. 3, M, N, F, G, O, L). It would not be a good strategy for the parasitoid to respond to plant volatile compounds that lead them to toxic hosts. These strategies must be interacting in a different way if they are to evolve. One of these ways is to consider that these interactions are beneficial to all players.

TRI-TROPHIC INTERACTIONS, COOPERATION

In addition to the better-understood cooperative and beneficial pollinator insect-plant interaction, many other plant-insect relationships are beneficial.

For example, ants are one of the most numerous of arthropods and play a significant role in seed dispersal (Gorb and Gorb, 1995; Rodgerson, 1998). Such seeds have an elaiosome that stimulates ants to pick up the seed and carry it back to the nest where only the elaiosome is consumed leaving the intact seed behind (Keeler, 1989; Mark and Olesen, 1996).

Plant defense also can involve arthropods. The best example is the bull horn acacia, *Acacia cornigera* L., which supports colonies of ants of the genus *Pseudomyrmex* that nest within domatia (hollow thorns) that are common on the plant. The adult ants are supplied carbohydrate and moisture from extrafloral nectaries that occur on each leaflet. However, the plant also provides proteins required for larval development and reproduction in the form of Beltian bodies that occur at the tip of each new leaflet (Rickson, 1980). Although the ants patrol the plant, they are not generally in evidence until the plant is disturbed, and then the ants swarm out over the plant, sting, and remove (if small enough) any intruder encountered. Arthropod herbivores, which are prevalent on the rare, undefended plants, are not generally consumed by the ants but killed and removed (Janzen, 1967). Thus, the plant maintains a defense at some cost (production of large domatia, proteinaceous Beltian bodies, nectaries and nectar secretion). The ants patrol and protect the plant, but at some cost (loss of venom, energy, and some workers).

Thus, like the ants protecting the *Acacia* plant, some parasitoids patrol select plants that may harbor their prospective host herbivores. Although extrafloral nectaries may provide some reward, for parasitoids the more significant rewards are hosts. One of the challenges for parasitoids is to locate a number of hosts (herbivores) to insure their reproductive success. This means that there must be suitable herbivores available in some numbers to insure the reproductive success of the parasitoids. These hosts could be viewed in a similar manor to the production of Beltian bodies by the *Acacia* plant. But this requires a very different view of herbivores than is conventionally considered.

Herbivores: What Is Their Relationship to the Plant?

Although the coevolutionary arms race (Berenbaum, 1983b) plays an important role in plant-insect herbivore evolution, there are other interpretations as discussed by Vinson (1999). Further, in considering the parasitoid-herbivore evolution, there are a number of questions that arise. Each question is discussed in the following paragraphs.

Why Have Plants Generally Not Won the Arms Race through the Production of Powerful Toxins?

Most plants contain a suite of related toxins. A number of authors (Gilbert, 1971; Rathcke and Poole, 1975) suggest that the occurrence of a suite of plant defensive compounds is due to the plant making a series of modifications to their defensive chemistry as the herbivore becomes resistant to the last modification. In contrast, many microorganisms and animals produce toxins that are very toxic to insects (Hardegsee and Tu, 1988; Edstrom, 1992) and this has even lead to a search to locate the genetic basis of these compounds so they can be engineered into plants. But, is this a good idea? Instead of expressing serious toxins, plants have evolved to express digestibility reducers, molting inhibitors, reproductive suppressants, and nutrient competitors, all of which tend to reduce the fitness of the herbivore (Feeny, 1970) but generally do not eliminate the herbivore population. Is this saying something?

Why Are So Many Defensive Compounds Employed by the Plant Structurally Similar?

Considering the speed at which insects become resistant to insecticides and the importance of using compounds with different modes of action to suppress the development of resistance (Denholm and Rowland, 1992), it is surprising that plants have not either accumulated a number of very different defensive compounds, or different compounds occur in even closely related species. Instead, plants often possess a group of related defensive compounds. For example, the presence of heliocides and their condensed form, gossypol, in cotton (Stipanovic et al., 1986), the hydroxycoumarins and angular furanocoumarius of plants attacked by various species of *Papilio* (Berenbaum, 1983a), or the cardiac glycosides of the Asclepidaceae (Roeske et al., 1976). In effect, the slow and often accumulative changes in the defensive chemistry of the plant over time, even among related species, would appear to be an ideal way to insure the changes can be tracked by a specialist herbivore and would also help insure their presence.

Why Do Plants Release So Many Volatile Compounds If Being Unapparent Is Advantageous?

Most plants release a great variety of compounds. For example, Yamaguchi and Shibamoto (1980) reported forty-six volatiles were released from chestnut flowers, *Castanea crenata* Sieb. and Zucc., and Averill et al. (1988) reported sixty volatiles have been identified from apples. Such volatile release is not limited to the more "apparent" plants, but includes the so-called unapparent species. For example, Buttery et al. (1980) and Flath et al. (1978) have reported sixty-eight volatiles released by corn silk and corn tassels, yet these plants do not depend on insects for either pollination or seed dispersal and must be considered to be a short-lived, unapparent plant. While the blend of volatiles released is dependent on the plant species, for example cotton releases over thirty compounds (Hedin et al., 1976), it also depends on the time of day, developmental age, and physiological state of the plant and can be influenced not only by the type of damage but also by the species of organism inflicting the damage (Turlings et al., 1998). These data would suggest that plants have not evolved a very effective way of being unapparent. In fact, I would suggest they have evolved to advertise, but in a way that is selective. Part of this selectivity is the release by the plant of a blend made up of a limited group of related compounds. The other side of the issue is whether herbivores can respond selectively to blends of plant volatiles. Metcalf and Metcalf (1992) reported that over 300 insect species in five insect orders have been shown to be attracted to one or several of over seventy different volatiles. However, herbivores optimally respond to a specific blend of compounds (Fein et al., 1982) that constitute only a fraction of the volatiles released. For example, from a mixture of over twenty-three volatiles released by carrot, only a mixture of trans-asarone, trans-2-hexanol, heptanal, and hexanal attract the carrot rust fly *Psila rosae* (Fabricius) (Guerin et al., 1983). Out of a mixture of over thirty compounds from cotton only a mixture of β-bisobolal, β-caryophyllene, β-caryophyllene oxide and limonene is attractive to the boll weevil *Anthonomus grandis* Boheman (Hedin et al., 1976). Thus, these specialist herbivores are adapted to efficiently locate a specific species of plant by responding to a specific blend of compounds released by the plant (Fraenkel, 1969; Minyard et al., 1969; Harborne, 1988). As a result, even the more r-selected herbaceous successional plants appear to be more readily located and colonized by their specialist herbivores than would be predicted if the plant and herbivore were in an adversarial relationship.

Why Have So Many Herbivores Evolved a Dependency on the Plant That They Are "At War" With?

The list of factors that herbivores have evolved to require from plants on which they feed is long and diverse, including biting incitants, phagostimulants, precursors to the insect's hormones, required accessory growth factors, egg production stimulants, ovipositional stimulants, precursors to or use of secondary plant compounds as defensive factors, and even factors involved in sexual selection (Metcalf and Metcalf, 1992). Such a dependency would argue against a competitive or adversarial hypothesis. An alternative hypothesis is that the plant-herbivore relationship is a tolerated-parasitic relationship that is less arguable. However, the evolution of a true parasitic relationship is toward a mutualist relationship, as it benefits the parasite if the host's fitness is maximized.

Is it possible that the plant has evolved to maintain a few herbivores? The idea of plant-herbivore mutualism is not new (Kimmins, 1971; Mattson and Addy, 1975; Owen, 1980). As pointed out by Owen (1980) many grasses have evolved to depend on grazers or fire, but are there other benefits?

Herbivores May Benefit the Host Plant

A plant maintaining a specialist herbivore may benefit by the reduction in colonization rate of the plant by opportunistic species because no two species can occupy the same niche (Gause, 1934). This is known as the "competitive exclusion principle" (Hardin, 1960; Krebs, 1985). Diamond (1978) suggested that competition for a niche is a major driving force of natural selection. Thus, the presence of specialist herbivores, which as I argue is under the plant's control, would reduce the attack and possible establishment of generalist or other specialist herbivores that are not under the plant's control. Although Lawton and Strong (1981) and Strong et al. (1984) have suggested an alternative to the "competition" hypothesis, Wiens (1977) argues that competition does occur, but is intermittent or rare. While this controversy is unresolved (Denno et al., 1995), the scenario outlined here would support Wiens's views. If competition exists, then the plant would benefit by encouraging those herbivores it could manage. It would be advantageous for the plant to provide cues to attract the "managed" herbivore and it would also be advantageous not to kill it. However, there would be an advantage to force the herbivore to remain on the plant as long as possible to deter any potentially less manageable herbivores. This would result in reducing competition to a less common event. This also does not mean that a plant may not support a diversity of herbivores, as a plant may represent a number of niches. This is particularly true of the more long-lived "apparent" plants.

Another benefit to the plant of having herbivores is that a mutualistic relationship between a plant and select specialist parasitoid could occur. For example, if plants have evolved in a tri-trophic environment to tolerate and encourage the presence of herbivores, this would also provide hosts for carnivores. As a result several plant-herbivore relationships could be predicted. One is that plants would evolve to invest some resource into producing and maintaining hosts, as occurs in the production of Beltian bodies by the *Acacia* plant. Such an investment in a herbivore by a plant would most likely be expressed in the plant's tolerance or compensation. Tolerance does not exert any selection pressure on the pest population (Horber, 1980), and a wide range of plants tolerates or compensates for a degree of herbivore damage. In effect, the plant can suffer some herbivore damage without a reduction in fitness (Crawley, 1987; Paige and Whitham, 1987). Further, while herbivore outbreaks can severely impact plants, such effects can be much less serious than hail storms or severe wind damage, which plants must have also evolved to handle through compensation. However, another way of looking at the ability of a plant to tolerate or even compensate for a degree of herbivory is that herbivore feeding can be considered as an investment in maintaining a population of herbivores as a resource for parasitoids without, as pointed out, suffering a loss in fitness.

A second tri-trophic consideration is that the plant should evolve to maintain the "proper" species of herbivore that can serve as a host for the parasitoid. Most plants, whether apparent or unapparent, do have a select guild of specialist herbivores that attack them (Bernays and Graham, 1988). To insure this situation the herbivore should evolve to be a specialist in locating a particular species of plant and the plant should evolve to release cues that a particular species of specialist herbivore can track. As discussed, plants do release a species-specific blend of volatiles and these blends do attract a particular species-specific specialist herbivore. Bernays (1997b) suggested that the reason there are so many specialist herbivores is that the specialist herbivore spends much less time and neural capacity to make a food plant choice. In effect the specialist herbivore can focus its attention on vigilance and escaping predation (Bernays, 1996). In contrast, the generalist must spend time searching for, sampling for suitability and changing plants to reduce the accumulation of toxins (Levins and MacArthur, 1969), which reduces vigilance (Lima and Dill, 1990).

I would suggest that plants have evolved to accommodate their specialist herbivores and that many of the specific interactions can be viewed as beneficial to the plant and to the herbivore. If specialist herbivores are part of the plant system, then the plant can evolve as a part of a tri-trophic system that provides several ways to benefit the plant and several ways to insure that the herbivore does not cause serious damage. These are discussed here:

1. The plant can evolve with a herbivore in a way that reduces the plants vulnerability to interloping herbivores the plant may be unable to manage. In effect the plant evolves to insure that a specialist herbivore is present on most plants in the population. This involves three factors: (a) the release of a blend of volatiles to which a specialist herbivore has evolved to respond to; (b) the evolution of a tolerance to the specialist herbivore; and (c) providing factors that result in the herbivores dependence on the plant.

2. The plant can evolve to attract a select guild of parasitoids to patrol the plant and help protect the plant from an overabundance of herbivores and to provide rewards, in addition to herbivores, such as extrafloral nectaries.

3. It is also important that plants produce secondary defensive compounds that do not kill the third trophic level, but allow the third trophic level to develop, such as compounds that just reduce herbivore fitness. This certainly is possible (Barbosa, 1988) and is likely much more common than indicated in the present literature (Vinson, 1999).

4. Assuming a dependent herbivore, the plant is provided three very different ways to manage the herbivore population. These are discussed in the next section.

Herbivore Management by Plants

There are several ways that a plant can manage its habitual associated herbivores. One is a plant's attractiveness to its habitual herbivore, a second is through the plants' expression of secondary compounds (herbivore suppressants), and a third is through the manipulation of parasitoids. Plants vary in the volatiles they release. Although much of this variability is inherent, some variability is due to induction by herbivores, the time of day, age, physiological state, and environmental influences. Similarly, herbivores vary in their inherent response to specific plant volatile cues that are influenced by their age, physiological state, time of day, and environment. This allows for a few plants to escape low volatile output, without any undo selection pressure on the herbivores. Herbivores are not forced to become extremely efficient in plant location, which allows some plants to escape detection, and all plants do not become extremely efficient in hiding, which allows some plants to be found. This helps maintain an optimal attractiveness for the plant to herbivores and an optimal sensitivity in the herbivore's ability to respond to the plant's cues.

Herbivore-suppressive plant secondary compounds (toxins) have been investigated more thoroughly than any of the other herbivore plant defenses (Harborne, 1988). Some of the basic aspects are shown to the left of the line marked "increasing plant cost" (the figure's "stem") in the center of figure 14.4.

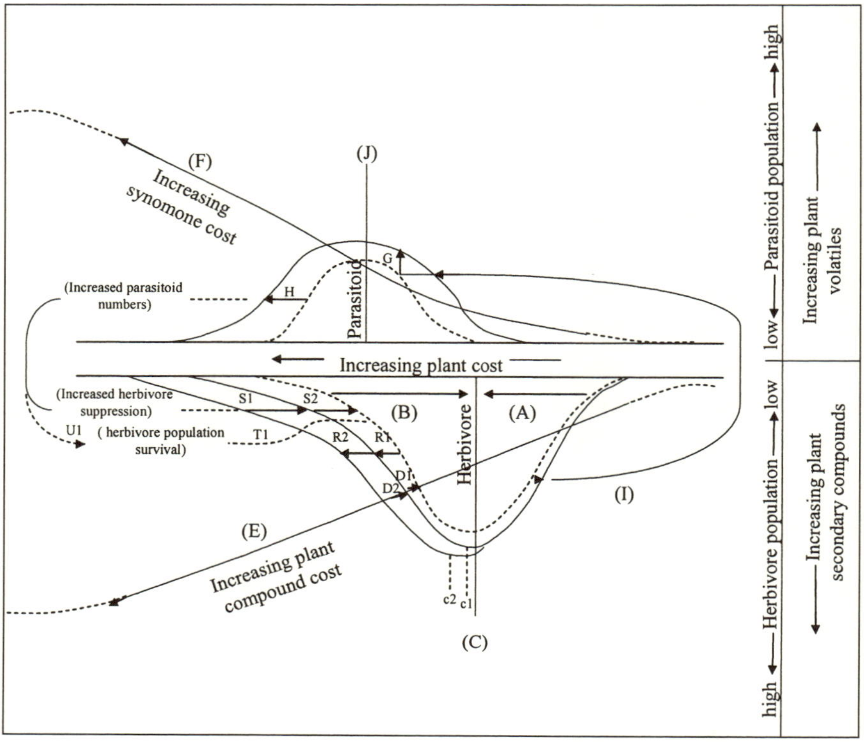

Figure 14.4. This diagrammatic tri-trophic interaction has the ordinate in the middle that states "increasing concentration of secondary compounds in plant," which increases to the right. This should be viewed as the stem of the plant. The abscissa shows "increasing cost to plant" in both directions. Above the stem shows a population of herbivores that require some "optimal level" of plant secondary compounds in their diet (A), but also are susceptible to increasing toxicity of secondary plant compounds (B). This results in herbivores that are most fit near some optimal level of secondary compounds. If they become more resistant to the defensive secondary compounds (T1) their population would increase to some new optimal level (upper dotted line). The plant may counter with some increased level of defensive compounds (D1) at some increased cost (represented by the inclining solid line—C). However, the increased level of the defensive compound may be countered by the herbivore adapting to the defensive compound (N1), leading to an "arms race." However, the arms race can be disrupted by the plant as shown by the parasitoid population below the stem. The plant releases some amount of a volatile compound that attracts parasitoids (P). The parasitoids attack, suppressing the herbivore population (S1). Because the increased herbivore population consists of more plant secondary defensive compound resistant individuals, the parasitoid removes more resistant individuals, pushing the herbivore back. Although the herbivore may counter by mounting an immune response (R1), the parasitoid counters with an immune suppressant. More importantly the parasitoid marks hosts and when encountering a number of marked hosts the parasitoid leaves, which insures that not all herbivores are attacked (M1). This is particularly important to the next generation. When herbivores are abundant, more parasitoids are produced (E1), resulting in a high parasitoid population. When the herbivore population is low, fewer parasitoids are produced (E2), resulting in a reduced parasitoid population. Regardless, host marking prevents the over-exploitation of hosts.

A solid ascending line (E) on the left shows an increase in the titer and cost of the plant secondary compounds. As shown by the dashed line bell curve with a line running through the middle (Figure 14.4, C), the herbivores' fitness clusters around some optimal level of plant secondary compounds, some of which the herbivore requires (Figure 14.4, A) and some of which become increasingly toxic as the titer increases (Figure 14.4, B). As a result the herbivore is dependent on the plant within some narrow limits. If the herbivore becomes resistant or tolerant to the toxic compounds (Figure 14.4, R1, R2) the herbivore may increase its population to a new level represented by the solid line above the two solid bell-shaped curves (Figure 14.4, c1, c2). The plant counters by increasing the level of the toxic compound (Figure 14.4, D1, D2) at some increased cost (Figure 14.4, E). However, the plant is limited in reducing the herbivore population back to its earlier level (Figure 14.4 left dotted-line bell-shaped curve) by several factors. These factors include the increased pressure to adapt by the herbivore due to the plant's increased selection of the more resistant individuals and the removal of the more susceptible individuals, which increases the herbivores pressure to shift up, thus increasing the plant cost. This shift up increases the herbivore's secondary substance need, and the pressure for this shift can be increased if the required substance and herbivore suppressant are the same, as shown by Williams et al. (1987). As a result there is the potential for an arms race. The arms race can be reduced if the herbivore's required plant substance is different from the suppressant, but since both substances are genetically linked, the arms race is not prevented. However, the arms race is prevented by the plant's investment in a third trophic level defensive system (Vinson, 1999).

Although the involvement of the third trophic level in plant defense is recognized (Price et al., 1980; Turlings and Benrey, 1998), it is generally considered fortuitous (Harris, 1980; Bottrell et al., 1998). However, Vinson (1993; 1999) suggested that in tri-trophic relationships involving specialist herbivores and their habitual parasitoids, the parasitoids are part of the plant defensive system. This is depicted to the right of the "stem" of figure 14.4, which shows a population of parasitoids that develop from herbivores around a median (Figure 14.4, J) so their population reflects the host population (Figure 14.4, I). As the herbivore population expands, so does the parasitoid (Figure 14.4, G), thus there are more parasitoids to attack the herbivores (Figure 14.4, H). The plant's cost would appear to be lower than the cost of the herbivore, as suggested by data demonstrating that in most systems, the parasitized host causes less damage (Vinson, 1972; Strand et al., 1997) and is reflected in the shorter bell curve of the parasitoid population compared to the one for the herbivore

(Figure 14.4). Like the herbivores, usually each new generation of adult parasitoids must be attracted at some cost (Figure 14.4, F). This cost is probably less than the cost of the plant secondary substances, as volatiles are smaller molecules. As a result of being attracted, the parasitoid attacks and suppresses the herbivore population (Figure 14.4, S1, S2). As suggested earlier, the more successful herbivore patches, which represent a herbivore population shifting slightly to the right, may be more attractive and the parasitoids may be retained longer, resulting in a disproportionate reduction of the more abundant resistant herbivores. This pressure is against an arms race (Figure 14.4, S1 and S2 work in concert with D1 and D2 against R1 and R2). Further, S1 and S2 are not directly linked with R1 and R2.

Although there is some resistance to attack on the part of herbivores (Figure 14.4, T1), the parasitoid's ability to reduce the herbivore population to levels lower than the original population (Figure 14.4, dotted herbivore population curve) or even to extinction, is reduced by the parasitoids phenomena of host marking (Figure 14.4, U1). Host marking is important to the parasitoid population, as a low herbivore population results in a low future parasitoid population (Figure 14.4, I), but host marking is also important to the herbivore (they are not eliminated) and indirectly to the plant population.

Cooperative Tri-Trophic Interactions

There has been a lot of discussion in regard to whether there are top-down insect-plant effects or bottom-up effects. In either case the discussion is from a herbivore-centric point of view. If instead we examine the relationships with the plant at the base of a triangle supporting both the parasitoid and herbivore (Figure 14.5), instead of the herbivore in the middle, a different view can be developed. This view is not meant to replace others or to apply to all plant-insect tri-trophic situations. Whether this view has any merit beyond a few select situations will remain to be seen. But if it has some merit then it may serve as a foundation to further improvements.

The flow of the interactions between the plant, herbivore, and parasitoid is shown in a more representative tri-trophic sense in figure 14.5. Most plants release a blend of volatile compounds (Metcalf and Metcalf, 1992), and these are attractive to their habitual herbivores (Figure 14.5, 1). This suggests that the plant is not trying to hide. In fact this may benefit the plant because the niche represented by the plant is filled by those herbivores (habitual herbivores) that share a long evolutionary history with the plant. With filled niches, establishment of the more threatening interlopers is reduced thus benefiting

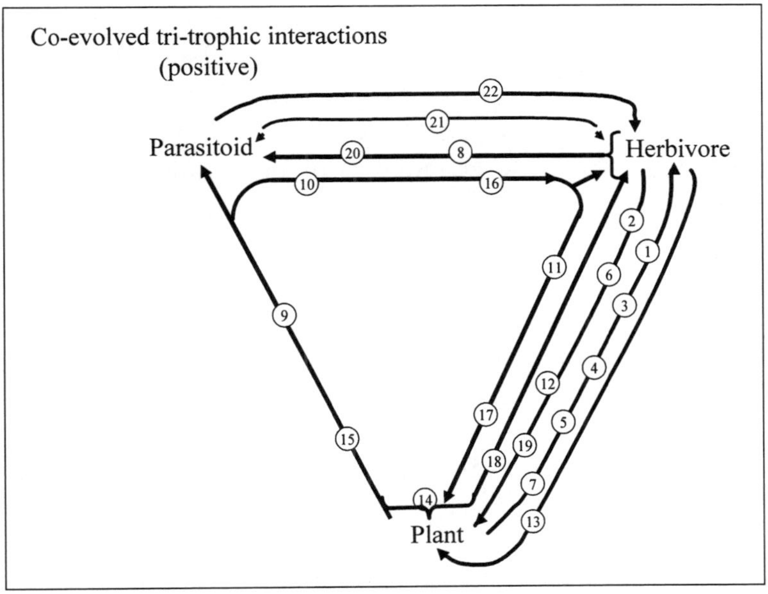

Figure 14.5. A positive (mutual) tri-trophic coevolved system is proposed in which the parasitoid and herbivore are linked together with the plant. (See text for discussion.)

the plant (Figure 14.5, 2). The herbivore also benefits by orienting to a specific volatile blend of compounds released by the host plant it habitually attacks. This ensures the host plant is rapidly located (Figure 14.5, 3) thus reducing potential predation (Bernays, 1997b). The herbivore not only obtains nutrients from its host plant (Figure 14.5, 4) but also various secondary compounds or precursors often essential to its survival (Figure 14.5, 5). While herbivores do feed on the plants, these can tolerate some degree of damage (Figure 14.5, 6). Further, some of the secondary plant compounds may be toxic to herbivores at certain concentrations. This provides opportunities for the plant to manipulate herbivore population by releasing more or less of a particular toxic secondary compound or essential micronutrient. Toxic secondary compounds may eliminate less fit individuals from habitual herbivore populations and may slow down the herbivore's growth and development (Figure 14.5, 7). With increased development time there are more chances for the herbivore to be found by parasitoids, and a prolonged occupation of the plant's niche by the habitual herbivore, thus reducing interloper colonization and allowing the plant to develop a long-term relationship with specialist parasitoids. Plant secondary compounds function in a way that does not reduce the suitability of the herbivore to serve as a host for the parasitoid (Figure 14.5, 8).

Parasitoids are also attracted to certain plants (Figure 14.5, 9) (Elzen et al.,

1983; McAuslane et al., 1991). This attraction occurs regardless of the presence or absence of hosts (Figure 14.5, 10). Once the plant is located, parasitoids can respond to herbivore-released short-range cues and locate hosts (Figure 14.5, 11), which benefits the plant. Since the parasitized herbivores are not removed from the plant (Vinson and Iwantisch, 1980), the plant niche remains filled, thus reducing colonization by interlopers (Figure 14.5, 12). Parasitism also reduces herbivore energy needs; parasitized hosts often feed less than healthy ones and cause less damage (Vinson and Barbosa, 1987). Host marking may be used by the parasitoid to assess impact on the herbivore population. This benefits both parasitoid and herbivore populations (Figure 14.5, 21).

An interesting observation by Reed et al. (1970) suggested that parasitoids were attracted to herbivore-damaged plants. This has been confirmed (Vinson and Williams, 1991; Turlings et al., 1998). More recently it has been demonstrated that herbivores produce substances (Figure 14.5, 13) that can induce the plant to produce and release (Figure 14.5, 15) either more or novel parasitoid attractive volatiles (DeMoraes et al., 1998). As a result parasitoids are attracted to plants that harbor their habitual herbivores (Figure 14.5, 16), which benefit the plant (Figure 14.5, 17). This, however, does alter the same aspects of the "detectability-reliability" paradigm (Vet et al., 1995) since the long-range cues released by damaged plants are not only highly detectable but also highly reliable. The feeding activity of the herbivore also induces the plant (Figure 14.5, 14) to produce more and sometimes unique secondary substances that further suppress the growth and development of the herbivore (Figure 14.5, 18). The overall result is, whether parasitized or not, the herbivore is on the plant longer and consumes less plant tissue (Figure 14.5, 19). At the same time these herbivores can still serve as hosts for their habitual parasitoids (Figure 14.5, 20).

What about the herbivore and parasitoid? To reduce complexity, most of the details of this interaction are not graphed on figure 14.5. Whether the parasitoid develops in a particular host herbivore depends on a series of adversarial interactions. Once the parasitoid reaches the microhabitat of the herbivore in response to plant attractants, it must search (Figure 14.5, 10 or 16) for the herbivore (Vinson, 1998). The herbivore tries to hide and defend itself, but this is often not successful and the parasitoid attacks, laying an egg on or in a herbivore. The herbivore has an internal immune defense (Strand and Peck, 1995) that is also often not successful when attacked by a habitual parasitoid (Vinson, 1990). As a result, parasitoids reproduce (Figure 14.5, 8 + 20) at the expense of the herbivore, which benefits the plant (Figure 14.5, 19). In many cases the herbivore, in spite of the evolution of various defenses, is vulnerable to the parasitoid. This vulnerability could lead to extinction of the herbivore population but is prevented by parasitoid through the phenomena of host marking (Figure 14.5, 21). Host marking provides the parasitoid with a mechanism to insure that it does not over-exploit the herbivore population. They abandon searching when the encounter rate of marked hosts reaches a certain level. This level will depend on a balance of all mortality factors for both the parasitoid and herbivore insuring the replacement of both. Thus, host marking is of benefit to the parasitoid as well as the herbivore (Figure 14.5, 21 arrows in both directions).

There are several aspects of this model that are different from a herbivore-centric model, but there are some aspects that are the same, interpreted differently. While the model remains a "bottom-up" tri-trophic relationship, it is more of a "bottom-up partnership," as opposed to "bottom-up herbivore centric control." Although some of the competitive relationships appear to remain, when the larger context is considered the competition is lost. For example, when we consider the herbivore feeding on the plant (Figure 14.5, 2) and the plant defending itself (Figure 14.5, 3), or the parasitoid attacking a herbivore (Figure 14.5, 10, 16), and the herbivore defending itself in the context of the tri-trophic relationship, these interactions are counterbalanced by benefits. Thus competition is very weak. For instance, it is beneficial for the plant to attract and maintain both specialist herbivores and parasitoids and not to subject herbivores to toxins that lead to countermeasures by the herbivore or cause the death of parasitoids. In fact, all relationships can be viewed as beneficial to the maintenance of the tri-trophic system.

Selection leading to evolution does not occur in a vacuum. Selection of the fittest depends on the context in which the selection takes place and the context depends on the environment that develops due to the presence of associated organisms. If two organisms are brought together in some way, a new association (Holkkanen and Pimentel, 1989), and they interact in a positive way, there is the opportunity for the genes that are responsible and already present in the population to result in a distinct advantage to both organisms. Such traits would rapidly spread through the population. The new relationship would then be subject to selection in various new directions that could involve a third organism. Over time this could lead to a complex of interrelated activities.

Assuming the positive plant-parasitoid-herbivore interaction involving specialists occurs, the opportunity for cheats, subversion, and exploitation exists. One of the exploiters of the plant tri-trophic system has been people. We have basically focused on the elimination of herbivores and in so doing we have often ignored the third level. As a result we have created some pest problems and we have polluted our environment. Yet pests remain a problem.

There may be a lot at stake if we take the wrong genetic approach to plant health. I believe we should consider that there may be alternatives to the present herbivore-centric view and our drive to create herbivore-free environments. We need to focus on developing a better understanding of biological relationships and the evolution of systems, particularly the plant system on which our survival still depends. Competition leading to the survival of the fittest is not the only evolutionary route. The evolution of multicellular organisms such as humans, or the social bees, ants, and termites, attest to this. We will not ask the right questions if we presume that herbivores are pests that are detrimental to plants. In fact the whole bottom-up versus top-down control of herbivores is predicated largely on the view that herbivores are a detriment to plants (Polis and Winemiller, 1996). For example, there are several studies that have assumed that parasitoid attack should result in a reduction in herbivory and as a result should show a reduction in plant fitness (Fraser and Grime, 1998). However, in the model presented here a reduction in plant fitness through increased herbivore attack would not be predicted.

What does all of this mean to agriculture? We stand at a threshold of new approaches. One is to produce a plant that does not tolerate its specialist herbivore, but eliminates it. If the intolerant plant was extensively produced but still attractive to the specialist parasitoid, the parasitoid could be eliminated, even if the herbivore had a refuge. The parasitoid elimination could be due to hosts that were poison to the progeny of the parasitoid or no hosts are present to produce a second generation. Once a specialist parasitoid is eliminated it cannot be resurrected. Thus, if the herbivore later emerges from a refuge there may be no natural way to bring it into balance. Or, since the niche once occupied by the specialist herbivore is vacated, an interloper with different characteristics and even resistant to the engineered toxin could adapt to occupy the niche and not under the plant's control. Both of these scenarios could have devastating consequences.

However, there are other possibilities. We have altered many plants very drastically and some do not even resemble the wild-type. It is possible that the specialist herbivore has already been eliminated along with its parasitoid, and the plant is dealing with an interloper that has become a semi-specialist. Further, some of our crops are highly modified and pressed to their productive limit through a high input of fertilizers, water, and the elimination of any competition. They may have lost their tolerance to herbivory, their ability to express appropriate defensive compounds, and lost their specialist parasitoids and possibly their specialist herbivores. Such plants only survive through the input of the agricultural community. In these cases the engineering of exotic toxins into these plants may be the best course, but for the majority of plants this may be the wrong approach.

REFERENCES CITED

Altieri, M. A., S. Annamalai, K. P. Katiyai, and R. A. Flath. 1982. Effects of Plant Extracts on the Rates of Parasitization of *Anagasta kuehniella* (Lep.: Pyralidae) Eggs by *Trichogramma pretiosum* (Hym.: Trichogrammatidae) under Greenhouse Conditions. *Entomophaga* 27: 431–38.

Altieri, M. A., W. J. Lewis, D. A. Nordlund, R. C. Gueldner, and J. W. Todd. 1981. Chemical Interactions between Plants and *Trichogramma* sp. Wasps in Georgia Soybean Fields. *Protection Ecology* 3: 259–63.

Averill, A. L., W. H. Reissig, and W. L. Roelofs. 1988. Specificity of Olfactory Response in the Tephritid Fruit Fly *Rhagoletis pomonella. Entomologia Experimentalis et Applicata* 47: 211–22.

Barbosa, P. 1988. Natural Enemies and Herbivore-Plant Interactions: Influence of Plant Allelochemicals and Host Specificity. In *Novel Aspects of Insect-Plant Interactions,* ed. P. Barbosa and D. Letourneau, 201–10. New York: John Wiley and Sons.

Berenbaum, M. R. 1983a. Effects of Tannins on Growth and Digestion in Two Species of Papilionids. *Entomologia Experimentalis et Applicata* 34: 245–50.

———. 1983b. Coumarins and Caterpillars: A Case for Coevolution. *Evolution* 37: 163–79.

Bernays, E. A. 1996. Selective Attention and the Evolution of Diet Breadth in Insects. In *Proceedings of the 9th International Symposium on Insect-Plant Interactions,* ed. E. Städler, M. Rowell-Rahler, and R. Baur, 125–31. Dordrecht, Netherlands: Kluwer.

———. 1997a. Feeding by Lepidopteran Larvae Is Dangerous. *Ecological Entomology* 22: 121–23.

———. 1997b. Why Are There So Many Specialists among Insect Herbivores? In Zesde voordracht gehouden ter herdenking van Prof. dr. J. de Wilde, hoogleraar in de entomologie aan de Landbouwuniversitent Wageningen (1954–1982) op vrijdag 29 augustus 1997, 1–20. Wageningen, Netherlands.

Bernays, E. A., and M. Graham. 1988. On the Evolution of Host Specificity in Phytophagous Arthropods. *Ecology* 69: 886–92.

Bohart, G. E., and N. N. Youssef. 1976. Biology and Behavior of *Evylaeus galipinsiae* Cockerell (Hymenoptera: Halicitidae). *The Wasmann Journal of Biology* 34: 185–234.

Bottrell, D. G., P. Barbosa, and F. Gould. 1998. Manipulating Natural Enemies by Plant Variety Selection and Modification: A Realistic Strategy? *Annual Review of Entomology* 43: 347–67.

Brattsten, L. B., C. F. Wilkinson, and T. Eisner. 1977. Herbivore-Plant Interactions: Mixed Function Oxidases and Secondary Plant Substances. *Science* 196: 1349–52.

Bruin, J., A. T. Groot, M. W. Sabelis, and M. Dicke. 1992. Mite Herbivory Causes Better Protection in Downwind Uninfested Plants. In *Proceedings of the 8th International Symposium on Insect-Plant Relationships,* ed. S. B. J. Menken, J. H. Visser, and P. Harrewijn, 357–58. Boston: Kluwer Academic.

Bruin, J., M. W. Sabelis, J. Takabayashi, and M. Dicke. 1991. Uninfested Plants Profit from Their Infested Neighbors. *Proceedings of the Section Experimental and Applied Entomology of the Netherlands Entomological Society* 2: 103–108.

Buttery, R. G., L. Ling, and R. Teraneshi. 1980. Volatiles of Corn Tassels: Possible Corn Earworm Attractants. *Journal of Agricultural and Food Chemistry* 28: 771–74.

Caprio, M. A., and B. E. Tabashnik. 1992. Evolution of Resistance to Plant Defensive Chemicals in Insects. In *Insect Chemical Ecology,* ed. B. D. Roitberg and M. B. Isman, 179–215. New York: Chapman and Hall.

Cates, R. G., and D. Rhoades. 1977. Patterns in the Production of Antiherbivore Chemical Defenses in Plant Communities. *Biochemistry and Systematic Ecology* 5: 185–93.

Crawley, M. J. 1987. Benevolent Herbivores? *Trends in Ecology and Evolution* 2: 167–68.

DeBach, P., and D. Rosen. 1991. *Biological Control by Natural Enemies.* New York: Cambridge University Press.

DeMoraes, C. M., W. J. Lewis, P. W. Pare, H. T. Alborn, and J. H. Tumlinson. 1998. Herbivore Infested Plants Selectively Attract Parasitoids. *Nature* 393: 570–73.

Denholm, I., and M. W. Rowland. 1992. Tactics for Managing Pesticide Resistance in Arthropods: Theory and Practice. *Annual Review of Entomology* 37: 91–112.

Denno, R. F., M. S. McClure, and J. R. Ott. 1995. Interspecific Interactions in Phytophagous Insects: Competition Reexamined and Resurrected. *Annual Review of Entomology* 40: 297–331.

Diamond, J. M. 1978. Niche Shifts and the Rediscovery of Interspecific Competition. *American Scientist* 66: 322–31.

Duffey, S. S., K. A. Bloem, and B. C. Campbell. 1986. Consequences of Sequestration of Plant Natural Products in Plant-Insect-Parasitoid Interactions. In *Interactions of Plant Resistance and Parasitoids and Predators of Insects,* ed. D. J. Boethel and R. D. Eikenbary, 31–60. New York: John Wiley and Sons.

Edstrom, A. 1992. *Venomous and Poisonous Animals.* Malabar, Fla.: Krieger Press.

Ehrlich, P. R., and P. H. Raven. 1964. Butterflies and Plants: A Study in Coevolution. *Evolution* 18: 586–608.

Elzen, G. W., H. J. Williams, and S. B. Vinson. 1983. Response by the Parasitoid *Campoletis sonorensis* to Chemicals (Synamones) in Plants: Implications for Habitat Location. *Environmental Entomology* 12: 1873–76.

Feeny, P. 1970. Seasonal Changes in Oak Leaf Tannins and Nutrients as a Cause of Spring Feeding by Winter Moth Caterpillars. *Ecology* 51: 565–81.

———. 1976. Plant Apparency and Chemical Defense. *Recent Advances in Phytochemistry* 10: 1–40.

Fein, B. L., W. H. Reissig, and W. L. Roelofs. 1982. Identification of Apple Volatiles Attractive to the Apple Maggot, *Rhagoletis pomonella. Journal of Chemical Ecology* 8: 1473–87.

Flath, R. A., R. R. Forrey, J. O. John, and B. C. Chan. 1978. Volatile Components of Corn Silk: Possible Corn Earworm Attractants. *Journal of Agricultural and Food Chemistry* 26: 1290–93.

Fraenkel, G. 1969. Evaluation of Our Thoughts on Secondary Plant Substances. *Entomologia Experimentalis et Applicata* 12: 473–86.

Fraser, L. H., and J. P. Grime. 1998. Top-Down Control and Its Effect on the Biomass and Composition of Three Grasses at High and Low Soil Fertility in Outdoor Microcosms. *Oecologia* 113: 239–46.

Futuyma, D. J., and F. Gould. 1979. Associations of Plants and Insects in a Deciduous Forest. *Ecological Monographs* 49: 33–50.

Gause, G. F. 1934. *The Struggle for Existence.* New York: Macmillan / Hofner Press.

Georghiou, G., and T. Saito. 1982. *Pest Resistance to Pesticides.* New York: Plenum Press.

Gilbert, L. E. 1971. Butterfly-Plant Coevolution: Has *Passiflora adenopoda* Won the Selectional Race with Heliconiine Butterflies? *Science* 172: 585–86.

Gorb, S. N., and E. V. Gorb. 1995. Removal Rates of Seeds of Five Myrmecochorous Plants by the Ant *Formica polyctena* (Hymenoptera: Formicidae). *Oikos* 73: 367–74.

Gross, P. 1993. Insect Behavioral and Morphological Defenses against Parasitoids. *Annual Review of Entomology* 38: 251–73.

Guerin, P. M., E. Städler, and H. R. Buser. 1983. Identification of Host Plant Attractants for the Carrot Fly *Psila rosae. Journal of Chemical Ecology* 9: 843–61.

Harborne, J. B. 1988. *Introduction to Ecological Biochemistry.* London: Academic Press.

Hardin, G. 1960. The Competitive Exclusion Principle. *Science* 131: 1292–97

Harris, M. O., ed. 1980. *Biology and Breeding for Resistance to Arthropods and Pathogens in Agricultural Plants.* College Station: Texas Agricultural Experiment Station, MP-1451.

Hedin, P. A., A. C. Thompson, and R. C. Gueldner. 1976. Cotton Plant and Insect Constituents That Control Boll Weevil Behavior and Development. *Recent Advances in Phytochemistry* 10: 271–350.

Herdegree, M. C., and A. T. Tu, eds. 1988. *Bacterial toxins,* vol. 4. New York: Marcel Dekker.

Holkkanen, H. M. T., and D. Pimentel. 1989. New Associations in Biological Control. *Canadian Entomologist* 121: 829–40.

Horber, E. 1980. Type and Classification of Resistance. In *Breeding Plants Resistant to Insects,* ed. F. G. Maxwell and P. R. Jennings, 15–21. New York: John Wiley and Sons.

Janzen, D. H. 1967. Interaction of the Bull's Horn Acacia (*Acacia cornigera*) with an Ant Inhabitant (*Pseudomyrmex ferruginea*) in Eastern Mexico. *Scientific Bulletin of the University of Kansas* 47: 315–558.

Keeler, K. H. 1989. Ant-Plant Interactions. In *Plant-Animal Interactions,* ed. W. G. Abrahamson, 207–42. New York: McGraw-Hill.

Kimmins, J. P. 1971. Variations in the Foliar Amino Acid Composition of Flowering Balsam Fir *abies balsames* (L.) Mioll. and White Spruce *Picea glauca* (Moench) Voss in Relation to Outbreaks of the Spruce Budworm *Choristoneura fumiferana* (Clen.). *Canadian Journal of Zoology* 49: 1005–11.

Kogan, M., and D. C. Fischer. 1991. Inducible Defenses in Soybean against Herbivorous Insects. In *Phytochemical Induction by Herbivores,* ed. D. W. Tallamy and M. J. Raupp, 347–78. New York: John Wiley and Sons.

Krebs, C. J. 1985. *Ecology: The Experimental Analysis of Distribution.* New York: Harper and Row.

Kuc, J. 1982. Induced Immunity to Plant Disease. *Bioscience* 32: 854–60.

Lawton, J. H., and D. R. Strong. 1981. Community Patterns and Competition in Folivorous Insects. *American Naturalist* 118: 317–38.

Levins, R., and R. H. MacArthur. 1969. An Hypothesis to Explain the Incidence of Monophagy. *Ecology* 50: 910–11.

Lima, S. L., and L. M. Dill. 1990. Behavioral Decisions Made under Risk of Predation: Review and Prospectus. *Canadian Journal of Zoology* 68: 619–40.

Mark, S., and J. M. Olesen. 1996. Importance of Elaiosome Size to Removal of Ant-Dispersed Seeds. *Oecologia* 107: 95–101.

Mattson, W. J., and N. D. Addy. 1975. Phytophagous Insects as Regulators of Forest Primary Production. *Science* 190: 515–21.

McAuslane, H. J. 1994. Interactions between Host-Plant Resistance and Biological Control Agents. In *Pest Management in the Subtropics: Biological Control—A Florida Perspective,* ed. D. Rosen, F. D. Bennett ,and T. L. Capinera, 681–708. Andover, United Kingdom: Intercept.

McAuslane, H. J., S. B. Vinson, and H. J. Williams. 1991. Influence of Adult Experience on Host Microhabitat Location by the Generalist Parasitoid, *Campoletis sonorensis* (Hymenoptera: Ichneumonidae). *Journal of Insect Behavior* 4: 101–13.

Metcalf, R. L., and E. R. Metcalf. 1992. *Plant Kairomones in Insect Ecology and Control.* New York: Chapman and Hall.

Minyard, J. P., D. D. Hardee, R. C. Gueldner, A. C. Thompson, G. Wiygul, and P. A. Hedin. 1969. Constituents of the Cotton Bud: Compounds Attractive to the Boll Weevil. *Journal of Agricultural and Food Chemistry* 17: 1093–97.

Morgan, A. C. 1910. *Observations Recorded at the 236th Regular Meeting of the Entomological Society of Washington* 12: 72.

Olff, H., V. K. Brown, and R. H. Drent, eds. 1999. *Herbivores: Between Plants and Predators.* Oxford: Blackwell Scientific.

Orr, D. B., D. J. Boethell, and W. A. Jones. 1985. Biology of *Telenomus chloropus* (Hymenoptera: Scelionidae) from Eggs of *Nezara viridula* (Hemiptera: Pentatomidae) Reared on Resistant and Susceptible Soybean Genotypes. *Canadian Entomologist* 117: 1137–42.

Owen, D. 1980. *Camouflage and Mimicry.* Chicago: University of Chicago Press.

Paige, K. N., and T. G. Whitham. 1987. Overcompensation in Response to Mammalian Herbivory: The Advantage of Being Eaten. *American Naturalist* 129: 407–16.

Panda, N., and G. S. Khush. 1995. *Host Plant Resistance to Insects.* Wallingford, United Kingdom: CAB Press International.

Plapp, F. W. 1977. Biochemical Genetics of Insecticide Resistance. *Annual Review of Entomology* 22: 179–97.

Polis, G. A., and K. O. Winemiller. 1996. *Food Webs: Integration of Patterns and Dynamics.* New York: Chapman and Hall.

Price, P. W. 1991. Evolutionary Theory of Host and Parasitoid Interactions. *Biological Control* 1: 83–93.

Price, P. W., C. E. Bouton, P. Gross, B. A. McPheron, J. A. Thompson, and A. E. Weis. 1980. Interactions among Three Trophic Levels: Influence of Plants on Interactions between Insect Herbivores and Natural Enemies. *Annual Review of Ecology and Systematics* 11: 41–65.

Rathcke, B. J., and R. W. Poole. 1975. Coevolutionary Race Continues: Butterfly Larval Adaptation to Plant Trichomes. *Science* 187: 175–76.

Reed, D. P., P. P. Feeny, and B. B. Root. 1970. Habitat Selection by the Aphid Parasite *Diaretiella rapae. Canadian Entomologist* 102: 1567–78.

Reese, J. C. 1977. The Effects of Plant Biochemicals on Insect Growth and Nutritional Physiology. In *Host Plant Resistance to Pests,* ed. P. A. Hedin, 129–52. Washington, D.C.: American Chemical Society.

Rhoades, D. F., and R. G. Cates. 1976. Toward a General Theory of Plant Antiherbivore Chemistry. *Recent Advances in Phytochemistry* 10: 168.

Rickson, F. R. 1980. Developmental Anatomy and Ultrastructure of the Ant-Food Bodies (Beccariian Bodies) of *Macaranga triloba* and *M. hypoleuca* (Euphorbiaceae). *American Journal of Botany* 67: 285–92.

Rodgerson, L. 1998. Mechanical Defense in Seeds Adapted for Ant Dispersal. *Ecology* 79: 1669–77.

Roeske, C. N., J. N. Seiber, P. L. Brower, and C. M. Moffitt. 1976. Milkweed Cardenolides and Their Comparative Processing by Monarch Butterflies. *Recent Advances in Phytochemistry* 10: 93–167.

Roitberg, B. D., and M. B. Isman, eds. 1992. *Insect Chemical Ecology: an Evolutionary Approach.* New York: Chapman and Hall.

Rosenthal, G. A., and D. H. Janzen. 1979. *Herbivores: Their Interactions with Secondary Plant Metabolites.* New York: Academic Press.

Rosenthal, G. A., and E. R. Berenbaum, eds. 1991. *Herbivores: Their Interaction with Secondary Plant Metabolites.* New York: Academic Press.

Ruberson, J. R., M. J. Tauber, C. A. Tauber, and W. M. Tingey. 1989. Interactions at Three Trophic Levels: *Edovum puttleri* Grissell (Hymenoptera: Eulophidae), the Colorado Potato Beetle, and Insect-Resistant Potatoes. *Canadian Entomologist* 121: 841–51.

Smith, J. M. 1957. Effects of the Food of the California Red Scale, *Aonidiella aurantii* (Mask.) on Reproduction of Its Hymenopterous Parasites. *Canadian Entomologist* 89: 219–30.

Stipanovic, R. D., H. J. Williams, and L. A. Smith. 1986. Cotton Terpenoid Inhibition of *Heliothis virscens* Development. In *Role of Allelochemicals,* ed. M. A. Green and P. Hedin, 79–94. American Chemical Society Symposium Series 296. Washington, D.C.: American Chemical Society.

Strand, M. R., and L. L. Pech. 1995. Immunological Basis for Compatibility in Parasitoid-Host Relationships. *Annual Review of Entomology* 40: 31–56.

Strand, M. R., R. Sequeira, and M. Otto. 1997. Growth and Development in Parasitoid Wasps: Adaptation to Variable Host Resources. *Vertical Food Web Interactions* 130: 191–203.

Strong, D. R., J. H. Lawton, and S. R. Southwood. 1984. *Insects on Plants: Community Patterns and Mechanisms.* Cambridge, Mass.: Harvard University Press.

Turlings, T. C. J., and B. Benrey 1998. Effects of Plant Metabolites on the Behavior and Development of Parasitic Wasps. *Ecoscience* 5: 321–33.

Turlings, T. C. J., and J. H. Tumlinson. 1991. Do Parasitoids Use Herbivore-Induced Plant Chemical Defenses to Locate Hosts? *Florida Entomologist* 74: 42–50.

Turlings, T. C. J., M. Bernasconi, R. Bertossa, F. Bigler, G. Caloz, and S. Dorn. 1998. The Induction of Volatile Emissions in Maize by Three Herbivore Species with Different Feeding Habits: Possible Consequences for Their Natural Enemies. *Biological Control* 11: 122–29.

Udayagiri, S., and R. L. Jones. 1992. Flight Behavior of *Macrocentrus grandii* Goidanich (Hymenoptera: Braconidae), a Specialist Parasitoid of European Corn Borer (Lepidoptera: Pyralidae)—Factors Influencing Response to Corn Volatiles. *Environmental Entomology* 21: 1448–56.

van Dam, N. M., and S. K. Bhairo-Marhe. 1992. Induced Chemical Defence in *Cynoglossum officinale. Proceedings of the 8th International Symposium on Insect-Plant Relationships* 1: 79–82.

van den Bosch, R. 1978. *The Pesticide Conspiracy.* New York: Doubleday.

van Emden, H. F. 1986. The Interaction of Plant Resistance and Natural Enemies: Effects on Populations of Sucking Insects. In *Interactions of Plant Resistance and Parasitoids and Predators of Insects,* ed. D. J. Boethe and R. D. Eikenbary, 138–50. New York: John Wiley and Sons.

van Emden, H. F., and M. J. Way. 1972. Host Plants in the Population Dynamics of Insects. In *Insect-Plant Relationships,* ed. H. F. van Emden, 81–199. London: Royal Entomological Society.

Vet, L. E. M., W. J. Lewis, and R. T. Carde. 1995. Parasitoid Foraging and Learning. In *Chemical Ecology of Insects,* vol. 2, ed. R. T. Carde and W. J. Bell, 65–101. New York: Chapman and Hall.

Vinson, S. B. 1972. Effect of the Parasitoid, *Campoletis sonorensis,* on the Growth of Its Host, *Heliothis virscens. Journal of Insect Physiology* 18: 1509–14.

———. 1975. Biochemical Coevolution between Parasitoids and Their Hosts. In *Evolutionary Strategies of Parasitic Insects and Mites,* ed. P. Price, 14–48. New York: Plenum.

———. 1981. Habitat Location. In *Semiochemicals: Their Role in Pest Control,* ed. D. A. Nordlund, R. L. Jones, and W. J. Lewis, 51–77. New York: John Wiley and Sons.

———. 1985. The Behavior of Parasitoids. In *Comprehensive Insect Physiology, Biochemistry, and Pharmacology,* ed. F. A. Kerkut and L. I. Gilbert, 417–69. New York: Pergamon Press.

———. 1990a. How Parasitoids Deal with the Immune System of Their Host: An Overview. *Archives of Insect Biochemistry and Physiology* 13: 3–27.

———. 1990b. Immunosuppression. In *New Directions in Biological Control: Alternatives for Suppressing Agricultural Pests and Diseases,* ed. R. R. Baker and P. E. Dunn, 517–35. New York: A. R. Liss.

———. 1990c. Potential Impact of Microbial Insecticides on Beneficial Arthropods in the Terrestrial Environment. In *Safety of Microbial Insecticides,* ed. M. Laird, A. Lacey, and E. W. Davidson, 43–64. Boca Raton, Fla.: CRC Press.

———. 1991. Chemical Signals Used by Insect Parasitoids. *Redia, Geornale de Zoologie* 124: 15–42.

———. 1998. The General Host Selection Behavior of Parasitoid Hymenoptera and a Comparison of Initial Strategies Utilized by Larvaphagous and Oophagous Species. *Biological Control* 11: 79–96.

———. 1999. Parasitoid Manipulation as a Plant Defense Strategy. *Annals of the Entomological Society of America* 92: 812–28.

Vinson, S. B., and G. F. Iwantisch. 1980. Host Suitability for Insect Parasitoids. *Annual Review of Entomology* 25: 397–419.

Vinson, S. B., and P. Barbosa. 1987. Interrelationship of Nutritional Ecology of Parasitoids. In *Nutritional Ecology of Insects, Mites, and Spiders,* ed. F. Stansky Jr. and J. G. Rodriguez, 673–93. New York: John Wiley and Sons.
Vinson, S. B., and H. J. Williams. 1991. Host Selection Behavior of *Campoletis sonorensis:* A Model System. *Biological Control* 1: 107–17.
Vinson, S. B., G. W. Frankie, and J. Barthell. 1993. Threats to the Diversity of Solitary Bees in a Neotropical Dry Forest in Central America. In *Hymenoptera and Biodiversity,* ed. J. LaSalle and E. Gould, 53–81. Wallingford, United Kingdom: CAB.
Waage, J. K. 1983. Aggregation in Field Parasitoid Populations: Foraging Time Allocation by a Population of *Diadegma* (Hymenoptera: Ichneumonidae). *Ecological Entomology* 8: 447–53.
Waeckers, F. L., E. Mitter, and S. Dorn. 1998. Vibrational Sounding by the Pupal Parasitoid *Pimpla (Coccygominmus) turionellae:* An Additional Solution to the Reliability-Detectability Problem. *Biological Control* 11: 141–46.
Whitman, D. W. 1988. Allelochemical Interactions among Plants, Herbivores, and Their Predators. In *Novel Aspects of Insect-Plant Interactions,* ed. P. Barbosa and D. K. Letourneau, 11–64. New York: John Wiley and Sons.
Whitman, T. G. 1981. Individual Trees as Heterogeneous Environments: Adaptation to Herbivory or Noise? In *Insect Life History Patterns: Habitat and Geographic Variations,* ed. R. F. Denno and H. Dingle, 9–27. New York: Springer-Verlag.
Wiens, J. A. 1977. On Competition and Variable Environments. *American Scientist* 65: 590–97.
Williams, H. J., R. D. Stipanovic, L. A. Smith, S. B. Vinson, P. O. Darnell, R. Montandon, D. L. Begin, G. W. Elzen, H. Gunasenan, and A. A. Bell. 1987. Effects of Gossypol and Other Cotton Terpenoids on *Heliothis virscens* Development. *Revista Latinoamericana de Quimica* 18: 119–31.
Yamaguchi, K., and T. Shibamoto. 1980. Volatile Constituents of the Chestnut Flower. *Journal of Agricultural and Food Chemistry* 28: 82–84.
Zucker, W. V. 1983. Tannins: Does Structure Determine Function? An Ecological Perspective. *American Naturalist* 121: 335–36.

Part V

Entomology:
A Science for All Society

Chapter 15

Celluloid Entomology

MAY R. BERENBAUM
Department of Entomology
University of Illinois at Urbana-Champaign
RICHARD J. LESKOSKY
Unit for Cinema Studies
University of Illinois at Urbana-Champaign

As a science, entomology is about the pursuit of knowledge; in conducting their science, entomologists are daily involved in ascertaining facts about insects. But entomology is not just for scientists and, in reality, never has been just for scientists. Insect biology has been the inspiration for all manner of human pursuits, including art, literature, music, sports, and many other nonscientific endeavors. Through popular culture, the public constructs its own reality about insects. Because entomologists must function within the limits and according to the dictates of society at large, it is instructive to examine these alternate realities and become familiar with them.

But which alternate reality? What are the drivers of popular culture? If there is one common experience that defines American culture, it is, in all probability, going to the movies. This fascination with film was clearly illustrated during the recent impeachment hearings, a historic occasion that arguably could affect the very future of the American presidency. During these history-making hearings, lawyers on both sides of the issue repeatedly quoted dialogue from or made reference to popular films. Congressman Henry Hyde alone quoted from at least two film sources—*A Man for all Seasons* and *Henry V*—in making a point, and the House Managers found themselves in dispute with White House lawyer Nicole Seligman as to which Charles Laughton role, Sir Wilfred Robarts in *Witness for the Prosecution* or Inspector Javert in *Les Miserables,* was most applicable to the proceedings. Every year, the broadcast of the Academy Awards ceremony garners enormous ratings, rivaled in magnitude only by airings of that other bastion of popular culture, football playoff games. Movies are among the principal legal forms of escaping the stress of everyday life.

The twin cities of Urbana and Champaign, in Illinois, provide a case in point. These towns have a population of about 100,000 and are the geographic home for the University of Illinois (UIUC), flagship campus for the state university system. In this college town, there are over forty movie screens (sixteen at one multiplex alone), not to mention dozens of outlets for video rentals, including grocery stores and gas stations, and a cable system with over a dozen channels dedicated to showing movies all of the time. Far fewer venues exist for obtaining information about any scientific discipline, including entomology. For the majority of students at the University of Illinois (and at most major universities), science classes comprise a remarkably small part of their undergraduate experience (only six of 120 hours are required at UIUC).

Avoidance of science is oftentimes attributed to the perception of science teaching as boring. Many studies have revealed that students find the manner in which science is taught in school to be insufferably dull. High school students polled by Harvard's Educational Technology Center agreed overwhelmingly that science is "important but boring" (Office of Technology Assessment, 1988), and, in a recent study investigating why undergraduates switch majors in college, Seymour and Hewitt (1997) found that the proportion of undergraduates remaining in science majors is substantially lower than the proportion remaining in social science or humanities majors; the principal reason these science students switched majors (mostly out of science altogether) was that they were "turned off" or bored by science.

Brustein (1958) argued that movies reflect society's view of science; it can also be argued that movies help to shape society's view of science, particularly in the absence of other inputs. The influence of formal education on attitudes toward science is limited by time; few individuals pursue studies of science after college, and opportunities for obtaining such information are restricted to technical books, a few science-oriented magazines, some forms of television programming, an occasional public lecture, and Internet Web sites of wildly variable quality. In general, negative attitudes toward science persist after formal education ends and are directed against these very sources of science facts. In stark contrast, movies are readily accessible to all (or, to those with the price of admission) and have an extraordinary ability to convey information, regardless of accuracy, through entertainment. They're effective because they are in many ways the antithesis of the lecture experience; they have what lectures generally lack (characters, plot, drama, and, all too often, humor). Because of their power, movies can create an alternate reality and because of their widespread appeal, that alternate reality becomes familiar to a significant segment of society. Take aliens, for example; everyone more or less recognizes what aliens look like, despite the fact that they do not exist. This "reality" was wholly created by the popular media (Skal, 1998).

Frequent appearances by insects in live action films are a relatively recent phenomenon in the history of film. For many years, the technical challenges of filming very small, largely untrainable, fast-moving creatures proved a disincentive for incorporating them into films. Jack Arnold, one experienced director of science fiction films of the 1950s who was successful in incorporating arthropods into feature films, explained the frustrations of the experience in filming *The Incredible Shrinking Man:* "It's very hard to direct a spider, I used spurts of air, as I had done in *Tarantula,* to prod him in the direction I wanted him to move. We flew in sixty Panamanian tarantulas because the domestic ones were too small and we couldn't keep a sharp focus on them. They were tremendous beasts—six inches in diameter! We used so many during the shooting because we had to light to such a high intensity they cooked" (Brosnan, 1991).

The pioneering efforts of special effects genius Willis O'Brien, starting in the 1930s, and of his protégé Ray Harryhausen, as well as technical advances in the production of film stock and traveling matte techniques gradually made the incorporation of insect images in film economically attractive. Moreover, competition for audiences, particularly with the rise of television, led the major film studios to increase investment in hitherto minor genres such as science fiction. With bigger budgets, more elaborate effects became feasible (Thompson and Bordwell, 1994). Today, insect images in film (as well as in television and print media) are everywhere in evidence; their role in cinema is firmly established. In fact, their impact on culture has been so strong that references to insect movies serve as punch lines in jokes and cartoons (e.g., Larson, 1984; Larson, 1987).

METHODS

Not surprisingly, insect images are most frequently encountered in science fiction films; because entomology is a biological subdiscipline, it follows logically that subjects of scientific study are rightly subjects of science fiction. In order to define the "alternate reality" of insect biology created by Hollywood, it is an interesting exercise to examine insect biology as it is collectively depicted in science fiction films. For this analysis, we compiled information on over sixty feature-length live action films released between 1938 and 1998 that involve insects either as central figures or essential plot devices (Figure 15.1). These films were all viewed at least once; details about cast, crew, plot, and production history were obtained in some cases from published sources (e.g., Warren, 1982; 1986). Taxonomic identity of arthropods in films (to order) was easily established, by virtue of both name and standard morphological characteristics, in virtually all films. Animated films were omitted from this analysis because animation rules for the depiction of reality differ from rules governing live action films (generally, humor derives from an inversion of normality, often made possible by the animation process) (Wells, 1998; Leskosky and Berenbaum, 1988).

RESULTS AND DISCUSSION

Movie arthropods differ in taxonomic distribution from entomological arthropods. Certain orders are conspicuous by their absence; Coleoptera, for example, the largest order of insects (indeed, of any organism), have rarely if ever been featured as characters in a live action feature-length film (in *The Hephaestus Plague,* the novel on which *Bug* is based, beetles are crossed with cockroaches by a careless scientist who inadvertently creates carbon-eating explosive hybrids that can spell out death threats on the wall with their bodies, but these were played by Madagascar hissing cockroaches in the film version). But, remarkably, the differences aren't quite as dramatic as might be expected. Two of the largest orders of real-life arthropods, Hymenoptera and Diptera, are in fact the most frequently depicted in insect-fear films. There may be several reasons for this proportional similarity—among them the fact that bees are relatively easily manipulated for the camera in comparison with other insects, and that both bees and flies can be reared in enormous numbers with comparative ease. Perhaps an even more important factor, however, is their familiarity to the audience. Encounters with bees, ants, wasps, and flies are part of the normal course of life for most moviegoers. Such an explanation also can account for the proliferation of films involving cockroaches (Berenbaum, 1995), although these, too, share the practical advantage of ease of rearing in enormous numbers. While it's true that other large orders (e.g., Hemiptera, Homoptera) are unrepresented or (Odonata, Orthoptera) underrepresented, it is also true that the smaller orders appear infrequently. No films have been

Table 15.1. **Live Action Feature Films with Insects as Major Components.**

Year	*Film Title*	*Year*	*Film Title*
1938	Yellow Jack	1975	Bug
1944	Sherlock Holmes vs. the Spider Woman	1975	The Giant Spider Invasion
1944	Once upon a Time	1976	The Savage Bees (TVM)
1953	Mesa of Lost Women	1977	Empire of the Ants
1954	Them	1977	Exorcist II—The Heretic
1954	Naked Jungle	1977	Ants: It Happened at Lakewood Manor (TVM)
1955	Tarantula	1977	Terror Out of the Sky (TVM)
1955	Panther Girl of the Kongo	1978	Tarantulas: The Deadly Cargo (TVM)
1957	The Black Scorpion	1978	The Bees
1957	Beginning of the End	1978	The Swarm
1957	Earth vs. the Spider	1978	Curse of the Black Widow (TVM)
1957	The Deadly Mantis	1980	Island Claws (Night of the Claw)
1957	Monster from Green Hell	1982	Creepshow
1958	The Cosmic Monsters	1986	The Fly
1958	She Devil	1987	Blue Monkey
1958	The Fly	1987	The Nest
1959	Return of the Fly	1987	Deep Space
1959	The Brain Eaters	1989	The Fly II
1959	Wasp Woman	1990	Arachnophobia
1962	Mothra	1991	Meet the Applegates
1964	Godzilla vs. The Thing	1993	Ticks
1965	Horrors of Spider Island	1994	Skeeters
1968	Destroy All Monsters!	1995	Mosquito (Nightswarm)
1969	The Blood Beast Terror (TheVampire Beast Craves Blood)	1996	Angels and Insects
1970	Flesh Feast	1996	Wax
1971	The Hellstrom Chronicles	1996	Joe's Apartment
1971	The Legend of Spider Forest	1996	Wasp Woman (TVM)
1972	Kiss of the Tarantula (Shudders)	1997	Starship Troopers
1973	Invasion of the Bee Girls	1997	Men in Black
1974	Phase IV	1997	Mimic
1974	Locusts (TVM)	1998	X-Files: The Movie
1974	The Killer Bees (TVM)		

made about orders containing fewer than 5,000 species (Ephemeroptera, Plecoptera, Isoptera, Thysanoptera, Phthiraptera) (*Autobiography of a Flea,* a pornographic film, is a conspicuous exception to the pattern). That such is the case is consistent with the need on the part of filmmakers to invoke images that are familiar. Insects with life histories unknown to the majority of viewers have no symbolic value.

The status of noninsect arthropods in films is less clear. Indeed, in the taxonomy of Hollywood, jointed-legged segmented organisms with exoskeletons are generally classified as insects, regardless of how many legs or how few antennae they possess. For example, the character of Sherlock Holmes, in *Sherlock Holmes and the Spider Woman,* identifies a spider (used by the nefarious Spider Woman to dispatch her victims) as *Lycosa carnivora* from the Obongo River in Africa, not only an "ugly-looking brute" but also the "deadliest insect known to science" (proving himself to be a better detective than taxonomist). Taxonomic hierarchies are obscured in cinema as well; in the film *Tarantula,* the artificial nutrient-enhanced giant spider is identified as being "from a species called Arachnida—a tarantula to be exact." In *Mimic,* world-class entomologist Dr. Gates, the mentor of Dr. Susan Tyler, a young scientist engaged in genetic engineering experiments with cockroaches, makes reference during a conversation with his former student to the "Phylum Insecta."

Insect morphology in the movies is also clearly different from entomological insect morphology. Mouthparts in particular function in all kinds of ways that are unfamiliar to the entomologist. In *Mosquito,* what appears to be a hypopharynx is green and oozes slime, and in *Skeeters* the mosquito proboscis takes on the complexity of a Swiss army knife. In a promotional poster, Mothra's mouthparts are more hamsterlike than lepidopteran. Compound eyes also are cause for some confusion; many films present an "insect-eye" view of a particular scene (usually a soon-to-be-victim) through a Fresnel lens to simulate what is imagined to be the image created by compound eyes (e.g., in *Empire of the Ants*). In reality, these images appear more like mosaics than repeated images. Even external morphology is cause for some confusion; in *Monster from Green Hell,* the compound eyes of the cosmic wasps roll in their sockets.

But, all told, there is remarkable congruence between cinematic and entomological arthropods. Almost all movie insects have six legs (when filmmakers have bothered to construct an entire model), whereas movie arachnids have eight (even when they're felt-covered PVC tubes attached to a

Volkswagen, as they were in the low-budget *The Giant Spider Invasion*). Almost without exception, movie insects have the three hallmark body regions—head, thorax, and abdomen—that differentiate them from other arthropods. Even at the ordinal level, there is remarkable fidelity. Movie mantids have raptorial forelegs (e.g., *The Deadly Mantis*), and movie dipterans have two wings (advertising posters for the film *Mosquito* notwithstanding), whereas movie arthropods from other orders have four. Mothra, the giant moth that attacked Tokyo in a series of Japanese films from the 1960s, clearly possesses scales. Large flat objects the size of pie plates scattered around the countryside (eventually identified as oversized scales) provide clues to the presence of an enormous moth in *The Blood Beast Terror.*

The physiology of movie arthropods bears only a passing resemblance to the physiology of entomological arthropods. This is a point frequently made by movie entomologists. Witness the statement, in *Blue Monkey,* by Dr. Elliot Jacobs, the entomologist called in to assist with an outbreak of genetically engineered mutant cockroaches in a hospital: "Insects aren't like humans or animals. They're 80 percent water and muscle. They have very few internal organs." A recurring conceit is the familiar violation of the surface area to volume ratio constraint—movie arthropods routinely grow to the size of mobile homes and suffer no ill effects, while their entomological counterparts are restricted by constraints of molting and breathing via tracheae to toy-truck size at a maximum. Even so, there are physiological attributes of film arthropods that should be more than faintly familiar to entomologists. Metamorphosis is a characteristic of a substantial proportion of movie arthropods, although, admittedly, the process takes place in a markedly different manner on screen from in an entomology laboratory. The transformation most frequently depicted in films is not larva to pupa or pupa to adult but rather insect to human or human to insect. This transformation generally follows voluntary exchange of body fluids—"*Drosophila* serum" in the case of *She Devil,* which allows the patient to transform herself at will from brunette to blonde; "spider hormones" in *Mesa of Lost Women,* which creates a race of women with long fingers and diaphanous gowns; "royal jelly" in *Wasp Woman,* which provides cosmetics magnate Janice Starlin with a return to youth accompanied by an uncontrollable desire to kill; and "DNA" in the 1986 remake of *The Fly.* It is likely true that insects appear frequently in films involving metamorphosis because of the shock value—what worse fate could befall a human than to become transformed into a life form that is so radically different in appearance? But their appearance may also be a subconscious recognition of the universality of metamorphosis within the class. As a general rule, transformations of humans into other animal forms more commonly involves magic or reincarnation (*The Shaggy Dog, The Shaggy D.A., Oh, Heavenly Dog, Lucky Dog*) or some form of genetic predisposition (*Teen Wolf* and its sequel, *The Howling, Cat People*) than it does mediation by hormones (conspicuous exceptions to this pattern are the early films of Bela Lugosi, including *The Ape Man* and *Return of the Ape Man,* that involve serum exchanges between humans and apes).

Yet another, rather obscure, aspect of insect metamorphosis is depicted rather effectively, if possibly inadvertently, in *The Fly II.* In this film, Martin Brundle, the son of Seth Brundle, unfortunate victim of a matter transfer experiment with a fly gone awry, experiences a fantastically accelerated but otherwise normal youth. Only at puberty is the fly DNA inherited from his father activated; the unfortunate teenager then begins to grow dipteran structures. In general, the process depicted is analogous to that by which cells of imaginal disks in holometabolous insects are activated only upon pupation.

Along with insect hormones, insect pheromones figure prominently in insect films (although they aren't always identified as such; in *The Bees* they're called "pherones"). In *Empire of the Ants,* giant ants use pheromones to enslave the local human population and to compel the humans to operate a sugar factory for them. The explanation provided for the response is that a pheromone "causes an obligatory response—did you hear that? Obligatory. It's a mind-bending substance that forces obedience." Although they have long been documented to exist in a wide range of organisms (including humans), pheromones rarely appear in science fiction films outside an entomological context.

Insect ecology takes on different dimensions in the movies, as well. Life cycles are unorthodox by entomological standards, to say the least. In *Mosquito,* for example, mutated mosquitoes, the offspring of normal mosquitoes that consumed the blood of aliens in a crash-landed UFO, have a life cycle consisting only of egg and adult stages. In *Ticks,* full-grown ticks eclose from what appears to be a cocoon. And population dynamics differ as well. For a number of movie arthropods, population size appears to be one (as evidenced by the titles—e.g., *Tarantula, The Deadly Mantis*), and reproduction does not appear to occur (at least over the two-hour span of the movie) (Berenbaum and Leskosky, 1992). At the other extreme, populations often build up to enormous sizes without depletion of any apparent food source. Bees blacken the sky in *The Bees* and *The Swarm* in a remarkably short period of time with no superabundance of nectar sources in evidence (although, in one film, a massive swarm of bees does attack the Rose Bowl parade, a logical place for bees to go to find nectar sources in abundance). It must be assumed that food utilization efficiencies of virtually all film arthropods are far higher than they are in real life because in no film in this survey do arthropods, giant or otherwise, appear to produce any frass (in *Beginning of the End,* this is despite giant grasshoppers consuming several tons of wheat in a three-month period). In *Starship Troopers,* it's not clear what the giant arthropods living on a planet bereft of other life forms eat in order to attain their large size (but they're *alien* insects, so terrestrial biological standards may not necessarily be applicable).

Adding to the general difficulty of finding sufficient food to survive is the other handicap that filmmakers impose on cinematic insects—inept foraging techniques. Screen insect predators and herbivores alike almost without exception announce their intention to strike with an ear-piercing stridulating sound (e.g., *Them, The Deadly Mantis, Beginning of the End, Empire of the Ants,* and so on), which in reality would be sure to alert prey to danger and induce them to initiate escape or defensive behavior (which, on the part of humans in many of these films, involves machine guns and bazooka fire directed at the insect). For example, in *Beginning of the End,* a television newscaster updates viewers in Chicago on the Illinois National Guard's efforts against hordes of gigantic radiation-induced mutant grasshoppers descending on the city, reassuring them that "the one advantage our forces hold over the enemy is that they ALWAYS reveal their intention to attack. Before every attack the locusts send forth this warning in the form of a high-pitched screech. Now, this screech increases in intensity until it reaches ear-shattering proportions. And it's when this screech reaches its full intensity that the locust attacks." It is difficult to conceive of a selection regime that would favor the evolution of this sort of behavior.

In contrast with food web dynamics, other aspects of ecology are depicted fairly accurately, at least in principle. Biomagnification, accumulation of toxins up a food chain, is a recurrent theme. In *Kingdom of the Spiders,* tarantulas take over a town and start consuming livestock and eventually veterinarian Dr. Robert "Rack" Hansen (portrayed by William Shatner) because "DDT" destroyed the food chain and deprived them of their normal prey. Other films depicting altered food web dynamics as a result of pollution (radioactive and/or toxic waste) include *Skeeters* and *Empire of the Ants.* In *Ticks,* fertilizers and other agrochemicals used by illegal marijuana growers are encountered by ticks, which grow to enormous size and terrorize a group of inner city teens in the woods on a wilderness survival trip.

Another ecological phenomenon of concern both in the movies and in real life is the accidental introduction of alien species (although in the movies these are more likely to be real aliens, from outer space, not just Europe or South America). *Arachnophobia* depicts what happens when a South American spider species accidentally is transported to the Pacific Northwest. Any one of the many killer bee movies pointedly makes reference to the dangers of accidental importation of strains of bees into new habitats (although in *The Bees* their introduction is no accident; greedy cosmetics magnates bring over killer bees in the hope of producing large amounts of very saleable royal jelly).

Insect genetics figure prominently in insect-fear films. Genetic engineering first appears in 1987 in *The Nest,* within five years of Rubin and Spradling's (1982) prediction that species other than *Drosophila* might soon be transformed. In real life, nondrosophilids were not successfully genetically transformed until 1995 (*Ceratitis capitata*—Zwiebel et al., 1995) and 1998 (*Aedes aegypt*—Jasinskiene et al., 1998). By 1997, however, on screen in *Mimic,* entomologist Susan Tyler (Mira Sorvino) is able to incorporate termite and mantid DNA into cockroaches, with the goal of creating a "Judas bug" to bring contagion to the cockroach vectors of a human illness but instead unleashing a plague of six-foot-tall man-eating cockroaches in the subway system of New York City.

So the two realities—cinema and entomology—of insect life are not really as disparate as they may first appear. There are commonalities no doubt because entomologists are part of society over and above their status as scientists. Entomologists get no special dispensation from the insect world when it comes to serving as food for parasites, nor are entomologists exempt from concerns about accidental importation of alien species or food web contamination by pesticides or radioactive waste. Entomologists can pursue their profession only because society at large deems their pursuit worthwhile; tax dollars support the enterprise.

Is it important for entomologists to acknowledge the other reality? It is essential because doing so allows entomologists to work more efficiently within society's boundaries. This is not to say that the next grant proposal that goes to the USDA should focus on developing pest management programs for extraterrestrial insects. But movies reflect social concerns and it is important for entomologists to recognize and acknowledge those concerns. In this context it is perhaps most interesting to note the greatest disparity between entomological and cinematic "realities." Although the ecology, behavior, morphology, and even taxonomy of cinematic arthropods are at least suggestive of reality, pest management in the movies is almost unrecognizable. Biological control is almost unheard of in insect-fear films; so, surprisingly, is chemical control. Host plant resistance, competitive control, and use of pathogens simply do not exist. Control of insects in movies is almost entirely by physical or mechanical means. Fire and/or explosives, napalm, or incendiary devices are a common theme, both in older films (*Mesa of Lost Women, Them, The Naked Jungle, Tarantula, The Swarm*) and in more recent films (*Blue Monkey, Arachnophobia, Mimic, Mosquito*). Guns, too, are frequently used, albeit with variable success (they effectively dispatched Edgar the alien cockroach in *Men in Black* as well as the alien arachnoids in *Starship Troopers* but were useless against giant grasshoppers in *Beginning of the End*). Other mechanical means of destroying insects of a more idiosyncratic nature include an electric harpoon (*The Black Scorpion*), a hydraulic press (*The Fly*), a chainsaw (*Mosquito*), and a "neutron initiator" (*The Giant Spider Invasion*). There are no pheromone traps in the movies (although the use of sound lures to bring insects to a central location to drown or incinerate them does appear in *The Beginning of the End* as well as *The Swarm*), nor are there any examples of sterile male release (a technique that might prove problematic given the frequency with which parthenogenetic reproduction seems to appear) (Berenbaum and Leskosky, 1991).

It is easy to imagine why insects in movies die in such spectacular fashion; death by flamethrower or incendiary device is more dramatic than death by fungal infection. However, given the frequency with which moviegoers have been exposed to such approaches, it is not at all difficult to understand why undramatic management techniques may be such a hard sell. Consumers are all too willing to believe spectacular and unsubstantiated claims of purveyors of mechanical and other devices intended to destroy insects in a spectacular, rapid, and highly conspicuous manner. Despite repeated demonstrations of lack of efficacy against target species (e.g., Fradin, 1998), bug zappers continue to sell by the millions; tabloid newspapers with accounts of alien insects bent on human destruction sell millions of copies. While entomologists may not view themselves in competition with Hollywood filmmakers, they must nonetheless deal with the consequences of unequal access to the public. Accordingly, entomologists must make an effort to make optimal use of whatever access they have and learn to make their salient points effectively, clearly, and compellingly in order to do their part in shaping popular conceptions more to fit the entomological reality.

ACKNOWLEDGMENTS

We thank Arthur Zangerl for providing assistance with the table.

REFERENCES CITED

Berenbaum, M. R. 1995. *Bugs in the System: Insects and Their Impact on Human Affairs.* Reading, Mass.: Addison Wesley.

Berenbaum, M. R., and R. J. Leskosky. 1991. Mosquitoes in the Movies. *Vector Control Bulletin of the North Central States* 1: 94–98.

———. 1992. Life History Strategies and Population Biology in Science Fiction Films. *Bulletin of the Ecological Society of America* 73: 236–40.

Brosnan, J. 1991. The Metaphor That Ate Tokyo: Monster Movies of the Fifties. In *The Primal Screen: A History of Science Fiction Films,* chap. 6. New York: Little Brown.

Brustein, R. 1958. Reflections on Horror Movies. *Partisan Review* 25: 288–96.

Dubeck, L. W., S. E. Moshier, and J. E. Boss. 1994. *Fantastic Voyages: Learning Science through Science Fiction Films.* New York: American Institute of Physics Press.

Fradin, M. S. 1998. Mosquitoes and Mosquito Repellents: A Clinician's Guide. *Annals of Internal Medicine* 128: 93140.

Jasinskiene, N., C. J. Coates, M. Q. Benedict, A. J. Cornel, C. S. Rafferty, A. A. James, and F. H. Collins. 1998. Stable Transformation of the Yellow Fever Mosquito, *Aedes aegypti,* with the Hermes Element from the Housefly. *Proceedings of the National Academy of Sciences USA* 95: 3743–47.

Larson, G. 1984. *In Search of the Far Side,* 48. Kansas City, Mo.: Andrews and McMeel.

———. 1987. *The Far Side Observer,* 40. Kansas City, Mo.: Andrews and McMeel.

Leskosky, R. J., and M. R. Berenbaum. 1988. Insects in Animated Films; or, Not All "Bugs" Are Bunnies. *Bulletin of the Entomological Society of America* 34: 55–63.

Office of Technology Assessment. 1988. *Educating Scientists and Engineers: Grade School to Grad School.* Washington, D.C.: U.S. Government Printing Office.

Rubin, G. M., and A. C. Spradling. 1982. Genetic Transformation of *Drosophila* with Transposable Element Vectors. *Science* 218: 348–53.

Seymour, E., and N. M. Hewitt. 1997. *Talking about Leaving: Why Undergraduates Leave the Sciences.* Boulder, Colo.: Westview Press.

Skal, D. J. 1998. *Screams of Reason: Mad Science and Modern Culture.* New York: W. W. Norton.

Thompson, K., and D. Bordwell. 1994. *Film History: An Introduction.* New York: McGraw-Hill.

Warren, W. 1982. *Keep Watching the Skies!* Jefferson, N.C.: McFarland.

———. 1986. *Keep Watching the Skies! II.* Jefferson, N.C.: McFarland.

Wells, P. 1998. *Understanding Animation.* London: Routledge.

Zwiebel, L. J., G. Saccone, A. Zacharopoulou, N. J. Besansky, G. Favia, F. H. Collins, C. Louis, and F. C. Kafatos. 1995. The White Gene of *Ceratitis capitata:* A Phenotypic Marker for Germline Transformation. *Science* 270: 2005–08.

Chapter 16

Our War with the Insects

Analysis of Lost Battles

MARSHALL W. JOHNSON
UC Kearney Agricultural Center
University of California–Riverside

Military metaphors are often used to describe our attempts to suppress or manage arthropod populations. Titles of both lay and scientific articles relating to arthropod suppression often include military terms such as *battle* (Bueckert, 1992), *campaign* (Daniel, 1990), *defense* (Anonymous, 1991), *enemy* (Hoffman, 1992), and *war* (Stevens, 1993). L. O. Howard was probably the best-known entomologist to commonly use military metaphors (Russell III, 1999). Recently, in the book titled *Nature Wars: People vs. Pests,* Winston (1997) stated that pest "management has become a modern war against nature, but it is time to reconsider the terms of engagement." However, no obvious attempts were made in that publication or others to use ancient or current military strategies (Sun-tzu, 1993; Hart, 1967; von Clausewitz, 1976) in developing Integrated Pest Management (IPM) programs. Although entomologists have borrowed specific ideas or products from the military for use in pest management, such as sequential sampling (Wald, 1943, 1945; Waters, 1955), toxins used in pesticides (Ware, 1994), aerial pesticide applications (Akesson and Yates, 1974), radar (Wolf, 1979; Dunnigan and Nofi, 1994; Devereux, 1991), and global positioning systems (GPS) (Herring, 1996), perhaps we have truly neglected a gold mine of information in the form of strategy and tactics that could be applied to solve many challenges we face. This has been employed in other endeavors such as business (Musashi, 1982; Luecke, 1994; Krause, 1995) and medicine (Hartman, 1993) to enhance efficiency and productivity.

In this chapter I examine aspects of pest management programs that are analogous to military campaigns. Concepts developed for analysis of military failures will be used to elucidate common, but perhaps not obvious, reasons for unsuccessful management of pests in agroecosystems. Using methods suggested by Cohen and Gooch (1990), I analyze a "lost battle" in pest management using a selected case history. Lastly, I discuss the potential of this analysis technique to provide insights into pest management failures.

Few would argue against the notion that humans compete with insects and mites for food and fiber products (Metcalf et al., 1962). But do these arthropods truly behave like a wartime enemy? In contrast to civil disasters (fire, flood, hurricanes, etc.) which do not "react" to actions taken to lessen their impact, opponents in military campaigns (Cohen and Gooch, 1990) and arthropods targeted in pest management programs do react to offensive and defensive efforts, often negating efforts to overcome them. In addition, many arthropod pests are practically impossible to eradicate once firmly established and widely distributed in a new area (Dahlsten et al., 1989). Evolution is the major process by which arthropod species react to and overcome man's management efforts (Scott, 1990; De Benedictis and Granett, 1992; Gould, 1983). In both war and pest management, "resistance" to commonly used tactics often develops and new technologies have to be developed.

Cohen and Gooch (1990) presented a method for analyzing overwhelming military defeats or routs (i.e., disorganized retreats), as well as numerous military case histories where opponents were similar in force as well as ability and a less decisive outcome was expected. This methodology may be applied to IPM systems in which critical control components were available but not implemented for some reason. However, there are certainly situations in which new pest problems develop due to unusual scenarios (i.e., establishment of some exotic pests) and the IPM community is caught unprepared. In the following discussion I will examine pest control failures in systems where all necessary components for control existed initially and growers should have been able to manage the pests discussed.

Control failures can be examined at many levels: locally, area-wide, regionally, or nationally. Many different reasons exist for overall control failures. Relative to pests these include, but are not limited to, pesticide resistance development, broadening or switching of host plant range, hormoligosis, destruction or impairment of effective natural enemies, poor timing of pest suppression tactics, and establishment of an exotic species. Other reasons not directly related to changes in pests include the introduction of an arthropod-transmitted plant pathogen, changes in farming practices or gate value of a farm product, and demands for pest-free farm products.

FAILURES IN PEST MANAGEMENT

Military Campaigns versus Pest Management Programs

Although few would propose that an act of war is equal to our efforts to manage arthropods, numerous analogies may be drawn between military campaigns and pest management programs. Warfare often has the defensive goal of preventing an enemy from taking a valued commodity or occupying a given land area (Peckham, 1964; Keegan, 1993; Diamond, 1997; Milton, 1999); similarly, IPM strives to prevent arthropods from reducing food and fiber production (Strickland, 1970). Human warfare tends to flow from poor to rich lands (Keegan, 1994), just as many arthropod pests tend to concentrate in resource rich crop monocultures versus polycultures (Risch et al., 1983). In both warfare and IPM, definite organizational "command levels" are commonly in place that serve as a conduit for information and decision-making (Table 16.1). Actions taken in both arenas may also be affected by political and special interests (e.g., business). Additionally, successful outcomes in both areas typically require large amounts of current and accurate information on conditions that may rapidly change (Cohen and Gooch, 1990; Pedigo and Buntin, 1994; National Research Council, 1996). With respect to the logistics of conducting a military campaign or pest management program, several parameters are always limited. These include funding, time, and personnel as

Table 16.1. **Analogies between Military Campaigns and Pest Management Programs Relative to Information Flow.**

Military campaign	*Pest management program*
High command	National and regional research funding programs; regulatory agencies, state and federal control programs, industry
Theatre command	University, federal, state, private, and industry research
Regional command	Cooperative extension, industry
Intermediate command	Consultants
Field command	Grower

well as availability of strategies and tactics that may be employed to achieve specific goals (Keegan, 1994; U.S. Congress, 1995). Climate, weather, and terrain significantly impact execution of military operations (Keegan, 1994) as well as implementation of pest suppression tactics (e.g., application of pesticides, Matthews, 1979; triggering of epizootics, Benz, 1987; natural enemy efficacy, Varley et al., 1974; Messenger, 1970). Even the offensive and defensive strategies and tactics are operationally similar (Table 16.2).

Concepts of Military Failure Applied to IPM

Cohen and Gooch (1990) stated that there are generally three types of failures that occur in military campaigns that can result in total defeat or a rout of an opposing force. These are failures to "learn," "anticipate," and "adapt." Failures also occur at different levels in the chain of command or in the flow of information and decisions. These failures are discussed here relative to their application to IPM programs.

Failure to Learn

The failure to absorb readily accessible lessons from recent history is the basis of the "failure to learn" concept (Cohen and Gooch, 1990). Pest management researchers, extension personnel, and practitioners often learn about management strategies and tactics by consulting the scientific literature, participating in classes or workshops, interacting with recognized experts, or consulting their colleagues in various ways (e.g., the Internet). Failure to learn valuable information and apply it to the decision-making process can result in complete destruction of crops by arthropod pests. Some of the simplest lessons to learn in pest management are the recognition and identification of arthropod species (and their various life stages) commonly found in the field. Knowing the difference between a pest and a natural enemy (or a healthy pest vs. a parasitized one) can either cost or save a grower the price of one or more pesticide applications. Similarly, knowledge of the correct manner to sample for pests and natural enemies, the economic thresholds for common pest species, as well as the effectiveness and optimum tactics for managing specific pests, all play an important role in effective management. However, as Cohen and Gooch (1990) point out, failure to learn can be due to many reasons other than the "learning" abilities of the decision-maker in question. For example, if no extension education programs are available for growers to learn about solutions to specific pest problems (e.g., biological control, U.S. Congress, 1995), then the root of the failure is not at the grower level, but at the

Table 16.2. **Analogies between Military Campaigns and Pest Management Programs Relative to Strategies, Tactics, and Other Actions.**

Actions	*Military campaign*	*Pest management program*
Defensive	Deception (an attempt to mislead)	Trap crops, bait sprays, sterile male release, etc.
	Fortification of lines and facilities	Host plant resistance (e.g., antixenosis)
	Misdirection (attempt to provide misleading data)	Pheromone confusion technique, plastic mulch, etc.
	Monitoring of enemy movements (i.e., reconnaissance, surveillance)	Surveys and monitoring of pest populations
Offensive	Biological warfare	Commercial preparations of insect pathogens
	Blitzkrieg	Overwhelming pest migrations
	Concealment (hiding of preparatory moves)	Inability to reliably detect insects in the field
	Intelligence gathering on enemy's weak points	Research on pest's biological weak points
	Development of new tactics	Development of new controls
	Saturation bombing	Broad spectrum sprays
	Strategic bombing	Specific placement of pesticides
	Monitoring of enemy movements (i.e., reconnaissance, surveillance)	Surveys and monitoring of pest populations
Other	Allies	Natural enemies
	Anti-personnel mine (booby trap)	Insecticide baits
	Collateral damage[a]	Nontarget impacts (environmental pollution, destruction of desired species, etc.)
	Mortality from "friendly fire"	Natural enemy death / impairment after pesticide applications[a]
	Trojan horse	Pests concealed in import products

[a]Pest resurgences and secondary pest upsets are products of this occurrence.

extension/education level. Likewise, if no solutions have been developed then the extension worker has nothing to extend.

Failure to Anticipate

Central to this area of failure is the inability to foresee and take appropriate measures to deal with an opponent's action, or a likely response of your opponent to your own action (Cohen and Gooch, 1990). In pest management, the ability to predict the actions of a pest population is usually based on our skill at monitoring its populations via routine scouting of plantings, combined with an understanding of its biology and ecology. However, additional information can be gathered by monitoring the pesticide resistance levels in local populations in a region, allowing us to predict possible pockets of resistant individuals and take action to reduce the impact of these pockets (i.e., pesticide resistance management, Croft, 1990). Predictive modeling may also contribute to our abilities to foresee population changes. Phenomena that growers commonly fail to anticipate are: seasonal variation in arthropod populations; dispersal of pests from senescing crops; fluctuations in arthropod numbers due to biological control; development of pesticide resistance; and pest resurgences and secondary pest upsets. Cohen and Gooch (1990) emphasize that the essence of failure to anticipate is the failure to take reasonable precautions against a known hazard. Just as the chief function of military intelligence is to provide a warning (predicting what an opponent is about to do) and information on the order of battle (who, what, and where of an enemy's forces), scouting in IPM allows a consultant or grower to make similar decisions about the arthropod populations in crop plantings.

It is also important to note that in military campaigns and pest management programs alike, failure to anticipate may not result from poor data analysis, but from poor data sources. An excellent example of this is illustrated by sampling *Liriomyza* leafminers in watermelon in Hawaii. Lynch and Johnson (1987) demonstrated that variation in leafminer larval counts was reduced by monitoring densities only in the medium sized leaves (12–20 cm midrib length) versus small (< 12 cm midrib length) or large leaves (> 20 cm midrib length). Prior to full establishment of the watermelon foliar canopy, higher leafminer densities were recorded near the base of the plant as compared with the distal end. Thus, sampling stratification with respect to the distance from the plant base resulted in greater precision of leafminer foliar estimates. Individuals without this knowledge and who routinely sampled leaves near the plant base would most likely apply more pesticides to a planting than one who sampled leaves from the middle areas of the plant.

Failure to Adapt

The root of failure here is an inability to cope with unfolding events (Cohen and Gooch, 1990). This is quite obvious in military situations when a protagonist overruns or flanks an enemy's position. In pest management, the most destructive example of this may be the introduction of alien pests that are not problematic in their countries of origin but pose a significant management challenge in the area of introduction for a variety of reasons (e.g., lack of effective natural enemies). Fortunately, this is not a common problem in most localities. Perhaps the three most common events that present the greatest control challenges are the evolution of pesticide resistance, the dispersal of plant pathogens transmitted by arthropods, and the mass movement of insect pests into plantings from a highly infested local source (the "blitzkrieg" of the insect world). In each situation, the application of pesticides to the system rarely solves the problem, and the solutions often exist in taking proactive measures rather than reactive ones.

Levels of Failure

As Cohen and Gooch (1990) point out, battles are not always lost at the level of the "soldier in the field," but often at higher levels (see Table 16.1). Likewise, growers and consultants are not always responsible for the losses experienced in agricultural plantings. The failure may be made at a higher level such as the university, Agricultural Experiment Station, Cooperative Extension, federal and state departments of agriculture, or industry levels. Solutions may be available in the scientific literature but not accessible or understandable to growers. Funding for research to address evolving problems or to maintain the effectiveness of established IPM programs may not be allocated based on their priority in the eyes of the public or grant review panels.

Accumulate Failures

Frequently, failures leading to destruction in military operations are not of an individual nature or single type but have multiple origins (learning, anticipatory, and adapting failures). In addition, they may stem from various levels of command or information flow (high command to field command) (Cohen and Gooch, 1990). Likewise, the same patterns can be seen in the implementation of arthropod control practices where growers' failures to learn, anticipate, and adapt compound initial inadequate responses to pest problems. Cohen and Gooch (1990) apply the terms "aggregate failures" and "catastrophic failures" to the simultaneous occurrence of two and three failure events, respectively. Catastrophic failures usually require outside assistance to overcome the military opponent (or pest problem). In the case of arthropod management, this may mean that assistance either via research, extension, or industry (or a combination) is necessary to overcome the problem.

FAILURE ANALYSIS: A CASE HISTORY

To solve pest management problems it is inadequate to know only that failures can happen. One must identify them, determine the necessary solutions, and modify practices to achieve success. To identify failures in military campaigns, Cohen and Gooch (1990) suggest the use of a "failure matrix" (Table 16.3). First, one needs to identify the failure using a process that Cohen and Gooch call "counterfactual" analysis. To do this, one asks the question, what would have been required to reduce or eliminate the failure? After the failure is identified, one must determine, what critical task went unfulfilled because of the failure? Those failures having the greatest influence on the eventual outcome were referred to as "critical" failures. Next, a "layered" analysis is conducted (Table 16.3). This analysis examines the behaviors at the different organizational levels of and their relative contributions to the overall failure. From this step, one draws an analytical matrix, a simplified chart of failures that presents graphically the key problems leading to failure. From the chart produced, "pathways" to failure that lead to the larger cause of failure in question may be derived. To find the critical tasks that must be accomplished, one determines the ways in which success for specific objectives was almost achieved. Once the nature of the failure has been determined, one should look for the critical tasks that were not completed as well as the critical lapses or mistaken assumptions that led to the failure. Correction of these errors should provide management success, and unlike war, modifications to the system can sometimes be refined for optimum effects. Cohen and Gooch (1990) provide an in-depth explanation of this methodology.

Leafminers on Watermelon in Hawaii

Following the introduction of the agromyzid leafminer *Liriomyza trifolii* Burgess into Hawaii in the mid-1970s, it became a primary pest on watermelon in the Kahuku area on the island of Oahu, Hawaii (Robin, 1983; Johnson, 1987). Growers directed most of their pesticide applications toward this leafminer as well as *Liriomyza sativae* Blanchard, which tended to be easier to control and less common than pesticide resistant *L. trifolii* populations that

Table 16.3. **Failure Matrix Indicating Critical Management Components and Nature of Common Failures in Suppression of Arthropod Pests of Watermelon in Hawaii before Development of an Organized IPM Program.**

	Critical management components				
Level of development / implementation	*Biological control*	*Monitoring of pest populations*	*Proper timing of pesticide applications*	*Application of selective pesticides*	*Alternative control methods for aphid vectors of cucurbit viruses*
Research	Natural enemies established for leafminer biological control	*Critical Failure* [a]: Monitoring techniques not developed for all watermelon pests	*Critical Failure* [a]: Density Treatment Levels (DTLs) not established for all pests	*Critical Failure* [a]: Selective pesticides and application techniques not developed for leafminers, thrips, and aphids	*Critical Failure* [a]: Alternative control methods not developed for aphid vectors of cucurbit viruses
Education / Extension	Growers advised to conserve natural enemies by using monitoring and DTLs				
Grower	*Failure* [b]: Loss of potential effectiveness of natural enemies due to destruction by pesticides	*Failure:* Pesticides applied regardless of need	*Failure:* Overuse of pesticides promotes pesticide resistance in leafminers, and impairment and destruction of natural enemies	*Failure:* Broad spectrum pesticides used for all pests promotes pesticide resistance, elimination of leafminer natural enemies	*Failure:* Pesticides used to maintain aphids at zero densities, but virus spread not necessarily contained

[a] Anticipatory failure; [b] Learning failure.

dominated the heavily sprayed watermelon plantings. Some growers sprayed their watermelon plantings daily, others more frequently, to control the resistant *L. trifolii,* without much success (Robin, 1983). Although some parasitoids (e.g., *Diglyphus begini* [Ashmead], *Chrysocharis oscinidis* [Crawford]) that attacked *Liriomyza* species were present, the Hawaii Department of Agriculture introduced others to increase biological control (Nakao and Funasaki, 1979; Nakao et al., 1981). Of these, *Chrysonotomyia punctriventris* (Crawford) and *Ganaspidium utilis* Beardsley became established. Watermelon growers were advised to reduce their chemical treatments to promote biological control (L. Nakahara, personal communication). However, because growers were reluctant to risk crop losses, demonstration plots were established in the early 1980s to illustrate the effectiveness of the leafminer parasitoids (W. C. Mitchell, unpublished data). Although many growers were convinced of the ability of the parasitoids to control the leafminers, they still had several other major pest species in the cropping system. These included the primary direct pest of the watermelon fruit, the melon fly, *Dacus cucurbitae* Coquillett, and several indirect (foliar) pests, which were: melon aphid, *Aphis gossypii* Glover; melon thrips, *Thrips palmi* Karny; greenhouse whitefly, *Trialeurodes vaporariorum* (Westwood); and the carmine mite, *Tetranychus cinnabarinus* (Boisduval). The melon thrips was newly introduced and most chemicals applied for its control were ineffective (Johnson, 1986). Additionally, the melon aphid not only fed on plants and produced honeydew (resulting in sooty mold accumulation) but also served to transmit the mosaic viruses among plants it fed upon. Failure to control these pests could potentially reduce crop yields substantially, and in the case of the melon aphid, could lead to total crop destruction if mosaic viruses spread unchecked within a watermelon planting (and potentially to neighbor plantings). Attempts to control these pests with chemicals could disrupt the leafminers by impairment or destruction of their effective natural enemies. Thus growers found themselves on the "horns of a dilemma," and failures to effectively manage arthropod pests in watermelon continued into the 1980s.

Critical Management Components

Once effective leafminer parasitoids were established by the 1980s, successful management of the overall watermelon pest complex (leafminers, aphids, thrips, etc.) required specific control tactics (and strategies for their use) for each species that routinely infested the crop. These tactics may be referred to as the "critical management components" (Table 16.3) and in this example included the use of biological control for the leafminers; the need to monitor populations of indirect pests to identify potentially damaging densities; effective and timely applications of pesticides to suppress indirect pests; application of selective pesticides to reduce potential pest resurgences (i.e., melon thrips, spider mites) and secondary pest upsets (i.e., leafminers); and alternatives to unilateral chemical control for management of aphids and the associated mosaic viruses they transmitted. As stated, decisions and actions not only occur at the grower level but also at levels above the grower that are associated with information development and flow (i.e., extension and research components). Several failures occurred after the introduction of natural enemies to control the leafminers. Note that most of the critical failures did not originate at the grower level but had their roots at the research level due to the lack of information on protocols and methods for managing the arthropod pests (i.e., melon thrips, melon aphid, melon fly) that did not have highly effective natural enemies. The critical failures were anticipatory failures on the part of research that prevented extension personnel from having useful information and protocols to extend to the growers, which in turn forced the growers to fail in their attempts to manage the crop pests. For some growers, failure to learn to conserve the leafminer natural enemies also occurred. All these failures contributed to the inability of the growers to successfully manage (failure to adapt) the pesticide resistant leafminers due to the lack of effective compounds and the reduced effectiveness of the leafminer natural enemies, resulting in further crop losses. Given that there were at least three failures in sequence, these can be called catastrophic failures that required the need of researchers and extension personnel to come to the growers' aid.

The Cavalry Arrives

A combination research-extension demonstration program was initiated in 1984 to assist the watermelon growers with their pest problems. The program included the establishment of a weekly-scouting service supported with Smith-Lever Funds. The results of this three-year program were described in detail by Johnson (1986; 1987; 1991), Johnson et al. (1989), and Lynch and Johnson (1987). To reduce the amount of pesticides applied to the watermelon crop for melon fly control, growers were advised to grow corn borders beside the watermelon plantings and routinely treat these with malathion bait sprays that targeted resting females (Nishida and Bess, 1957). Sampling protocols for the watermelon pests were developed (and implemented) as well as density treatment levels (DTLs) (Table 16.4) to time pesticide applications. DTLs were nominal thresholds (Pedigo, 1996) based on scientific information on the pests' biologies and growers' experiences with crop losses induced by the

Table 16.4. **Failure Matrix Following Change in Management Practices for Arthropod Pests on Watermelon in Hawaii.**

	Critical management components				
Level of development / implementation	*Biological control*	*Monitoring of pest populations*	*Proper timing of pesticide applications*	*Application of selective pesticides*	*Alternative control methods for aphid vectors of cucurbit viruses*
Research	Natural enemies established for leafminer biological control	Monitoring techniques developed for all major watermelon pests	Density treatment Levels (DTLS) established	Selective pesticides application techniques developed or identified for aphids and thrips	Alternative control methode developed or identified for aphid vectors of curcurbit viruses
Education / Extension	Growers advised to conserve natural enemies by using monitoring and DTLs	Techniques used in WM IPM Project sponsored by CES[a]	Growers advised of established DTLs	Growers advised to apply malathion for aphid control and selectively place pesticides for thrips and melon fly[b]	Growers advised to use alternative control techniques[b]
Grower	Leafminer natural enemies conserved and control obtained usually without pesticides	Decisions to apply pesticides based on routine survey data and DTLs	Pesticides applied within forty-eight hours of perceived need for application	Increased aphid and thrips control with minimum destruction of leafminer natural enemies	*Failure*[c]: Growers reluctant rogue plants

[a] Cooperative Extension Service.

[b] Malathion bait sprays on corn trap crops for melon fly; plastic mulch and roguing for virus containment.

[c] Learning failure.

pests (Johnson et al., 1989). Growers received weekly information on the pest densities in their watermelon plantings within forty-eight hours of the crop being sampled. Advice on the need to treat plantings based on DTLs was provided along with information on natural enemies found and the pests they would target. To reduce pesticide applications to the entire foliage canopy, growers were instructed to direct chemical applications to the growing vine tips when melon thrips were only a problem on that part of the plant. To limit the spread of mosaic viruses in the crops, growers were instructed to rogue infected plants, however, roguing was not adequately practiced.

Elimination of the failures indicated in table 16.3 as shown in table 16.4 resulted in successful pest control in watermelon plantings (Johnson et al., 1989; Johnson, 1991). However, one failure was identified among the growers, a learning failure related to inadequately roguing mosaic virus–infected plants that provided an in-field source for the mosaic virus.

INSIGHTS INTO MANAGEMENT FAILURES

Application of the concepts proposed by Cohen and Gooch (1990) to pest management has the potential to advance our understanding of arthropod control by identifying weak points within complicated systems. During the last decade, IPM has been scrutinized for not living up to its potential effectiveness and benefits. Thus, new names have been suggested for this management practice (e.g., biologically intensive IPM, Frisbie and Smith, 1991; ecologically based pest management, National Research Council, 1996), specifying greater focus on specific management tactics and strategies. However, I support the view of Kogan (1999) that the original concepts and methodologies of IPM were solid, but over time its implementation has fallen short of the initial ideas. Regardless of the term applied to IPM, analysis of failures would provide valuable learning opportunities and insights into a given management problem, thereby revealing numerous weaknesses in IPM programs that have been overlooked. Growers may more easily comprehend why their control efforts are unsuccessful if they can be shown how their actions, or lack thereof, impact the ultimate level of control and its sustainability. Additionally, military analogies may enhance our abilities to communicate to nonscientists (i.e., growers) the importance of activities like sampling and monitoring as well as the need to only take action when necessary. This is somewhat supported by the frequent use of military metaphors in the pest management literature.

Most students rarely receive instruction on methodologies to critically evaluate failing pest management systems. In fact, failures are rarely discussed in the scientific literature. In contrast, military campaigns are subject to numerous post hoc analyses by professionals and amateurs alike, contributing to advances in the art of war. In pest management, emphasis is placed on studying successful case histories or development of pest management programs. Although ecologists have modeled numerous pest and natural enemy biologies and population dynamics (see Tummala et al., 1976), these models have not typically included the "human decision" elements in pest management systems except for decisions to apply chemical applications.

We may view recurring pesticide resistance in some species (e.g., diamondback moth, Colorado potato beetle, Georghiou and Lagunes-Tejeda, 1991) differently when seen as our failure to learn that simply replacing insecticides as resistance develops is a "doomed" strategy. Along these same lines, the idea of using genetically engineered plants with *Bacillus thuringiensis* endotoxin genes becomes a less innovative concept given the high potential for failure based on past history with conventional insecticides and *Bacillus thuringiensis*–based products (Georghiou and Lagunes-Tejeda, 1991; Tabashnik, 1994). Military strategists have long known (> 2,000 years) that repeated use of the same tactics eventually leads to total ineffectiveness (Sun-tzu, 1993). However, this practice is commonly relied upon by less creative and generally unsuccessful decision-makers (Sun-tzu, 1993; Hart, 1967) who tend not to be risk takers. Also in the arena of learning failures is the need for more education of extension personnel and growers about the effective use of biological control agents (U.S. Congress, 1995). When seen in the context of a failure matrix, this becomes quite obvious. Unfortunately, those individuals who fail to learn history many times repeat it in the field of IPM as well as war.

This chapter has focused on analysis of failure, with little reference to application of military concepts and strategies to pest management. Most biologists are totally unfamiliar with the latter area of thought. Many military concepts may exist that would improve our pest management efforts. However, just as pest management has many strategists, so does military science (Sun-tzu, 1993; Hart, 1967; von Clausewitz, 1976). Some concepts will be applicable and others will not. I believe that it would be worthwhile to examine the potential of these strategies for improving the efficacy of pest management.

ACKNOWLEDGMENTS

I thank Mark Carder, Randy R. Gaugler, Thomas O. Holtzer, Cerruti R. R. Hooks, Jian J. Duan, Lynn M. LeBeck, Ian Macrae, Michael J. Oraze, Jay A. Rosenheim, Nelson R. Powers, Ronald J. Prokopy, T. Eugene Reagan, Bruce E. Tabashnik, Nick C. Toscano, and Starker Wright for suggestions and comments that improved the manuscript. Appreciation is extended to Kevin M. Heinz and the Texas A&M graduate students for inviting me as a contributing speaker to the Texas A&M Entomology Centennial Lecture Series. This is Journal Series No. 4462 of the College of Tropical Agriculture and Human Resources, University of Hawaii at Manoa, Honolulu.

REFERENCES CITED

Akesson, N. B., and W. E. Yates. 1974. The Use of Aircraft in Agriculture. Food and Agriculture Organization Agricultural Development Paper, No. 94.

Anonymous. 1991. Cockroaches: A Multilateral Strategic Defense. University of California, *Berkeley Wellness Letter* 7 (12): 7.

Benz, G. 1987. Environment. In *Epizootiology of Insect Diseases,* ed. J. R. Fuxa and Y. Tanada, 177–214. New York: John Wiley and Sons.

Bueckert, D. 1992. The Battle against Superpests. *Canadian Geographic* 112 (6): 104–109.

Clausewitz, C. V. 1976. *On War,* trans. M. Howard and P. Paret. Princeton, N.J.: Princeton University Press.

Cohen, E. A., and J. Gooch. 1990. *Military Misfortunes: The Anatomy of Failure in War.* New York: Free Press.

Croft, B. A. 1990. Developing a Philosophy and Program of Pesticide Resistance Management. In *Pesticide Resistance in Arthropods,* ed. R. T. Roush and B. E. Tabashnik, 277–96, New York: Chapman and Hall.

Dahlsten, D. L., R. Garcia, and H. Lorraine, eds. 1989. Eradication as a Pest Management Tool: Concepts and Contexts. In *Eradication of Exotic Pests: Analysis with Case Histories,* 3–15. New Haven, Conn.: Yale University Press.

Daniel, P. 1990. A Rogue Bureaucracy: The USDA Fire Ant Campaign of the Late 1950s. *Agricultural History* 64 (2): 99–114.

De Benedictis, J. A., and J. Granett. 1992. Variability of Responses of Grape Phylloxera (Homoptera: Phylloxeridae) to Bioassays That Discriminate between California Biotypes. *Journal of Economic Entomology* 85: 1527–34.

Devereux, T. 1991. *Messenger Gods of Battle: Radio, Radar, Sonar: The Story of Electronics in War.* London: Brassey's.

Diamond, J. 1997. *Guns, Germs, and Steel: The Fates of Human Societies.* New York: W. W. Norton.

Dunnigan, J. F., and A. A. Nofi. 1994. *Dirty Little Secrets of World War II.* New York: Quill / William Morrow.

Frisbie, R. E., and J. W. Smith Jr. 1991. Biologically Intensive Integrated Pest Management: The Future. In Progress and Perspectives for the 21st Century, ed. J. J. Menn and A. L. Steinhauer, 151–64. Entomological Society of America Centennial Symposium. Lanham, Md.: ESA.

Georghiou, G. P., and A. Lagunes-Tejeda. 1991. *The Occurrence of Resistance to Pesticides in Arthropods.* Rome: Food and Agriculture Organization.

Gould, F. 1983. Genetics of Plant-Herbivore Systems: Interactions between Applied and Basic Studies. In *Variable Plants and Herbivores in Natural and Managed Systems,* ed. R. F. Denno and M. S. McClure, 599–653. New York: Academic Press.

Hart, B. H. L. 1967. *Strategy,* 2d rev. ed. New York: Praeger.

Hartman, A. L. 1993. The Role of Military Strategies in Medical Thinking. *Perspectives in Biology and Medicine* 36: 177–83.

Herring, T. A. 1996. The Global Positioning System. *Scientific American* 274 (2): 44–50.

Hoffmann, M. 1992. The Enemy of My Enemy is My Friend. *American Scientist* 80: 535–36.

Johnson, M. W. 1986. Population Trends of a Newly Introduced Species, *Thrips palmi* Karny, in Commercial Watermelon Plantings in Hawaii. *Journal of Economic Entomology* 79: 718–20.

———. 1987. Parasitization of *Liriomyza* spp. (Diptera: Agromyzidae) Infesting Commercial Watermelon Plantings in Hawaii. *Journal of Economic Entomology* 80: 56–61.

———. 1991. Integrated Pest Management: An Integral Component of Sustainable Agriculture in the Pacific Basin. In *Proceedings of the International Conference on Agriculture for the 21st Century,* 56–62. M. Okada International Association, October 12–14, 1990, Maui, Hawaii. New York: MOA Foundation.

Johnson, M. W., R. F. L. Mau, A. P. Martinez, and S. Fukuda. 1989. Foliar Pests of Watermelon in Hawaii. *Tropical Pest Management* 35: 90–96.

Keegan, J. 1993. *A History of Warfare.* New York: Vintage Books.

Kogan, M. 1999. Integrated Pest Management: Constructive Criticism or Revisionism? *Phytoparasitica* 27: 93–96.

Krause, D. G. 1995. *The Art of War for Executives.* New York: Berkeley Publishing Group.

Luecke, R. 1994. *Scuttle Your Ships before Advancing and Other Lessons from History on Leadership and Change for Today's Managers.* New York: Oxford University Press.

Lynch, J. A., and M. W. Johnson. 1987. Stratified Sampling of *Liriomyza* spp. (Diptera: Agromyzidae) and Associated Hymenopterous Parasites on Watermelon. *Journal of Economic Entomology* 80: 1254–61.

Matthews, G. A. 1979. *Pesticide Application Methods.* New York: Longman.

Messenger, P. S. 1970. Bioclimatic Inputs to Biological Control and Pest Management Programs. In *Concepts of Pest Management,* ed. R. L. Rabb and F. E. Guthrie, 84–102. Proceedings, North Carolina State University, Raleigh, N.C., March 25–27, 1970.

Milton, G. 1999. *Nathaniel's Nutmeg; or, The True and Incredible Adventures of the Spice Trader Who Changed the Course of History.* New York: Farrar, Straus and Giroux.

Metcalf, C. L., W. P. Flint, and R. L. Metcalf. 1962. *Destructive and Useful Insects.* New York: McGraw-Hill.

Musashi, M. 1982. *A Book of Five Rings.* Woodstock, N.Y.: Overlook Press.

Nakao, H. K., and G. Y. Funasaki. 1979. Introductions for Biological Control in Hawaii: 1975–1976. *Proceedings of the Hawaiian Entomological Society* 23: 125–28.

Nakao, H. K., G. Y. Funasaki, S. Y. Higa, and P. Y. Lai. 1981. Introductions for Biological Control in Hawaii, 1977–1978. *Proceedings of the Hawaiian Entomological Society* 23: 425–30.

National Research Council. 1996. *Ecologically Based Pest Management: New Solutions for a New Century.* Washington, D.C.: National Academy Press.

Nishida, T., and H. A. Bess. 1957. Studies on the Ecology and Control of the Melon Fly, *Dacus* (Strurneta) *cucurbitae* Coquillett (Diptera: Tephritidae). *Hawaii Agricultural Experiment Station Bulletin,* No. 34.

Peckham, H. H. 1964. *The Colonial Wars, 1689–1762.* Chicago: University of Chicago Press.

Pedigo, L. P. 1996. General Models of Economic Thresholds. In *Economic Thresholds for Integrated Pest Management,* ed. L. G. Higley and L. P. Pedigo, 41–57. Lincoln: University of Nebraska Press.

Pedigo, L. P., and G. D. Buntin, eds. 1994. *Handbook of Sampling Methods for Arthropods in Agriculture.* Boca Raton, Fla.: CRC Press.

Risch, S. J., D. A. Andow, and M. Altieri. 1983. Agroecosystem Diversity and Pest Control: Data, Tentative Conclusions, and New Research Directions. *Environmental Entomology* 12: 625–29.

Robin, M. R. 1983. Sampling the Leafminers, *Liriomyza sativae* Blanchard, and *Liriomyza trifolii* Burgess (Diptera: Agromyzidae), and Their Associated Hymenopterous Parasites in Watermelon. Master's thesis, University of Hawaii at Manoa, Honolulu.

Russell III, E. P. 1999. L. O. Howard Promoted War Metaphors as a Rallying Cry for Economic Entomology. *American Entomologist* 45: 74–78.

Scott, J. G. 1990. Investigating Mechanisms of Insecticide Resistance: Methods, Strategies, and Pitfalls. In *Pesticide Resistance in Arthropods,* ed. R. T. Roush and B. E. Tabashnik, 39–57. New York: Chapman and Hall.

Stevens, J. 1993. Winning the War against Mealybugs. *Technology Review* 96 (8): 17–18.

Strickland, A. H. 1970. Some Economic Principles of Pest Management. In *Concepts of Pest Management,* ed. R. L. Rabb and F. E. Guthrie, 30–44. Proceedings, North Carolina State University, Raleigh, N.C., March 25–27, 1970.

Sun-tzu [Sunzi]. 1993. *Sun-tzu: The Art of Warfare.* The first English translation incorporating the recently discovered Yin-Ch'üeh-Shan Texts, trans. R. T. Ames. New York: Ballantine.

Tabashnik, B. E. 1994. Evolution of Resistance to *Bacillus thuringiensis. Annual Review of Entomology* 39: 47–79.

Tummala, R. L., D. L. Haynes, and B. A. Croft, eds. 1976. *Modeling for Pest Management: Concepts, Techniques, and Applications, U.S.A./USSR.* East Lansing: Michigan State University.

U.S. Congress. 1995. *Biologically Based Technologies for Pest Control.* Office of Technology Assessment, U.S. Congress, OTA-ENV-636, Washington, D.C.: U.S. Government Printing Office.

Varley, G. C., G. R. Gradwell, and M. P. Hassell. 1974. *Insect Population Ecology.* Berkeley: University of California Press.

Wald, A. 1943. *Sequential Analysis of Statistical Data: Theory.* Columbia University Statistical Research Group, Report 75, and Office of Scientific Research and Development, Report 1998.

———. 1945. Sequential Tests of Statistical Hypotheses. *Annals of Mathematical Statistics* 16: 117–86.

Ware, G. W. 1994. *The Pesticide Book,* 4th ed. Fresno, Calif.: Thomson Publications.

Waters, W. E. 1955. Sequential Sampling in Forest Insect Surveys. *Forest Science* 1: 68–79.

Winston, M. L. 1997. *Nature Wars: People vs. Pests.* Cambridge, Mass.: Harvard University Press.

Wolf, W. W. 1979. Entomological Radar Studies in the United States. In *Movement of Highly Mobile Insects: Concepts and Methodology in Research,* ed. R. L. Rabb and G. G. Kennedy, 263–66. Raleigh: University Graphics, North Carolina State University.

Chapter 17

Entomology Courses

Conduits for Educating the Public about Science

DONALD W. HALL
Department of Entomology
University of Florida–Gainesville

Entomologists are uniquely positioned to teach the public about science for a number of obvious reasons. In terms of numbers of species, insects are the largest group of animals by far. They also are among the most abundant in terms of sheer numbers. Because of the great numbers and often-spectacular appearance of insects, everyone has some familiarity with them. Most people have some opinion, either positive or negative, about insects.

Insects have permeated every aspect of human culture from ancient history to the present (Hogue, 1987). They are found abundantly in our literature (both religious and secular), art, music, movies, and television commercials. Insects are particularly "hot" now. It is difficult to go into a shopping mall without seeing insect jewelry and fabrics with insect prints. Currently there are a number of television commercials that utilize insects. Two of the most popular movies of late 1998 and early 1999 are PDI/Dreamworks' *ANTZ* and Pixar/Disney's *A Bug's Life*—a masterpiece of computer animation. Even the U.S. Postal Service is getting in on the act. In October of 1999, a sheet of thirty-three-cent postage stamps will be issued that features sixteen species of insects and four species of spiders.

Insects are of great importance economically both as beneficial and destructive organisms, and as six-legged "guinea pigs" for scientific research. Because of their great diversity, insects can be used to teach a variety of scientific concepts. Entomologists, particularly those of us in academia, are uniquely positioned to take advantage of this great opportunity provided by insects. Two of the missions (teaching and extension) of land grant universities are directed specifically at education. We merely need to find the optimal approaches to take advantage of this great opportunity to educate the public.

EDUCATING THE PUBLIC

It is critical that the general public and especially those who hold public office have an appreciation and understanding of science. Legislators and other politicians influence science policy through funding decisions. It is essential that they have an appreciation for science and a basic understanding of the nature of the scientific endeavor. Most legislators were not science majors in college. Therefore, it is important for us as scientists to spend the same careful attention to quality education for nonscience majors as we do for majors. In fact, most nonscience majors have a fear of science and negative attitudes toward science courses. Considerably more effort may be required to first capture their interest in order to optimize their learning. Perhaps the most valuable thing we can do is to work to change the public's attitude toward science.

There is a considerable body of evidence primarily from the environmental education literature (Iozzi, 1989) that attitudes are formed mostly during elementary school years. Attitudes toward science in general likely follow the same pattern. Environmental attitudes of eighth graders were not significantly different from those of adults (Miller, 1975). The message is that if we want to impact attitudes toward science, the optimal time for doing that is while children are in elementary school. It becomes progressively harder to do as they get older (Simpson and Oliver, 1985; Talton and Simpson, 1985). There is also evidence that the childhood years are the critical years for development of interest in insects even by most adult amateur entomologists (Matthews, 1992). This also is likely true for professional entomologists. Alaimo and Doran (1980) found that children obtain most of their information on science from the media and from school.

Because educating the public is such an enormous task, scientists must leverage their efforts by first educating others who in turn can educate the masses. Because elementary school children are most receptive to changes in attitude, education of pre-service and in-service teachers provides an efficient approach to reaching the children. Some entomology departments, including Texas A&M's, are now offering formal courses for teachers.

Another mechanism for leveraging our efforts is to offer large enrollment introductory science courses with insects as the examples to nonscience majors. Many entomology departments are now teaching this type of course. At some universities, it is possible for students to meet part of the general education science requirement with such courses. It is generally more difficult to change college students' attitudes toward science, but by making the courses sufficiently interesting, it is possible. There are two groups of students who should be specially targeted for recruitment for these courses. First, journalism and communications students should be recruited because they have the future potential to educate millions of people via print and electronic media. Second, business students and students in majors that commonly produce law students should be recruited. Some of these students ultimately will become politicians—including legislators. The leading prior occupations of members of the 105th Congress (1997) were law (172 representatives; 53 senators) and business or banking (181 representatives; 33 senators) (Ornstein et al., 199798). There are special pedagogical issues and approaches to consider when teaching lower division nonscience majors who may already have negative attitudes toward science. I will first give a brief discussion of some of these issues and how I have attempted to address them in a large enrollment course for nonmajors and in a summer field course for pre-service teachers.

COGNITIVE BASIS FOR TEACHING AND LEARNING

Before beginning a discussion of pedagogy, it may be helpful to look at current ideas about how we learn. Learning is believed to take place in two memory compartments—working memory and long-term memory. For learning to occur, new information must be taken into working memory, integrated with previously learned information retrieved from long-term memory and then itself be encoded into long-term memory—a process known as cognitive transformation (Farquhar and Surrey, 1995). Working memory or short-term memory has a very limited and short-term capacity. Long-term memory has almost unlimited capacity and memory items may be stored for long periods of time (probably for life) although we may have difficulty with retrieval of some of it. Information entering working memory must be reinforced continuously or encoded in long-term memory or it is rapidly lost. It is believed that a person's relative intelligence may depend on their capacity for manipulation of information in working memory (Wickelgren, 1997).

Chunking

George Miller (1956) published a seminal paper on the number of "chunks" of information that could be held in working memory for immediate recall and referred to this number as the "magical number seven, plus or minus two." Simon (1974) presented evidence that the chunk capacity of working memory is in the range of five to seven (probably closer to five). He gave the following working definition of a chunk: "a chunk of any kind of stimulus material is the quantity that short term memory will hold five of." For verbal information, Simon presented evidence that words are likely chunks (as opposed to syllables or phrases). It is interesting that we also can hold more digits of numerical information in working memory by "chunking" them. For example, our telephone numbers (e.g., 352–392–1901) and social security numbers are "chunked" into chunks of two to four digits, making them much easier to remember.

The principles of chunking theory have been demonstrated in different types of learning ranging from high level cognitive activities like chess (Chase and Simon, 1973) to athletic skills (Allard and Burnett, 1985). Chunking theory appears to have broad applicability, and our task is to find out how to exploit it in our entomological educational activities. The important thing to keep in mind is the very limited capacity of working memory. This can be readily demonstrated if one attempts a simple mental manipulative task like attempting to name a list of well-known items in alphabetical order (e.g., the days of the week). The days of the week are seven items that we all have ingrained in our long-term memory. However, we have great difficulty in bringing all seven names into working memory and manipulating them simultaneously. How much more difficult must it be for nonscience majors to process information made up of unfamiliar terms?

Working Memory

Studies suggest that most of the functions of working memory reside in the prefrontal cortex of the brain (Baddeley and Sala, 1996; Beardsley, 1997; D'Esposito and Grossman, 1996; Goldman-Rakic, 1996). Information to be encoded from working memory to long-term memory is dependent on intermediary processing in the hippocampus. Long-term memory is spread throughout various areas of the cerebral cortex depending on the nature of the information.

Working memory simultaneously stores and processes information. It is believed to be composed of three compartments: the Central Executive and two slave systems—the Phonological (articulatory) Loop and the Visuo-Spatial Sketch (Scratch) Pad (Baddeley, 1992). The Central Executive is an attention controller and task manager. The Phonological Loop is responsible for holding speech-based material (including digits) while the Visuo-Spatial Sketch Pad forms and manipulates images.

Baddeley (1995) suggested that one can get an idea of the function of the three components of working memory by counting the number of windows in her/his house from memory. Most people do this by forming images of the windows by mentally going from room to room or walking around the outside of the house. The Visuo-Spatial Sketch Pad is responsible for forming the images. The Phonological Loop does the subvocal counting, while the Central Executive maintains attention and organizes the overall process.

Cognitive Load

Because working memory has a very limited capacity, each additional cognitive process performed places additional demands on the system (D'Esposito and Grossman, 1996). The total load on working memory including learning and nonlearning events is known as "cognitive load." Farquhar and Surry (1995) have classified cognitive load into three categories: learning events, message comprehension events, and nonlearning events. They define a learning event as "an event in which the learner is actively involved and purposefully interacting with the material to be learned." Learning events are activities that assist in efficient encoding of material into long-term memory. Learning events listed by Farquhar and Surry (1995) include arranging a mental organization of information, developing mental hierarchies or categories, producing mental images, and performing mental repetition. High cognitive engagement results when cognitive load is composed primarily of learning events.

Message comprehension events comprise the part of the cognitive load expended to understand the message. Complex material, unfamiliar terminology, and too many simultaneous sensory stimuli impose on working memory and slow down encoding to long-term memory. Therefore, material to be learned should be presented in the simplest manner appropriate and possible. Some practical examples of message comprehension events and instructional formats for overcoming them are discussed by Chandler and Sweller (1991).

Nonlearning events include anything that distracts attention from the instructional presentation. Examples could include classmates talking, doors banging, daydreaming, personal problems, or a boring lecture. One of my colleagues, who requires the students in her large lecture class to write a paragraph card during each lecture, told of a student who complained on her card that she was distracted by the couple sitting next to her who "made out" whenever the lights were dimmed for the PowerPoint presentations. When message comprehension events and nonlearning events place significant load on working memory there is low cognitive engagement, and little encoding of material into long-term memory takes place.

COURSES FOR NONSCIENCE MAJORS

Large Enrollment Classes for Nonscience Majors

General Considerations

Teaching takes place in many formats, and teaching approaches should be appropriate to the type of class. Even for a particular classroom situation there are a variety of teaching approaches that can be successful. General information on effective teaching techniques may be found in a number of books including those by McKeachie (1994) and Newcomb et al. (1993). Large lecture classes present special problems—particularly for nonmajors. Carbone (1998) offers useful advice specifically for teaching large lecture classes.

It is particularly important for students in large classes to have a good idea of the instructor's objectives (what the successful student is expected to know or be able to do as a result of the course). Most of us have a tendency to give the students more information than they can possibly learn. This frustrates them, and some react by giving up. By giving students specific learning objectives, we encourage them to concentrate on the material we believe is most important. Also, the act of writing the learning objectives forces us as instructors to give serious thought to what we consider important in a course. Mager (1984) provides instructions for writing effective learning objectives. All or a significant percentage of questions on examinations should then be directed at students' achievement of the objectives.

Live Insects

The greatest resources we as entomologists have are real insects (preferably alive). Whenever possible, I use live insects even in large classes. When my class enrollment was fewer than 100 students, I attempted to have enough specimens so that each group of two or three students could have a specimen to handle (usually in a container). When the class size increased to over 300 students, this was no longer possible. I now bring live insects for projection under the document camera. For some reason, it is easier to keep students' attention with a live insect than with videotape even if the live insect is under a video camera. There is something about knowing that there is a live insect under the camera that holds their interest. By having live insects, there is the added advantage that students may see (and sometimes handle) the insects after class.

Multimedia

When it is impossible to use live insects, I have attempted to address the cognitive issues discussed earlier through the use of multimedia. In its simplest form, multimedia has come to mean computer presentation using Microsoft PowerPoint or similar presentation software. With today's high-powered LCD projectors, the computer (or document camera) images can be projected full screen in a fully lit room. Multimedia, by strict definition, is the use of three or more kinds of media. I have published brief discussions of my use of these multimedia techniques (Hall, 1996a, b). Initially, I used a multimedia authoring software program (Multimedia) that has its own high-level programming language. As standard presentation software programs (e.g., Microsoft PowerPoint) have become more powerful, I have converted most of my lectures to that format. Used inappropriately, computer presentation has the potential to be significantly worse than other presentation modes. However, properly used, it can be a powerful tool for effectively overcoming cognitive load and enhancing the learning process.

Possibly the most powerful effect of multimedia is that it forces the instructor to be better prepared and to organize her or his material more carefully. Think back to the working memory overload example of attempting to name the days of the week in alphabetical order. One can easily name the days of the week in chronological order. That is because the days were originally encoded into long-term memory in chronological order. This example suggests that we should attempt to present material to students in the structure or format in which they will most likely need to retrieve it from long-term memory.

Progressive Disclosure

One of the strengths of presentation software is the capability for progressive disclosure. Progressive disclosure is the equivalent of dragging a piece of paper down an overhead transparency as successive points are covered. However, with computer progressive disclosure, one has the capability to hide or dim previously covered points. The importance of hiding or dimming is evident when the importance of focusing attention and limiting the amount of information available to the student is considered in the context of cognitive load and the limits of working memory. Also, with most presentation software, it is possible to animate text by flying it in from different directions and adding various sound effects and such. This may have some benefit when used judiciously. Motion is known to be a powerful attention-getting mechanism. However, it also has significant potential for distraction. At best, it loses its effect after being used a few times, and students may concentrate on the motion and become distracted from the material to be learned. For that reason I prefer to have bulleted text appear without animation.

When preparing presentation slides it is again helpful to consider chunking and the limited capacity for working memory. As a general rule, the number of bulleted points per slide should be five or fewer, and each point should be composed of no more than four or five words. Simple graphics may be helpful when appropriate. It also is possible to use progressive disclosure to label diagrams (e.g., structures of insects) or to show sequential steps in an experimental procedure. Names of structures may be shown in sequence and hidden or dimmed after they are discussed.

Video, Animation, Slides, and Computer Graphics

The next best thing to a live insect is high quality video. Again, the motion is an effective way to gain and hold the students' attention. There are many videos available that contain insect material. Also, video is a great way to illustrate to students the prevalent place of insects and other arthropods in our culture. I show videotapes of television commercials and segments from a variety of movies to illustrate various aspects of insect biology. A particularly powerful format for video is laserdisc. With laserdiscs, one can generate bar-coded clips with special computer software. I tape these bar codes onto printouts of my PowerPoint presentations. When I am ready for a particular video sequence, I scan and play the bar code that cues to a specific frame on the laserdisc and plays the sequence of video frames specified by the bar code. The advantage of laserdiscs over videotape is random access to the desired frames without having to run forward or backward to search for them. The disadvantage is the limited amount of video that is available on laserdisc. Because of large file sizes, I don't use much digitized video. However, digital videodiscs will likely prove the method of choice for insect video in the near future.

If video is not available, animations may be used when motion is required or helpful. Very few insect animations are available commercially. Insect graphics for animations may be obtained by editing and using commercially available computer clip art or by drawing original insect art, scanning, and editing it. Simple animations can be created by using the animation functions of the presentation software, or more sophisticated programming-based animations can be created with multimedia authoring software or specialized animation software. The reader is referred to Hall (1996a) for a discussion of basic computer animation techniques. Slides (two-by-two or digitized) or computer graphics also can be effective components of multimedia-based lectures.

Multimedia techniques, properly used and accompanied by a liberal dose of instructor enthusiasm, are valuable methods to maintain and focus student attention, thereby reducing the "nonlearning events" compartment of cognitive load. Also, because of the visual nature of multimedia, it should assist students with formation of visual images in the Visuo-Spatial Sketch Pad of working memory.

Courses for Education Majors and In-Service Teachers

Matthews et al. (1997) have reviewed the use of insects as teaching tools in primary and secondary education. Their review also cites many of the most valuable resources for specific entomology-related activities and materials. Also, the National Science Teachers Association publishes the age-graded science education journals *Science and Children* and *The Science Teacher* that include teaching activities with insects. My students and I maintain a list of insect-related educational articles from these and other journals. Many of the published activities integrate art, music, and language arts with insect material.

The National Research Council (1996) has included inquiry as an important component of its National Science Education Standards. Unfortunately, most elementary school teachers and many teachers in the higher grades have inadequate training in science to feel comfortable even to teach science content let alone scientific inquiry. Therefore, it is important for scientists to assist with this part of teacher training. There are many valid contexts for providing this training. A couple of excellent sources for outdoor inquiry activities with insects and other organisms are publications funded by the Nongame Wildlife Program of the Florida Game and Freshwater

Fish Commission (Cronin-Jones, 1992; Feinsinger and Minno, n.d.). Because most children are now growing up spending most of their time watching television or on the computer, my preference has been to attempt to help teachers learn to take their students outdoors on the school grounds for hands-on activities with insects.

I teach a summer field course titled "Insect Field Biology" for science education majors and in-service teachers. The course is designed to provide the students with information and exercises that will be useful to them in their teaching. I have published a summary of the course as it was a few years ago (Hall, 1995), but the course is continuously evolving. There is considerable entomological content in the course. Students are required to learn 100 common Florida insects to common name and order. The insects were carefully selected to be representative of insects that the students and their future students are likely to encounter. Also, a major consideration was to select insects that are useful for teaching various biological concepts. Students also are required to learn twenty common plants that have interesting insect associations. A small but representative insect collection is required. The collection requirements are geared toward making the collection a "teaching" collection with emphasis on species that have various morphological adaptations or demonstrate important biological concepts.

The course meets for two whole mornings a week for six weeks. Emphasis is on insect communities. We begin with decomposer communities—leaf litter, dung, carrion, and logs. We look at insects that live on foliage and then at the insect faunas of various types of aquatic communities. There also are activities with ants, ant-extrafloral nectary associations, antlions and tiger beetle larvae, and insect pollination biology. The emphasis throughout these activities is on food webs and specific adaptations of insects to their environment and way of life.

While we are in the field, we are continually observing the structure and behaviors of insects. We then use the focus question, "how is a given structure or behavior adaptive?" This focus question provides a starting point for inquiry—generation of testable hypotheses, planning of experiments, predicted outcomes, etc.—the traditional "scientific method" (Hall, 1997). By going through this process repeatedly with hands-on activities with familiar insects, the students begin to develop a mind-set for inquiry activities that they will be able to use with their own students.

It is critical that courses for teachers be primarily hands-on activities. In fact, this is the most effective approach to teaching for any course when possible. This approach addresses most of the pedagogical issues discussed earlier. When the hands-on activities are outdoors with living insects, there is great potential for permanently changing students' attitudes toward insects, the environment, and science.

REFERENCES CITED

Alaimo, S., and R. Doran. 1980. Students' Perceptions of Environmental Problems and Sources of Environmental Information. *Journal of Environmental Education* 12: 17–21.

Allard, F., and N. Burnett. 1985. Skill in Sport. *Canadian Journal of Psychology* 39: 294–312.

Baddeley, A. 1992. Working Memory. *Science* 255: 559.

———. 1995. Memory. In *Cognitive Psychology*, ed. C. C. French and A. M. Colman. New York: Longman.

Baddeley, A., and S. D. Sala. 1996. Working Memory and Executive Control. *Philosophical Transactions of the Royal Society of London, Series B* 351: 1397–1404.

Beardsley, T. 1997. The Machinery of Thought. *Scientific American* 277: 78–83.

Carbone, E. 1998. *Teaching Large Classes: Tools and Strategies.* Thousand Oaks, Calif.: Sage.

Chandler, P., and J. Sweller. 1991. Cognitive Load Theory and the Format of Instruction. *Cognition and Instruction* 8: 293–332.

Chase, W. G., and H. A. Simon. 1973. Perception in Chess. *Cognitive Psychology* 4: 55–81.

Cronin-Jones, L. 1992. *The Schoolyard Wildlife Activity Guide.* Tallahassee: Florida Game and Freshwater Fish Commission.

D'Esposito, M., and M. Grossman. 1996. The Physiological Basis of Executive Function and Working Memory. *The Neuroscientist* 2: 345–52.

Farquhar, J. D., and D. W. Surrey. 1995. Reducing Impositions on Working Memory through Instructional Strategies. *Performance and Instruction* 34: 4–7.

Feinsinger, P., and M. Minno. n.d. *Handbook to Schoolyard Plants and Animals of North Central Florida.* Tallahassee: Florida Game and Freshwater Fish Commission.

Goldman-Rakic, P. S. 1996. Regional and Cellular Fractionation of Working Memory. *Proceedings of the National Academy of Sciences USA* 93: 13473–80.

Hall, D. W. 1995. Bringing Hands-On Experience to Teaching Insect Field Biology. *Journal of College Science Teaching* 24: 195–200.

———. 1996a. Creating Computer-Based Instructional Animations. *NACTA Journal* 40: 8–11.

———. 1996b. Multimedia in the Entomology Classroom. *American Entomologist* 42: 92–98.

———. 1997. Inquiry Learning: A Paradigm from an Entomology Course. *Journal of Natural Resources and Life Sciences Education* 26: 180–82.

Hogue, C. 1987. Cultural Entomology. *Annual Review of Entomology* 32: 181–99.

Iozzi, L. A. 1989. What Research Says to the Educator. Part One: Environmental Education and the Affective Domain. *Journal of Environmental Education* 20: 3–9.

Mager, R. F. 1984. *Preparing Instructional Objectives,* rev. 2d ed. Belmont, Calif.: Lake Publishing.

Matthews, J. R. 1992. Adult Amateur Experiences in Entomology: Breaking the Stereotypes. In *Insect Potpourri: Adventures in Entomology,* ed. J. Adams, 321–28. Gainesville, Fla.: Sandhill Crane Press.

Matthews, R. W., L. R. Flage, and J. R. Matthews. 1997. Insects as Teaching Tools in Primary and Secondary Education. *Annual Review of Entomology* 42: 269–89.

McKeachie, W. J. 1994. *Teaching Tips,* 9th ed. Lexington, Mass.: D. C. Heath.

Miller, G. A. 1956. The Magical Number Seven, Plus or Minus Two: Some Limits on Our Capacity for Processing Information. *Psychological Review* 63: 81–97.

Miller, J. 1975. The Development of Pre-Adult Attitudes toward Environmental Conservation. *School Science and Mathematics* 75: 729–37.
National Research Council. 1996. *National Science Education Standards.* Washington, D.C.: National Academy Press.
Newcomb, L. H., J. D. McCracken, and J. R. Warmbrod. 1993. *Methods of Teaching Agriculture,* 2d ed. Danville, Ill.: Interstate Publishers.
Ornstein, N. J., T. E. Mann, and M. J. Malbin. 1997–98. *Vital Statistics on Congress 1997–1998.* Washington, D.C.: Congressional Quarterly.
Simon, H. A. 1974. How Big Is a Chunk? *Science* 183: 482–88.
Simpson, R. D., and J. S. Oliver. 1985. Attitudes toward Science and Achievement Motivation Profiles of Male and Female Science Students in Grades Six through Ten. *Science Education* 69: 511–26.
Talton, E. L., and R. D. Simpson. 1985. Relationships between Peer and Individual Attitudes toward Science among Adolescent Students. *Science Education* 69: 19–24.
Wickelgren, I. 1997. Getting a Grasp on Working Memory. *Science* 275: 1580–82.

Part VI

Global Issues in Entomology: Human Health, Biodiversity, and Sustainability

Chapter 18

Viruses, Vectors, and Virus Vectors

BARRY BEATY
Arthropod-Borne and Infectious Diseases Laboratory
Department of Microbiology
Colorado State University–Fort Collins

Arthropod-borne diseases, such as malaria, filariasis, leishmaniasis, dengue, and yellow fever, are major causes of morbidity and mortality in humans, and new arthropod-borne diseases continue to emerge (Gubler, 1998; Gratz, 1999). The morbidity and mortality associated with these diseases can be staggering; each year, billions of humans are at risk and hundreds of millions of infections and millions of deaths occur each year (World Health Organization, 1996). Many of these arthropod-borne diseases remain intractable to control and are resurgent in many areas of the world. The reasons for this are multifactorial and include: development of pesticide resistance in vectors, drug resistance in parasites (e.g., chloroquine resistance in malaria), lack of effective vaccines for diseases such as malaria and dengue, decline of vector control programs, increased urbanization in the tropics, and reduction in numbers of trained personnel to control vector-borne diseases. Thus the situation is bleak and prospects for control or eradication of vector-borne diseases using conventional techniques are not encouraging.

Novel strategies for control of vector populations and vector-borne diseases may be derived from new information concerning the molecular biology of the vectors and of vector-pathogen interactions. Despite their central role in disease transmission, relatively little is known of the molecular biology of vectors. In contrast, studies of the molecular biology of *Drosophila melanogaster* L. have been virtually revolutionized by the use of the "P" element transposon to mutagenize and to express genes of interest in vivo (Beaty and Carlson, 1997; Carlson et al., 1995). Recent successes in transformation of *Aedes aegypti* L. mosquitoes are landmark contributions to the field of vector molecular biology (Jasinskiene, et al., 1998; Coates et al., 1999). However, transformation remains a lengthy and laborious practice for characterization of genes of interest in vectors. Clearly new tools for characterizing vector genes that are being identified and will be identified from the emerging field of vector genomics are desperately needed. In addition, new tools or approaches are necessary to identify genes that condition important vector phenotypes. Viruses may also offer potential in this regard. Viruses can be used to identify new genes of interest (GOIs), to characterize new genes, and potentially to deliver genes in novel and efficacious biocontrol strategies. Selected examples of the use of viruses as tools in vector biology studies at the Arthropod-Borne and Infectious Diseases Laboratory (AIDL) follow.

VIRUSES AS TOOLS IN VECTOR BIOLOGY

Cap Scavenging of Bunyaviruses as a Tool to Identify GOIs in Mosquitoes

Transovarial transmission (TOT) of LaCrosse virus (California serogroup, genus *Bunyavirus,* family Bunyaviridae) by its mosquito vector *Aedes triseriatus* (Say) is undoubtedly one of the most successful of host parasite relationships (Beaty and Bishop, 1988). TOT and filial infection rates of > 90 percent occur in *A. triseriatus* from endemic areas in the upper Midwest. This occurs despite the fact that viruses in the family Bunyaviridae are major causes of abortion and teratology when vertically transmitted in vertebrate hosts. We have been investigating the hypothesis that coregulation of arthropod host and virus transcription condition efficient TOT of LAC in *A. triseriatus.*

The specific mechanism that we have targeted in these studies is virus transcription. Viruses in the family Bunyaviridae are cap scavengers; for example, LAC virus contains a primer-stimulated RNA polymerase and a methylated cap-dependent endonuclease (Patterson et al., 1984). LAC primes transcription of its mRNA by using host cell mRNA 5′ caps plus adjacent oligonucleotide sequences as primers. These nonviral-coded 5′ terminal extensions are heterogeneous in infected mammalian cells. This mechanism could coregulate host and virus transcription and reduce viral virulence in vectors, thereby minimizing detrimental effects to the host during critical life stages, and for enhanced overwintering efficiency of viruses. Viral transcription could be reduced or regulated during diapause or follicular quiescence by reduction in or absence of synthesis of host mRNA. This would result in a reduced amount of capped mRNA available to provide primers for transcription of viral mRNA. We are also exploiting the unique replication strategy of viruses in the family Bunyaviridae to reveal host mRNA synthesis in cells.

We previously demonstrated that LAC virus and *A. triseriatus* transcription are coregulated in mosquito ovaries (Chandler et al., 1996). After ingestion of a blood meal and virus infection, replicative forms of the LAC virus were detected in the gonadotropically active ovaries. If the female did not ingest another blood meal, ovarian metabolic activity was greatly reduced and replicative forms of LAC disappeared. However, if the female ingested another blood meal that stimulated ovarian activity, replicative forms of LAC appeared in the ovaries. This suggested that the potential deleterious effect of virus replication was modulated in the inactive ovary. In addition, field studies to determine the effect of LAC virus infection on the success of overwintering of *A. triseriatus* revealed that virus infection did adversely affect overwintering, but the detrimental effects only occurred after the embryos had emerged from diapause and were once again metabolically active (McGaw et al., 1998).

To determine if cap-scavenging could function to promote survival, nucleotide sequences were determined for 5′ termini of LAC S segment mRNAs from persistently infected *A. triseriatus* mosquito embryos (Dobie et al., 1997). Amazingly, during *A. triseriatus* embryogenesis, one 5′ terminal sequence, 5′ CCACTCGCCACT, comprised 81 percent of the scavenged host oligonucleotides. As eggs emerged from diapause, the LAC mRNA 5′ termini were more variable; 33 percent had this sequence and the rest were heterogeneous. In post-diapausing eggs, 100 percent of viral mRNAs had the sequence. This message cap predominated in long-term, persistent infections. Molecular analyses suggest that at least one host message with this 5′ terminal sequence is very similar to an apoptosis regulating gene of *Drosophila* (unpublished data). Cap scavenging from this host mRNA may somehow perturb apoptosis regulation, thereby attenuating virus virulence in embryos.

A different host mRNA preferentially donated caps (5′ AGGAAAGATGGT) to prime virus transcription during deep diapause. Interestingly, this cap sequence was also detected late in embryogenesis, when the embryo would be fully formed, dormant, and awaiting hatching stimuli. After *A. triseriatus* embryos had been in diapause for several months, the predominant scavenged sequence was 5′ AGGAAAGATGGT, and sequence 1 became less prevalent. The protein encoded by this gene remains to be determined, but could be a dormancy or diapause associated protein. Thus, the virus can be used to reveal host transcriptional activity during critical life stages and to identify new GOIs.

VIRUS TRANSDUCTION SYSTEMS IN MOSQUITOES

Viruses have been exploited by molecular biologists to genetically manipulate cells and organisms and to reveal fundamental molecular biological processes. Transducing virus vectors are powerful tools for introduction of genetic constructs into cells. Arboviruses, such as Sindbis (SIN) virus, have proven to be particularly useful for achieving efficient transient expression in vertebrate cells and stable expression in invertebrate cells (Olson et al., 1999). Mosquito parvoviruses (e.g., AeDNV) are environmentally stable and provide the unique ability to transduce GOIs into vectors. Our group has pioneered the use of SIN and AeDNV vectors to transduce and to characterize GOIs in vector cells and vectors. Selected studies follow.

SIN Virus Expression Systems

Full-length infectious cDNA clones of viruses within the family Togaviridae, including SIN, Semliki Forest, and Venezuelan equine encephalitis viruses, have been described. This technology has allowed genetic manipulation of RNA genomes and discovery of viral determinants of host range, virulence, replication, assembly and packaging, and such. Infectious clone technology also has led to the development of new expression systems for use in both vertebrate and invertebrate organisms. At AIDL, we have utilized SIN virus-based gene expression systems that are derived from cDNA clones of the virus (Olson et al., 1999). These virus-transducing systems transiently express heterologous RNAs and proteins in mosquito cell culture and in mosquito larvae and adults. The SIN-based transducing system most frequently used at AIDL is termed the double subgenomic SIN (dsSIN) expression system. The most utilized dsSIN system has been pTE3′2J. This dsSIN plasmid consists of an SP6 bacteriophage promoter followed by the 5′ NCR of SIN virus, the nonstructural genes *nsP1, nsP2, nsP3,* and *nsP4,* the capsid gene, and the envelope genes (*PE2* and *E1*). A second internal initiation site has been placed downstream of the *E1* gene. A polylinker sequence is inserted downstream of the duplicated promoter to facilitate the introduction of heterologous genes and is followed by the 3′ NCR of SIN virus. Full-length RNA is transcribed in vitro from the SP6 promoter and the RNA is transfected into the appropriate cells. Within the transfected cell three dsSIN RNA species occur, the genomic RNA, the first subgenomic mRNA that is translated to form the virus structural proteins and a second subgenomic mRNA that can be translated to form the heterologous protein. Virions containing the recombinant RNA genome are generated that can be used to infect cells or mosquitoes.

We also utilize a second system, the SIN replicon, to express genes, especially those that are larger in size. This construct transcribes a self-replicating RNA, but the virus structural genes are replaced with the GOI. The RNA can be can be packaged when co-transfected with RNA encoding the structural genes. Secondary cells can be infected but the infection is self-limited.

SIN Expression of Genes of Interest in Mosquitoes and Mosquito Cells

We have exploited SIN expression systems for expression and characterization of numerous GOIs and sequences in cells. For example, we have expressed: CAT (Olson et al., 1994; Kamrud et al., 1995, 1997); scorpion toxin for insects (Higgs et al., 1995); green fluorescent protein (GFP) (Higgs et al., 1996); luciferase (Johnson et al., 1999); and numerous viral and viral antisense sequences (Powers et al., 1994, 1996; Olson et al., 1996; Gaines et al., 1996; Higgs et al., 1998; Kamrud et al., 1998). In general, these studies revealed that dsSIN transducing viruses were very important tools for expressing heterologous genes in mosquito cells and in adult mosquitoes. Once inoculated, the SIN expression viruses infected most arthropod tissues and expressed the genes or antisense sequences robustly and for a very long period of time (Rayms-Keller et al., 1995). Several examples follow.

SIN Expression of Heterologous Proteins in Mosquito Saliva

The recognition that arthropod saliva can promote infection of vertebrate hosts by vector-borne pathogens was a landmark discovery (Titus and Ribeiro, 1988). Techniques to characterize the antihemostasis and immunomodulatory components of vector saliva that determine transmission of pathogens are critical. We exploited dsSIN and SIN replicon systems to express GOIs in saliva.

The dsSIN virus, TE/3′2J, and the SIN replicon, rep5/26S, were engineered to express chloramphenicol acetyltransferase (CAT) protein in salivary glands and saliva of transduced female *Culex pipiens pipiens* L. (Kamrud, et al., 1997). Indirect immunoflourescence analysis revealed that both the TE/3′2J/CAT and rep5/CAT/26S viruses infected and expressed CAT in the salivary glands. Importantly, saliva collected from mosquitoes transduced with TE/3′2J/CAT contained a 25kDA protein, presumably the CAT protein. CAT activity assays revealed that saliva collected from mosquitoes transduced with either TE/3′2J/CAT or rep5/CAT/26S contained approximately 3.0×10^6 trimers of CAT. Thus, the heterologous gene was abundantly expressed in saliva.

We subsequently decided to exploit SIN replicon system to express genes from LAC in mosquito saliva (Kamrud et al., 1998). The intent was to determine if virus proteins could be expressed in biologically and immunologically relevant fashion. The long-term goal would be to see if vectors could deliver sufficient viral protein to induce a host immune response. SIN replicon constructs expressing G1 and G2 were engineered and demonstrated to express the two LAC virus envelope proteins in a biologically relevant fashion. In addition, expression of the two genes from different constructs resulted in appropriate trafficking in cells and co-localization in the Golgi. A truncated form of G1 was secreted from mosquito cells, which would be necessary for it to be expressed in saliva. Studies are in progress to determine if adequate protein can be secreted in saliva of SIN rep5/LAC G1 infected mosquitoes to induce an immune response in vertebrates. The ability to express specific immunogens in saliva provides an unprecedented opportunity to investigate host immune response in the context of the immunomodulatory effects of vector saliva.

SIN Silencing of a Mosquito Salivary Gene

We conducted studies to examine the potential for exploiting Alphavirus expression vectors to knock out endogenous genes in vectors (Johnson et al., 1999). In these studies we utilized a transgenic line of *A. aegypti* mosquitoes that express the luciferase (LUC) reporter gene from the mosquito apyrase promoter (Jasinskiene et al., 1998). LUC is expressed only in the adult female salivary glands, as is apyrase itself. We constructed a dsSIN virus TE/3′2J/anti-LUC virus that expresses RNA complementary to the 5′ end of the LUC mRNA. LUC activity was monitored in salivary glands of mosquitoes that were infected with the anti-LUC or control dsSIN virus. Mosquitoes infected with anti-LUC virus exhibited > 90 percent reduction in LUC compared to uninfected and control dsSIN virus-infected mosquitoes. Importantly, expression of other salivary gland genes was not perturbed by anti-LUC virus. Thus, the dsSIN antisense expression system could be an important tool for studying gene function in vivo in mosquitoes.

SIN Molecular Genetic Reduction of Vector Competence

Intracellular immunization or pathogen-derived resistance strategies have been utilized to induce resistance in cells to virus infections. Typically, the strategy involves the expression of effector RNA sequences and/or viral proteins in the host cell to make it resistant to infection with the target virus. We have exploited SIN systems to determine if a similar strategy would be efficacious in inducing the anti-arboviral state in mosquito cell cultures and in vivo in vectors.

Initial studies focused upon knock out LAC virus infection in mosquitoes using antisense and sense suppression approaches (Powers et al., 1994; Powers et al., 1996). Studies were conducted to assess the efficacy of numerous sequences and strategies from the M (codes for G1 and G2 proteins and a nonstructural protein) and S (codes for the N protein and a nonstructural protein) RNA segments. These studies quickly revealed that effector molecules based on the S RNA segment were generally efficacious, while those based on M RNA sequences were not. Indeed, antisense to the S RNA knocked out virus infection in cells and even in mosquitoes. It is important to note that the rapid assessment of the antiviral efficacy of different strategies and portions of the genome would probably still be ongoing if we would have had to utilize normal transgenesis studies to characterize the sequences.

A similar approach was promptly used to knock out dengue virus infections in cell culture and mosquitoes. The dsSIN system was engineered to express antisense sequence to the premembrane (PRM) protein-coding region of dengue-2 virus. Expression of the antisense sequence knocked out dengue-2 virus infection and replication in cell cultures (Gaines et al., 1996). Similar results were obtained in vivo in *A. aegypti* mosquitoes. Indeed, salivary glands infected with the dsSIN-anti-PRM virus were not infected with dengue virus. Most importantly, transmission of dengue-2 virus was inhibited in the mosquitoes infected with dsSIN, the first demonstration of molecular ablation of transmission of a vector-borne pathogen (Olson et al., 1996). These proof-of-concept studies suggest that similar approaches may be exploited for disrupting the transmission cycle of parasitic diseases at the pathogen-vector interface.

These landmark studies have now been expanded to include yellow fever virus (Higgs et al., 1998), and the power of the system has been used to characterize the antiviral potential for other regions of the flavivirus genome. These studies would not have been possible without the power of the dsSIN expression system.

Need for a Second Generation of dsSIN Expression Vectors

Although an incredible tool, the dsSIN system based up pTE3′2J does have limitations. The most important is that dsSIN virus generated poorly infected mosquitoes; the virus apparently does not infect and replicate in the midgut epithelial cells of vectors. In the studies related previously, recombinant dsSIN viruses were intrathoracically inoculated into vectors, thereby bypassing the midgut entirely. The midgut and midgut cells are critical determinants of the vector competence of vectors; the pathogen must negotiate the midgut in order to gain access to the hemocoel and subsequently salivary glands for transmission. Thus, we began studies to develop a dsSIN system the exhibited tropisms for midgut epithelial cells of mosquitoes. The surface glycoproteins of viruses typically condition tropisms. The virus that was cloned to prepare pTE3′2J was a highly mouse adapted and neurovirulent virus. We located a Malaysian (MRE16) SIN virus that efficiently infects midgut epithelial cells in *A. aegypti* and has a 95 percent dissemination rate when blood fed into the mosquito. The subgenomic (26S) RNA of MRE16 was cloned and sequenced (Seabaugh, et al., 1998). A full-length MRE16 26S cDNA was assembled and a chimeric SIN infectious cDNA, designated MRE1001, was generated that contains the structural gene cDNA of MRE16 virus genome in the background cDNA sequences of pTE3′2J. The MRE1001 cDNA infectious clone efficiently infects midgut cells, and > 90 percent of infected mosquitoes develop disseminated infections after fourteen days extrinsic incubation.

A new dsSIN infectious cDNA clone (MRE/3′2J) has been generated from the MRE1001 chimeric cDNA. We have engineered MRE/3′2J virus to express GFP and used it to characterize the tropisms of the new dsSIN transducing system. MRE/3′2J/GFP efficiently infects and expresses GFP in mosquito midguts following *per os* infection (Olson et al., 2000), and the virus disseminates efficiently to express GFP in secondary target tissues. Surprisingly, MRE/3′2J/GFP virus also can efficiently infect mosquito larvae when larvae ingest C6/36 cells that have been infected 48 h earlier with the virus (Higgs et al., 1999). Thus, we now have a system that orally infects and expresses GOIs in midguts of both juvenile and adult stages of *A. aegypti.*

This is very important for a number of reasons. For example, intrathoracic inoculation of vectors with dsSIN viruses induced the vector immune response because of physical trauma. Thus, it was impossible to study host immune function genes. Oral infection of *A. aegypti* with dsSIN does not induce the immune response (Carl Lowenberger, University of Wisconsin, personal communication); thus the new constructs can be used to probe vector immunity.

We are also currently building dsSIN viruses that efficiently infect and express GOIs in *Anopheles* spp. mosquitoes. There is great need for provision of virus transduction systems to investigate the molecular biology of vector species.

Aedes Densonucleosis Virus Expression Systems

Insect viruses are potential transducing vectors and are attractive candidates for use in biological control strategies (Carlson, et al., 1999). DNVs of mosquitoes are well characterized and have been exploited as transducing vectors. DNVs appear to be rather widespread in nature, but are relatively restricted in their host specificity. They have a nonenveloped icosahedral structure and are stable viruses in the environment. We developed the AeDNV as a gene vector that is able to transduce genes into mosquito cells (Afanasiev et al., 1994). This virus transducing system provides the exciting potential for introducing genes that increase the virulence of the virus or perhaps modify the life cycle or behavior of the mosquito into natural populations. Some mosquito DNVs have been reported to be transmitted vertically, which could potentially be exploited in the implementation of control strategies (Carlson, et al., 1999). We have made significant progress in adapting the system to infect larvae naturally.

Transduction of Aedes aegypti *Mosquitoes with Vectors Derived from* Aedes *Densovirus*

Studies to demonstrate the potential for *Aedes* densovirus (AeDNV)–based transduction of vectors are in progress and have been very successful (Afanasiev et al., 1999). For example, *Aedes* densovirus constructs that express GFP have been used to investigate the pathogenesis of AeDNV in *A. aegypti.* A construct was made from in which GFP was expressed robustly as a fusion protein to the C-terminus of *NS1*. The two-plasmid system was used to package the *NS1*-GFP construct, and the resulting transducing virus particles were infectious to *A. aegypti* larvae. GFP expression was detected as early as forty-eight hours postinfection. Amazingly, the principal portal of entry seemed to be the anal papillae, which were the most commonly infected organ. The nuclei throughout the syncitium lining the anal papillae abundantly expressed GFP. Since the anal papillae are contiguous with the hemolymph, expression of GOIs in the papillae would result in delivery of potential effector GOIs throughout the arthropod system.

Packaging of AeDNV-GFP Transducing Virus by Expression of Densovirus Structural Proteins from a SIN Virus Expression System

Genetic recombination resulting in the production of wild-type virus is an obstacle in the current system for producing densovirus transducing particles (Allen-Miura et al., 1999). In some applications for AeDNV transducing systems, production of wild-type virus, which can cause morbidity and mortality in infected vectors, would be contraindicated. This problem can be eliminated by exploiting the dsSIN virus, TE/3′2J, to express the structural proteins (VPs) of AeDNV. Thus, wild-type virus could not be generated. The AeDNV VPs were cloned behind the second subgenomic promoter of TE/3′2J/VP virus; appropriate expression of the VPs was demonstrated in infected cells by Northern analyses and immunoflourescence. TE/3′2J/VP virus was then used to infect mosquito cells transfected with p7NS1-GFP, a plasmid expressing the nonstructural genes of AeDNV and GFP as a reporter gene. AeDNV-GFP transducing virus was produced that was of equivalent titer to that produced by the two-plasmid method and was infectious to *A. aegypti* larvae. Such a strategy could be utilized to preclude production of infectious virus in biocontrol efforts to control vector populations.

SUMMARY

Mosquito-borne diseases remain intractable to control or are resurgent in many areas of the world. Novel strategies for control of these diseases are desperately needed. Modern molecular approaches to be developed and applied in these studies will undoubtedly provide considerable information about gene regulation and expression in vectors and consequently a much better understanding of the biology/molecular biology of vectors and vector-pathogen interactions. This would be most beneficial in removing the "flying syringe" view of the vector in terms of vector biology, vector-pathogen interactions, vector competence, and other components of vector-pathogen amplification and maintenance cycles in nature. Elucidation of the determinants of vector-pathogen interactions and vector competence may provide biomarkers of vector populations that pose undue risk for pathogen transmission. Assay for such biomarkers could allow public health authorities to focus limited resources for control of the most dangerous vector populations. In addition, these studies will provide considerable information about the feasibility of developing a new generation of biocontrol agents for control of vectors. Potent target-specific effector molecules to disrupt pathogen transmission or vectorial capacity could be theoretically delivered to vector populations in viruses, bacteria, or transposons, or delivered using conventional carriers to control these important diseases of humans and animals.

ACKNOWLEDGEMENTS

This research was supported by NIH grants AI 25629, AI 32543, AI 46753 and the John D. and Catherine T. MacArthur Foundation.

REFERENCES CITED

Afanasiev, B., Y. Kozlov, J. Carlson, and B. Beaty. 1994. Densovirus of *Aedes aegypti* as an Expression Vector in Mosquito Cells. *Experimental Parasitology* 79: 322–39.

Afanasiev, B. N., B. J. Beaty, and J. O. Carlson. 1999. Transduction of *Aedes aegypti* Mosquitoes with Vectors Derived from *Aedes* Densovirus. *Virology* 257: 62–72.

Allen-Miura, T., B. Afanasiev, K. Olson, B. Beaty, and J. O. Carlson. 1999. Packaging AeDNV-GFP Transducing Viruses by Expression of Densovirus Structural Proteins from a Sindbis Virus Expression System. *Virology* 257: 54–61.

Beaty, B. J., and J. O. Carlson. 1997. Molecular Manipulation of Mosquitoes. *Current Opinion in Infectious Disease* 10: 372–76.

Beaty, B. J., and D. H. L. Bishop. 1988. Bunyavirus-Vector Interactions. *Virus Research* 1C: 289–302.

Carlson, J., S. Higgs, K. Olson, and B. Beaty. 1995. Molecular Manipulation of Mosquitoes. *Annual Review of Entomology* 40: 359–88.

Carlson, J. O., B. N. Afanasiev, and E. Suchman. 1999. Densoviruses as Transducing Vectors for Insects. In *Insect Transgenesis.* ed. A. M. Handler and A. A. James, 139–60. Boca Raton, Fla.: CRC Press.

Chandler, L., L. Wasieloski, C. D. Blair, and B. J. Beaty. 1996. Analysis of LaCrosse Virus S-Segment RNA and Its Positive-Sense Transcripts in Persistently Infected Mosquito Tissues. *Journal of Virology* 70: 897276.

Coates, C. J., N. Jasinskiene, G. B. Pott., and A. A. James. 1999. Mariner Transposition and Transformation of the Yellow Fever Mosquito, *Aedes aegypti. Gene* 226: 317–25.

Dobie, D., C. D. Blair, L. Chandler, A. Rayms-Keller, L. P. Wasieloski, and B. J. Beaty. 1997. Analysis of LaCrosse Virus S mRNA 5' Termini in Infected *Aedes triseriatus* Mosquitoes and Mosquito Cells. *Journal of Virology* 71: 4395–99.

Gaines, P., K. E. Olson, S. Higgs, B. Beaty, and C. Blair. 1996. Pathogen Derived Resistance to Dengue-2 Virus in Mosquito Cells by Expression of the Premembrane Coding Region of the Viral Genome. *Journal of Virology* 70: 2132–37.

Gubler, D. J. 1998. Resurgent Vector-Borne Diseases as a Global Health Problem. *Emerging Infectious Diseases* 4: 442–49.

Gratz, N. G. 1999. Emerging and Resurging Vector-Borne Diseases. *Annual Review of Entomology* 44: 51–75.

Higgs, S., K. Olson, L. Klimowski, A. M. Powers, J. O. Carlson, R. D. Possee, and B. J. Beaty. 1995. Mosquito Sensitivity to a Scorpion Neurotoxin Expressed Using an Infectious Sindbis Virus Vector. *Insect Molecular Biology* 4: 97–103.

Higgs, S., D. Traul, B. Davis, B. Wilcox, and B. Beaty. 1996. Green Fluorescent Protein Expressed in Living Mosquitoes without the Requirement for Transformation. *Biotechniques* 21: 660–64.

Higgs, S., J. Rayner, K. Olson, B. Davis, B. Beaty, and C. Blair. 1998. Engineered Resistance in *Aedes aegypti* to a West African and South American Strain of Yellow Fever Virus. *American Journal of Tropical Medicine and Hygiene* 58: 663–70.

Higgs, S., C. T. Oray, K. Myles, K. E. Olson, and B. J. Beaty. 1999. Infecting Larval Arthropods with a Chimeric, Double Subgenomic Sindbis Virus Vector to Express Genes of Interest. *Biotechniques* 27: 908–11.

Jasinskiene, N., C. J. Coates, M. Q. Benedict, A. J. Cornel, C. S. Rafferty, A. A. James, and F. H. Collins. 1998. Stable Transformation of the Yellow Fever Mosquito, *Aedes aegypti*, with the Hermes Element from the Housefly. *Proceedings of the National Academy of Sciences USA* 95: 3743–47.

Johnson, B. W., K. E. Olson, T. Allen-Miura, A. Rayms-Keller, J. O. Carlson, C. J. Coates, N. Jasinskiene, A. A. James, B. J. Beaty, and S. Higgs. 1999. Inhibition of Luciferase Expression in Transgenic *Aedes aegypti* Mosquitoes by Sindbis Virus Expression of Antisense Luciferase RNA. *Proceedings of the National Academy of Sciences USA* 96: 13399–403.

Kamrud, K. I., A. M. Powers, S. Higgs, K. E. Olson, C. D. Blair, J. O. Carlson, and B. J. Beaty. 1995. The Expression of Chloramphenicol Acetyltransferase in Mosquitoes and Mosquito Cells Using a Packaged Sindbis Replicon System. *Experimental Parasitology* 81: 394–403.

Kamrud, K. I., K. E. Olson, S. Higgs, J. O. Carlson, and B. J. Beaty. 1997. Secretion of a Heterologous Protein in the Saliva of *Culex pipiens* Mosquitoes. *Insect Biochemistry and Molecular Biology* 27: 423–29.

Kamrud, K., K. Olson, S. Higgs, J. Carlson, and B. Beaty. 1998. Use of the Sindbis Replicon System for Expression of LaCrosse Virus Envelope Proteins in Mosquito Cells. *Archives of Virology* 143: 1365–77.

McGaw, M., L. Chandler, L. Wasieloski, C. Blair, and B. Beaty. 1998. Effect of LaCrosse Virus Overwintering of *Aedes triseriatus. American Journal of Tropical Medicine and Hygiene* 58: 168–75.

Olson, K., S. Higgs, C. S. Hahn, C. M. Rice, J. O. Carlson, and B. J. Beaty. 1994. The Expression of Chloramphenicol Acetyltransferase in *Aedes albopictus* (C6/36) Cells and *Aedes triseriatus* Mosquitoes Using a Double Subgenomic Recombinant Sindbis Virus. *Insect Biochemistry and Molecular Biology* 24: 39–48.

Olson, K., S. Higgs, A. Powers, B. Davis, J. Carlson, C. Blair, and B. J. Beaty. 1996. Genetically Engineered Resistance in Mosquitoes to Dengue Virus Transmission. *Science* 272: 884–86.

Olson, K. E., B. Beaty, and S. Higgs. 1999. Sindbis Virus Expression Systems for the Manipulation of Insect Vectors. In *The Insect Viruses,* ed. L. K. Miller and L. A. Ball, 371–404. New York: Plenum Publishing.

Olson, K. E., K. M. Myles, R. C. Seabaugh, S. Higgs, J. O. Carlson, and B. J. Beaty. 2000. Development of a Sindbis Virus Expression System That Efficiently Expressed Green Fluorescent Protein in Midguts of *Aedes aegypti* Following Per Os Infection. *Insect Molecular Biology* 9: 57–65.

Patterson, J. L., B. Holloway, and D. Kolakofsky. 1984. La Crosse Virions Contain a Primer-Stimulated RNA Polymerase and a Methylated Cap-Dependent Endonuclease. *Journal of Virology* 52: 215–22.

Powers, A., K. Olson, S, Higgs, J. Carlson, and B. Beaty. 1994. Expression of the Antisense S RNA of La Crosse Virus Using a Sindbis Vector Induces Interference to Superinfection in Cells. *Virus Research* 32: 57–67.

Powers, A., K. Kamrud, K. Olson, S. Higgs, J. Carlson, and B. Beaty. 1996. Molecularly Engineered Resistance to California Serogroup Virus Replication in Mosquito Cells and Mosquitoes. *Proceedings of the National Academy of Sciences USA* 93: 4187–91.
Rayms-Keller, A., A. Powers, S. Higgs, K. Olson, K. I. Kamrud, J. O. Carlson, and B. J. Beaty. 1995. Replication and Expression of a Recombinant Sindbis Virus in Mosquitoes. *Insect Molecular Biology* 4: 245–51.
Seabaugh, R. C., K. E. Olson, S. Higgs, J. O. Carlson, and B. J. Beaty. 1998. Development of a Chimeric Sindbis Virus with Enhanced Per Os Infection of *Aedes aegypti. Virology* 243: 99–112.
Titus, R. G., and J. M. C. Ribeiro. 1988. Salivary Gland Lysates from the Sandfly *Lutzomyia longipalpis* Enhance *Leishmania* Infectivity. *Science* 239: 1306–08.
World Health Organization. 1996. *The World Health Report 1996: Fighting Disease, Fostering Development.* Geneva, Switzerland: World Health Organization, Office of World Health Reporting.

Chapter 19

Insects

Ambassadors for Biodiversity Conversation

MICHAEL J. SAMWAYS
Department of Entomology and Nematology
University of Stellenbosch
Matieland, South Africa

INSECT SUCCESS AND VARIETY

Insects are an immensely successful life form, with about 8 million species (Hammond, 1992). However, as so little is known about marine invertebrates, it is difficult to say what percentage of global species biodiversity this represents. A figure just over 56 percent is one estimate (Hammond, 1992). If just metazoans are considered, insects make up about four-fifths. Apart from a few marine insects such as the sea skaters *Halobates* spp., they are as a group terrestrial or aquatic. In consequence, their ambassadorial status for biodiversity conservation only pertains to the land and freshwater.

The bauplan of insects has been a successful one. Flight has been a hallmark, with the Carboniferous airspace being dominated firstly by insects. With pterodactyls and birds not yet having arisen, they had free evolutionary reign to reach an enormous size. The dragonfly *Meganeura monyi* Brongniart, for example, had a wingspan of 70 cm. Yet insects are limited in the size to which they can grow. An insect the size of a tortoise would have to have such a strong set of muscles to move that it would probably implode. A thicker cuticle would, in turn, be so cumbersome that all sensible mobility would be forsaken. So, insects remain small. But in doing so, they inhabit virtually every mineral and vegetable crack and crevice on the earth's surface. Such smallness also enables great abundance. Collembola, which, strictly speaking, are only honorary insects, can even occur at densities of between 10^4 and 10^5 per square meter (Hopkin, 1998).

We tend to think of insects as tough and as organisms that will survive anything. This is not entirely true. The groundbeetle *Africobatus harpaloides* Guen. of Central Africa for example, is so sensitive to microhabitat temperature changes that a difference of only 0.9°C stimulates it into dormancy (Greenwood, 1987).

Insects show an extensive array of polymorphisms. As for developmental polymorphism, the caterpillar for example is functionally quite a different animal from the butterfly. But insects also show many other types of polymorphisms, making them even more variable and various than straight species numbers would imply. Furthermore, genetic leakage resulting in hybridization can be very common, with a quarter of *Heliconius* butterfly species hybridizing in the wild (Mallet, 1997).

INSECT LOSS

Despite insects' great success, the huge unplanned experiment of worldwide landscape change is causing unprecedented widespread extirpation of insect populations and extinction of species, with perhaps thirty species disappearing each week (Mawdsley and Stork, 1995). Conservation strategies aim to mitigate this loss and to maintain as many and the most valuable genes, populations, and species as possible, along with their interactions within communities and ecosystems.

Currently there is an urgent need to establish some explicit guidelines for biodiversity conservation at the spatial scale of the landscape. Landscape ecological research is a relatively new field, and the interface between behavior and landscape form is newer still. This is especially so for insects, whether in wilderness, agro-forestry, or urban settings.

In late pre-agriculture times, insect extinctions appear to have been relatively rare (Coope, 1995). Indeed, this seems to have been the case for insect clades in general (Labandeira and Sepkoski, 1993). Today, however, through human geometrization, transformation, and fragmentation of landscapes, they are being shifted around on an enormous scale and at great intensity (Banaszak, 1996). Not that they have been passive responders. There are some indications of rapid genetic change in response to the changing land mosaic (Van Dyck and Matthysen, 1999).

Loss of insect populations, with perforation and retraction of ranges, has been particularly severe in Europe during the second half of the twentieth century (e.g., butterflies [Warren and Key, 1991; Dennis, 1993] and katydids [Samways, 1989]). Perhaps though, we have also just begun to be acutely aware of the decline that began over 100 years ago (Swinton, 1880). Certainly, there were early species extinctions (Rentz, 1993; Hambler and Speight, 1996) even of formerly widespread and abundant species (Lockwood and DeBrey, 1990), island species (Howarth and Ramsay, 1991), and certainly of those restricted to special habitats (Chapman, 1987). Yet ironically, narrow-range endemics are, in a way, preadapted to fragmentation, provided that the anthropogenic impact does not coincide with their locality. This seems to be the case with some highly localized, narrowly endemic damselflies in the Cape, where some populations were extirpated yet others appear relatively secure (Samways, 1999a). Similarly, the Cape Peninsula, with at least 111 documented endemic invertebrates, is a relatively safe haven for this extraordinarily unique fauna confined to a small mountain range completely surrounded by a city (Picker and Samways, 1996). But for some insects, such as damselflies, being in a reserve does not guarantee their survival (Samways, 1999a). Changing weather conditions, especially related to El Niño events, means that their populations must be spatially dynamic. Vagility becomes an important credential. On the positive side, a game reserve boundary is not the limit for the majority of insect species as it is for the mega-vertebrates (Rivers-Moore and Samways, 1996). For many insects, the landscape outside a reserve is just as habitable as that inside.

Although insect diversity loss is principally from landscape change and habitat loss, there are often interactive or synergistic factors (Samways, 1996). Among these are alien invasive plants, which reduce insect diversity principally at the species level rather than at the level of genera or families (Samways et al., 1996). Alien plants, however, are not necessarily all detrimental to insect diversity. Introduced aquatic and terrestrial plants can encourage some indigenous species. These tend to be the more abundant, widespread habitat-insensitive and highly mobile species, rather than the localized and rare specialists. This is the case for butterflies (Feltham, 1994) and for dragonflies (Samways, 1999a).

Insect mobility has been one of the major contributors to their survival, and some species have radically shifted their geographical range in response to retreating ice sheets and have not become extinct (Elias, 1994; Coope, 1995). Some insects are naturally invasive, and it is not unusual to see insects land on snowfields, colonize recently cooled lava flows, or travel on driftwood to arrive serendipitously on islands (Thornton, 1996). This intrinsic invasiveness means that some insects are themselves the enemy of biodiversity as they become alien invaders, taking advantage by stowing away on human transport and establishing in new lands (New, 1994; Samways, 1996).

INSECTS AND LANDSCAPE CONSERVATION

We do not know exactly what insect species, insect evolutionarily significant units (Moritz, 1994), or insect polymorphisms are needed under all environmental conditions. This means that viewing the higher spatial scale (i.e., landscape level) is sufficiently large to accommodate most of an insect's activity, yet at the same time it is just small enough for practical management. This may be considered as the "black box" approach, that is, a whole range of interaction types, interaction strengths, ecological processes, and biotic units are being conserved without us knowing the details. Besides, how can we know all possible interactions, when, for example, 1,000 species in the same community potentially produce 0.5 million interactions? In addition, strengths of those interactions vary, and hence outcomes. For insects, and much other components of biodiversity, this "coarse filter" landscape approach will achieve far more biodiversity conservation than the "fine filter" autecological approach. This is not to decry the role of behavior and the need for special single-species studies in special circumstances, especially when compiling the "Red List" (IUCN, 1996).

The disadvantage of the coarse filter approach is that it is oblivious to what and in which way the immensely complex contents are actually being conserved (Figure 19.1). Nevertheless, conserving black boxes (i.e., whole landscapes with high connectivity and minimal human devastation) is an answer in view of the magnitude of the biodiversity crisis and the shortage of time to conserve it.

In regard to insects as ambassadors for biodiversity conservation at the landscape level, there is a major note of caution. This applies equally to different insect groups as it does to insects as a whole and other taxa. Generally, there is a lack of congruence among taxa of various groups across landscapes (Prendergast et el., 1993; Samways, 1993; Lawton et al., 1998; Koch, 1999; Kotze and Samways, 1999). This means that it is important to undertake comprehensive species surveys (van Jaarsveld et al., 1998; Balmford and Gaston, 1999), yet with recognition of the limitations of species-only data for biodiversity conservation (Maddock and du Plessis, 1999).

The problem...

Black-box contents
The given complexity

1. Genetic variety (including polymorphisms, evolutionary significant units and species)
2. Interactions between phenotypes and their surroundings
3. Changes in interactions strengths, components and processes over time

The restraints...

Limitations to understanding
The complexity of the task

1. The taxonomic challenge
2. Knowing all the interactions
3. Tracking subtle changes over time
4. Sorting what is significant from what is not for conservation management
5. Time and money restraints

The solution...

Black-box conservation

1. Maintain wilderness
2. Restore naturalness
3. Create habitat webs with natural and indigenous heterogeneity and connectivity

(4. Additionally, include single-species recovery projects)

Figure 19.1. Biodiversity, especially the insect component, is immensely complex. Besides the huge number of species and genetic varieties, there is an even more complex set of interactions between insects and their surroundings. This imposes constraints on how to practically conserve global biodiversity. Nevertheless, a solution is to conserve entire landscapes, never quite knowing all the contents of these "black boxes." First, conservation of the last remaining tracts of the wilderness in crucial. Although financially expensive, morally disputable, and ecologically relatively unpredictable, restoration is nevertheless an important adjunct to maintaining wildlands. The creation of habitat webs either connects wildlands or forms an interconnected n/set of restored or semirestored habitats (see Figure 19.3). Finally, while this landscape approach is feasible, it is also blind to the needs of individual species. So a reasonable supplement is to build single-species plans on top of this landscape approach.

SOLUTIONS TO THE CURRENT INSECT DIVERSITY CRISIS

Although single-species conservation plays an important role in the protection of certain species under severe threat in the field (Thomas, 1991; Foucart and Lecoq, 1998; Hoekstra, 1998) and for reintroduction of species on the brink of extirpation or extinction (Oates and Warren, 1990; Pearce-Kelly et al., 1998; Sherley, 1998), it is not the mainstay of global insect diversity conservation. At the landscape level, a first premise is to conserve as much wilderness as possible, as it is irreplaceable (Rolston, 1994). Remaining wilderness is largely mountainous and often rich in local endemism (Howarth and Mull, 1992; Picker and Samways, 1996), or it is physically difficult to access, or both. Cousine Island, Seychelles, is a case in point, and with no alien invasive mammals it is possibly unique in the tropics (Samways, in press). This and other Seychelles islands are home to a unique fauna (Gerlach, 1997; Matyot, 1998).

The anthropogenic land mosaic is a differential filter (Ingham and Samways, 1996) (Figure 19.2), often placing restrictions on insect movement, as with butterflies in Africa (Wood and Samways, 1991). There is then likely to be loss of populations from small habitat fragments (Kindvall and Ahlén, 1992) and a proportional increase in habitat nonspecialists when the plant community is disturbed (Samways et al., 1996). Besides habitat disturbance, there are other survival hurdles to overcome. These include ensuring that there is genetic connectivity between isolated populations. Genetic connectivity can occur naturally through stepping-stone gene flow where in apparently sedentary species there is chronic long-distance dispersal or infrequent mass dispersal (Peterson, 1996).

Although conservation biology promotes landscape connectivity and natural heterogeneity, there are few definitive guidelines to follow. In South Africa, introduced pine trees can cause attrition of the grasshopper assemblage up to 30 m into the surrounding grassland (Samways and Moore, 1991), while thick riparian stands of *Acacia mearnsii* De Wild can locally blot out the local Odonata assemblage and endangers the survival of the highly threatened damselfly *Chlorolestes apricans* Wilmot (Samways, 1999a). Pine trees overhanging the water are similarly inhibiting, and it is recommended that the drip zones of trees are > 30 m from the river's edge (Kinvig and Samways, 2000).

For insects and other small ectotherms, small- and large-scale topography has a great influence on local distribution patterns. Hills, at least in Africa, allow insects to avoid cold-air drainage (Samways, 1990), and for strong fliers,

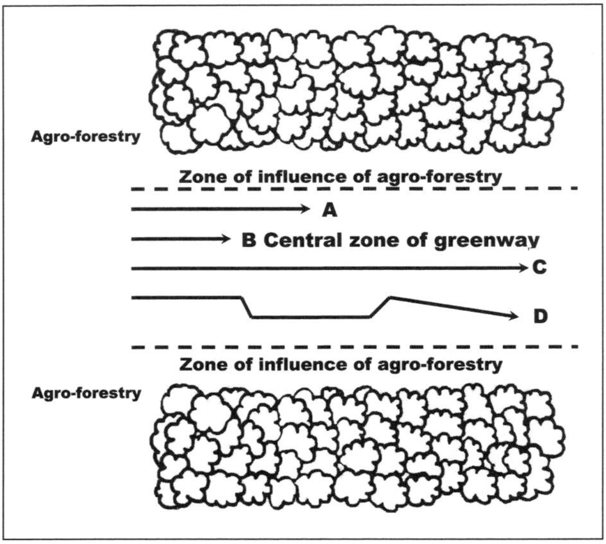

Figure. 19.2 The concept of the land differential filter. This is a section viewed from above of a habitat web composed of a grassland linkage or greenway between stands of agro-forestry. The effect of agro-forestry does not end at the edge, but because of shading, root, or other effects, spills over into the natural greenway. Indigenous species have different abilities to move through or along the greenway. The arrows (A to C) represent different dispersal abilities relative to that particular greenway. In reality, the movement paths are not straight (D). These lines are given as movement capabilities, but more meaningful for biodiversity conservation would be to use them to represent extent of gene flow. These lines are illustrative of both ecological and evolutionary time.

they are also meeting points for mating. Both burning (Chambers and Samways, 1998) and plant architecture (Lawrence and Samways, in preparation) strongly interact with topography, which in turn may determine the effectiveness of corridors or linkages.

Linkages (Harris, 1984) or greenways (Smith and Hellmund, 1993) have been known for a long time to be an ideal aim for biodiversity conservation. In regard to insects, however, there has been little definitive evidence showing that greenways or habitat webs work. Studies from metapopulations (Brakefield, 1991; Peterson, 1996) indicate that there does not necessarily have to be a continual flow of individuals along the linkages so long as there is enough genetic exchange to overcome the problems associated with severe and long-term genetic bottlenecks. This means that the linkage should provide a movement corridor at least occasionally. The problem with such movement corridors, however, is that there is little or no room for sideways movement. So the individuals are obliged to follow the one line of direction, stay still, or turn back. Inevitably this increases the filtration effect (Ingham and Samways,

Figure 19.3. An example of a habitat web (Goodhope Plantation, MONDI Forests near Pietermaritzburg, South Africa). Here the pine trees are planted sensitively on indigenous grasslands in a way that allows conservation within the boundaries of the plantation.

1996) of the landscape. Furthermore, edges of corridors are an added problem in that pines, for example, influence the adjacent grassland vegetation community and hence alter habitat quality. Another compounding factor is that large megaherbivores, whether domestic cattle or indigenous game, are funneled. When this happens there is extreme trampling, which may change the grasshopper (Gueguen, 1991) and the general insect assemblage (Rivers-Moore and Samways, 1996). This evidence suggests that for interconnected corridors (linkages) to be a viable option for biodiversity conservation in Africa they must be more than 60 m wide. For certain species a narrower corridor may suffice and for others only a wider corridor will do. Nevertheless, the ideal configuration is a habitat web (Figure 19.3), within which organisms live and move as they would in a natural, more expansive setting. We now need to test the effectiveness of such habitat webs for as wide a range of organisms as possible.

SUMMARY

Biodiversity is not simply what is visually obvious to us as humans, such as birds, mammals, and large plants. Indeed, small taxa, genetic variety, and ecological processes are the mainstay of biodiversity. Insects score highly on these three counts and yet are large enough to assess quantitatively. Insects historically have featured strongly in heuristic studies of systematics, genetics, physiology, behavior, and functional ecology. Now they are good ambassadors for biodiversity conservation.

ACKNOWLEDGMENTS

Special thanks to Kevin Heinz and Texas A&M University for the opportunity to present this chapter, and to Erika Börner, Melinda Hatton, and Pamela Sweet for preparation assistance.

REFERENCES CITED

Balmford, A., and K. J. C. Gaston. 1999. Why Biodiversity Surveys Are Good Value. *Nature* 398: 204–205.

Banaszak, J. 1996. Ecological Bases of Conservation of Wild Bees. In *The Conservation of Bees,* ed. A. Matheson, S. L. Buchmann, C. O'Toole, P. Westrich, and I. H. Williams, 55–62. London: Academic Press.

Brakefield, P. M. 1991. Genetics and the Conservation of Invertebrates, In *The Scientific Management of Temperate Communities for Conservation,* ed. I. F. Spellerberg, F. B. Goldsmith, and M. G. Morris, 45–79. Oxford, United Kingdom: Blackwell.

Chambers, B. Q., and M. J. Samways. 1998. Grasshopper Response to a 40-Year Experimental Burning and Mowing Regime, with Recommendations for Invertebrate Conservation Management. *Biodiversity and Conservation* 7: 985–1012.

Chapman, P. 1987. Extinction Faces the World's Cave Creatures. *IUCN Bulletin* 18: 7–9.

Coope, G. R. 1995. The Effects of Quaternary Climatic Changes in Insect Populations: Lessons from the Past. In *Insects in a Changing Environment,* ed. R. Harrington and N. E. Stork, 29–48. London: Academic Press.

Dennis, R. L. H. 1993. *Butterflies and Climatic Change.* Manchester, United Kingdom: Manchester University Press.

Elias, S. A. 1994. *Quaternary Insects and their Environments.* Washington, D.C.: Smithsonian Institution Press.

Feltham, N. F. 1994. *The Ecology of a Pietermaritzburg Butterfly Assemblage: A Conservation Perspective.* M.S. thesis, University of Natal, South Africa.

Foucart, A., and M. Lecoq. 1998. Major Threats to a Protected Grasshopper, *Prionotropis hystrix rhodanica* (Orthoptera: Pamphagidae, Akicerinae), Endemic to Southern France. *Journal of Insect Conservation* 2: 187–93.

Gerlach, J., ed. 1997. *Seychelles Red Data Book, 1997.* Mahé: The Nature Protection Trust of the Seychelles.

Greenwood, S. R. 1987. The Role of Insects in Tropical Forest Food Webs. *Ambio* 16: 267–70.

Gueguen, A. 1991. Effet du pâturage ovin sur la dynamique et la biomasse d'une population d' *Aeropus sibiricus* (L.) (Orthoptera: Acrididae) dans une alpage du sud de la France, 185–90. 4th Congrés International des Terres de Parcours, Montpellier, France.

Hambler, C., and M. R. Speight. 1996. Extinction Rates in British Nonmarine Invertebrates since 1900. *Conservation Biology* 10: 892–96.
Hammond, P. M. 1992. Species Inventory. In *Global Biodiversity: Status of the Earth's Living Resources,* ed. B. Groombridge, 17–39. London: Chapman and Hall.
Harris, L. D. 1984. *The Fragmented Forest.* Chicago: University of Chicago Press.
Hoekstra, J. M. 1998. Conserving Orthoptera in the Wild: Lessons from *Trimerotropis infantilis* (Oedipodinae). *Journal of Insect Conservation* 2: 179–85.
Hopkin, S. P. 1998. Collembola: The Most Abundant Insects on Earth. *Antenna* 22: 117–21.
Howarth, F. G., and W. P. Mull. 1992. *Hawaiian Insects and Their Kin.* Honolulu: University of Hawai'i Press.
Howarth, F. G., and G. Ramsay. 1991. The Conservation of Island Insects and Their Habitats. In *The Conservation of Insects and Their Habitats,* ed. N. M. Collins and J. A. Thomas, 71–119. London: Academic Press.
Ingham, D. S., and M. J. Samways. 1996. Application of Fragmentation and Variegation Models to Epigaeic Invertebrates in South Africa. *Conservation Biology* 10: 1353–58.
IUCN (International Union for Conservation of Nature and Natural Resources). 1996. *1996 IUCN Red List of Threatened Animals.* Gland, Switzerland: IUCN.
Kindvall, O., and I. Ahlén. 1992. Geometrical Factors and Metapopulation Dynamics of the Bush Cricket, *Metriopetra bicolor* Philippi (Orthoptera: Tettigoniidae). *Conservation Biology* 6: 520–29.
Kinvig, R. G., and M. J. Samways. 2000. Conserving Dragonflies (Odonata) along Streams Running through Commercial Forestry. *Odonatologica* 29: 195–208.
Koch, S. 1991. *The Distribution of Selected Scarabaeoid (Coleoptera) Taxa in South Africa and Namibia: Prospects for their Conservation.* M.S. thesis, University of Pretoria.
Kotze, D. J., and M. J. Samways. 1999. Support for the Multi-Taxa Approach in Biodiversity Assessment, as Shown by Epigaeic Invertebrates in an Afromontane Forest Archipelago. *Journal of Insect Conservation* 3: 125–43.
Labandeira, C. C., and J. J. Sepkoski Jr. 1993. Insect Diversity in the Fossil Record. *Science* 261: 310–15.
Lawton, J. H., D. E. Bignell, B. Bolton, G. F. Bloemers, P. Eggleton, P. M. Hammond, M. Hodda, R. D. Holt, T. B. Larsen, N. A. Mawdsley, N. E. Stork, D. S. Srivastava, and A. D. Watt. 1998. Biodiversity Inventories, Indicator Taxa and Effects of Habitat Modification in Tropical Forests. *Nature* 391: 72–76.
Lockwood, J. A., and L. D. DeBrey. 1990. A Solution for the Sudden and Unexplained Extinction of the Rocky Mountain Grasshopper (Orthoptera: Acrididae). *Environmental Entomology* 17: 1194–1205
Maddock, A., and M. A. du Plessis. 1999. Can Species Data Only Be Appropriately Used to Conserve Biodiversity? *Biodiversity and Conservation* 8: 603–15.
Mallet, J. 1997. A Species Definition for the Modern Synthesis. *Trends in Ecology and Evolution* 10: 294.
Matyot, P. 1998. The Orthopteroids of the Seychelles: A Threatened Island Fauna. *Journal of Insect Conservation* 2: 235–46.
Mawdsley, N. A., and N. E. Stork. 1995. Species Extinctions in Insects: Ecological and Biogeographical Considerations. In *Insects in a Changing Environment,* ed. R. Harrington and N. E. Stork, 321–69. London: Academic Press.
Moritz, C. 1994. Defining "Evolutionary Significant Units" for Conservation. *Trends in Ecology and Evolution* 9: 373–75.
New, T. R. 1994. *Exotic Insects in Australia.* Adelaide, Australia: Gleneagles.
Oates, M. R., and M. S. Warren. 1990. *A Review of Butterfly Introductions in Britain and Ireland.* Report for the Joint Committee for the Conservation of British Insects Founded by the World Wide Fund for Nature. Surrey, United Kingdom: Godalming.

Pearce-Kelly, P., R. Jones, D. Clarke, C. Walker, P. Atkin, and A. A. Cunningham. 1998. The Captive Rearing of Threatened Orthoptera: A Comparison of the Conservation Potential and Practical Considerations of the Two Species' Breeding Programmes at the Zoological Society of London. *Journal of Insect Conservation* 2: 201–10.

Peterson, M. A. 1996. Long-Distance Gene Flow in the Sedentary Butterfly, *Euphilotes enoptes* (Lepidoptera: Lycaenidae). *Evolution* 50: 1990–99.

Picker, M. D., and M. J. Samways. 1996. Faunal Diversity and Endemicity of the Cape Peninsula, South Africa: A First Assessment. *Biodiversity and Conservation* 5: 591–606.

Prendergast, J. R., R. M. Quinn, J. H. Lawton, B. C. Eversham, and D.W. Gibbons. 1993. Rare Species, the Coincidence of Diversity Hotspots and Conservation Strategies. *Nature* 365: 335–37.

Rentz, D. C. F. 1993. Orthopteroid Insects in Threatened Habitats in Australia. In *Perspectives on Insect Conservation,* ed. K. J. Gaston, T. R. New, and M. J. Samways, 125–38. Andover, United Kingdom: Intercept.

Rivers-Moore, N. A., and M. J. Samways. 1996. Game and Cattle Trampling, and Impacts of Human Dwellings on Arthropods at a Game Park Boundary. *Biodiversity and Conservation* 5: 1545–56.

Rolston, H. III. 1994. *Conserving Natural Value.* New York: Columbia University Press.

Samways, M. J. 1989. Insect Conservation and Landscape Ecology: A Case-History of Bush Crickets (Tettigoniidae) in Southern France. *Environmental Conservation* 16: 217–226.

———. 1990. Landforms and Winter Habitat Refugia in the Conservation of Montane Grasshoppers in Southern Africa. *Conservation Biology* 4: 375–82.

———. 1993. Dragonflies (Odonata) in Taxic Overlays and Biodiversity Conservation. In *Perspectives on Insect Conservation,* ed. K. J. Gaston, T. R. New, and M. J. Samways, 111–23. Andover, United Kingdom: Intercept.

———. 1996. Insects on the Brink of a Major Discontinuity. *Biodiversity and Conservation* 5: 1047–58.

———. 1996. Managing Insect Invasions by Watching Other Countries. In *Invasive Species and Biodiversity Management,* ed. O. T. Sandlund, P. J. Schei, and A. Viken, 295–304. Dordrecht, Netherlands: Kluwer.

———. 1999a. Diversity and Conservation Status of South African Dragonflies (Odonata). *Odonatologica* 28: 13–62.

———. 2000. A Conceptual Model of Ecosystem Restoration Triage Based on Experiences from Three Remote Oceanic Islands. *Biodiversity and Conservation* 9: 1073–83.

Samways, M. J., and S. D. Moore. 1991. Influence of Exotic Conifer Patches on Grasshopper (Orthoptera) Assemblages in a Grassland Matrix at a Recreational Resort, Natal, South Africa. *Biological Conservation* 57: 117–37.

Samways, M. J., P. M. Caldwell, and R. Osborn. 1996. Ground-Living Invertebrate Assemblages in Native, Planted, and Invasive Vegetation in South Africa. *Agriculture, Ecosystems and Environment* 59: 19–32.

Sherley, G. 1998. Translocating a Threatened New Zealand Giant Orthopteran: *Deinacrida* sp. (Stenopelmatidae): Some Lessons. *Journal of Insect Conservation* 2: 195–99.

Smith, D. S., and P. C. Hellmund, eds. 1993. *Ecology of Greenways: Design and Function of Linear Conservation Areas.* Minneapolis: University of Minnesota Press.

Swinton, A. H. 1880. *Insect Variety: Its Propagation and Distribution.* London: Cassell, Petter and Galpin.

Thornton, I. 1996. *Krakatau: The Destruction and Reassembly of an Island Ecosystem.* Cambridge, Mass.: Harvard University Press.

Thomas, J. A. 1991. Rare Species Conservation: Case Studies of European Butterflies. In *The Scientific Management of Temperate Communities for Conservation,* ed. I. F. Spellerberg, F. B. Goldsmith, and M. G. Morris, 149–97. Oxford, United Kingdom: Blackmore.

Van Dyck, H., and E. Matthysen. 1999. Habitat Fragmentation and Insect Flight: A Changing "Design" in a Changing Landscape? *Trends in Ecology and Evolution* 14: 172–74.

Van Jaarsveld, A. S., S. Freitag, S. H. Chown, C. Muller, S. Koch, H. Hull, C. Bellamy, M. Krüger, S. Endrödy-Younga, M. W. Mausell, and C. H. Scholze. 1998. Biodiversity Assessment and Conservation Strategies. *Science* 279: 2106–08.

Warren, M. S., and R. S. Key. 1991. Woodlands: Past, Present, and Potential for Insects. In *The Conservation of Insects and their Habitats,* ed. N. M. Collins and J. A. Thomas, 155–211. London: Academic Press.

Wodd, P. A., and M. J. Samways. 1991. Landscape Element Pattern and Continuity of Butterfly Flight Paths in an Ecologically Landscaped Botanic Garden, Natal, South Africa. *Biological Conservation* 58: 149–66.

CHAPTER 20

How Can Entomology Contribute to Sustainable Crop Protection?

J. C. VAN LENTEREN
Laboratory of Entomology
Department of Plant Sciences
Wageningen University and Research Centre
Wageningen, Netherlands

PESTS ARE A MINORITY WITHIN THE INSECTS

The word insect is often considered synonymous to the term "noxious organism," Insects are regularly mentioned as harmful, cruel, and dangerous creatures. Upon asking people to list insect names, they will commonly start with pestiferous species—mosquitoes, flies, wasps, locusts, and ants. Beneficial, beautiful, and benign insects do not usually make it to the top five of the list mentioned. There is a general lack of knowledge regarding insects and their importance.

The number of animal species is estimated to be between 20 and 50 million, and out of these, 1.7 million species have been identified and named to date. Eighty percent of all animal species are insects. Among the insects, plant eaters are our main competitors and they comprise about half of the insect world, that is, some millions of species. The other half consists of insects that live off other insects (predators and parasites), insects that live off other animals, and insects that feed on dead organic matter. Although the number of plant-eating insects is very large, the number of species regarded as pests is relatively small. Estimates of the number of insect pests (including medical, agricultural, structural, and nuisance pests) vary between 500 and 10,000 species worldwide (van Lenteren, 1993). Thus, a balanced view of the insect world should picture most insects as playing a positive role in nature, while only a minimal fraction of insect species—less than 0.01 percent—as causing problems for humans.

Due to its activities in controlling this small group of insect species with synthetic pesticides—a development that started only fifty-five years ago—the entomological profession has come under severe criticism. In this chapter I will illustrate how entomologists can help amend this criticism by developing environmentally safe and sustainable crop protection.

MOST INSECTS ARE USEFUL

Even the most conservative estimate of the number of pest species (10,000) clearly indicates that very few insects are problematic. Why is this number so small? Ecologists have ample evidence relevant to answer this question. First of all, not all plant-eating insects occur on the plants we grow for food, feed, or fiber. Ninety percent of our food comes from only 300 plant species. Many more insects feed on the other 300,000 plant species with little or no direct monetary value. Further, we rarely see these noncommercial plant species being completely defoliated or their populations severely affected by insects. In nature, only up to 10 percent of the plant biomass is normally eaten by insects and other herbivores. This contrasts strongly with the 20 to 100 percent defoliation by insects observed on crop species. What has gone wrong? Comparison of population dynamics of potential pest species in natural ecosystems and insect pests in agro-ecosystems points toward several explanations:

1. Wild plants possess resistance to a variety of insects, which slows insect population development or causes insects to avoid them.
2. In natural ecosystems, certain host plants are not easily discovered by most insect herbivores. Only a few highly host-specific species are able to locate these plants and reproduce successfully on them.
3. In natural ecosystems, a myriad of natural enemy species maintain plant-eating insects at low population densities and, even in agro-ecosystems, a number of potential pest species are regulated by natural enemies.

These combined factors generally result in low population densities of plant-eating insects in natural ecosystems. Agro-ecosystems, however, are characterized by monoculture of genetically identical crop species, are maintained in excellent growing conditions, and often regrown in the same fields season after season. This predictability in quantity and quality of resources benefits pestiferous species by maximizing their reproduction and survival. At the same time, many natural enemy species are not able to persist in ephemeral agro-ecosystems. Thus, man has created very favorable conditions for plant-eating insects through the provision of accessible and/or more nutritious food plants and through the decimation of their natural enemies (Huffaker et al., 1976).

Hundreds of thousands of insect species are beneficial. Besides natural enemies of pests (predators and parasites), we know of many other large groups of useful insects and their role in ecosystems functioning. These include pollinators, recyclers (insects eating dead organic material), and nectar-, silk-, and wax-producing insects (see review by Smith et al., 1973). Further, many insects are used as food, although this was more common in the past than it is today. Large orthopterans are still an important item of the human diet, especially for children, in certain parts of Africa (see review of entomophagy by DeFoliart, 1989). Many insects have also played an eminent role in culture and recreation (Southwood, 1977; Hogue, 1987). Clearly, entomologists need to change public perceptions regarding insects and their importance in our world.

THE ROLE OF CHEMICAL PEST CONTROL IS STRONGLY OVERRATED

The world market for chemical pesticides is estimated at U.S. $26 billion (Klassen, 1993). A generally accepted idea is that most insect pest populations can be kept under control using synthetic chemical pesticides. This is an incorrect assumption and new trends in crop protection are demonstrating the value of other pest control alternatives. Biologically based technologies, such as host-plant resistance to pests, already make up a significant part of the crop protection market (Table 20.1). The world market for seeds and propagules, most of which have been improved for resistance to harmful organisms, is about U.S. $6 billion. Quantities spent on biological control, behavior-modifying chemicals (e.g., pheromones), and botanical insecticides in the developed world amounted to U.S. $255 million in 1991 (van Lenteren, 2000). This is still a relatively modest amount compared to quantities spent on traditional synthetic pesticides. However, this figure is steadily increasing.

Biologically based pest control methods appear to play a minor role in crop protection worldwide. However, this perception is based on money spent on pest control. Thus, comparisons between pest control alternatives

Table 20.1. **Estimated Amount of Money Spent on Crop Protection (in Billions of $ U.S.).**

Control Agent	*U.S. $ billions*
Conventional synthetic pesticides	26.00
Plant material resistant to pests and diseases	6.00
Biological control	0.05
Bacterial and fungal-derived toxins	0.11
Botanical pesticides	0.09
Behavior modifying chemicals	0.05

may be inherently flawed. First of all, many potentially pestiferous species are kept at low population densities by host-plant resistance factors and by natural enemies (DeBach and Rosen, 1991). Their contribution to pest suppression is never accounted in monetary terms. Second, when long-term control is obtained by the release of natural enemies (classical and inoculative biological control), the annual benefits resulting from single natural enemy releases are not considered in cost to benefit calculations. Therefore, the importance of biologically based pest control methods is often seriously underscored in comparisons to chemical and other pest control methods. It is, of course, shortsighted and wrong to use the cost of a certain crop protection technique to indicate its importance. A correct means of evaluation would be to calculate cost to benefit ratios based on the costs to develop a method and the economic benefits obtained after its application. Conservative calculations indicate that biological control has cost to benefit ratios ten times better than those of chemical control (Table 20.2).

RESEARCH ON PEST CONTROL NEEDS TO BE REDIRECTED TO BIOLOGICALLY BASED, LONG-TERM, AND ENVIRONMENTALLY SAFE ALTERNATIVES

Some 200 chemicals in a vast array of formulations are used as insecticides worldwide. Insecticides are the most hazardous pesticides to humans because, unlike fungicides, for example, they target animal populations. The majority of insecticides can be characterized as broad spectrum, and pose risk to producers, applicators, consumers, and the environment. However, the main problem for farmers and the chemical industry today is the development of insecticide-resistance. The exponential increase of insecticide resistance has

Table 20.2. **Comparison of Aspects Related to the Development and Application of Chemical and Biological Control.**

Parameter	*Chemical control*	*Biological control*
Number of "ingredients" tested	> 1 million	5,500
Success ratio	1:30,000	1:20
Development costs	100 million US$	2 million US$
Development time	10 years	10 years
Benefit per unit of money invested	3:1	30:1
Risk of resistance development	great	none/small
Specificity	low	high
Harmful side-effects	several	none/few

caused a dramatic rise in human disease problems (e.g., insect-vectored malaria) and an increase in crop losses due to insect pests (Metcalf, 1980; Oerke et al., 1994). Furthermore, the development of new pesticides has become increasingly difficult. As more potential chemicals need to be screened, and more research on safety is required for registration, the overall pesticide development and production costs have become enormous. The rate at which insect populations have become resistant to common insecticides and even new and complex compounds has, however, not decreased comparatively. Populations of more than 500 insect species have been reported resistant to one or more pesticides (Georghiou and Lagunes-Tejeda, 1991). Chemical eradication of insect pests has generally failed; harmful insects have survived all chemical tactics invented to destroy them.

The rising costs of pesticide development have caused a strong decline in the number of new insecticides appearing on the market each year. As many as twenty new active ingredients per year were registered during the 1960s, but less than one is registered per year to date. The role of agricultural entomologists will have to change from studying opportunistic methods of (chemical) pest control whereby problems are awaited for, solutions are hoped for, and indicators of poor performance ignored to studying biologically based pest control methods. Research on conventional pesticide development, which is generally based on empiricism, is being replaced by research into new pest control methods that are based in biological mechanisms and relationships (Klassen, 1993). For an extensive review of developments in the area of biologically based technologies, I refer the reader to Lumsden and Vaughn (1993).

Many alternative methods are already available, and an important change in research approaches is taking place. There is an enormous and still largely untapped pool of biological and genetic resources that creates excellent possibilities for pest control methods to be developed and implemented in Integrated Pest Management (IPM) approaches. In my opinion, two cornerstones of IPM, host-plant resistance and biological control, should play a more important role in pest management in the future.

Although host-plant resistance has quite often been exploited for disease control, it can also be used in insect pest control to a much larger extent than it is used to date. Although the use of insect-resistant cultivars has been more common in North America than in Europe, recent investments in research and development of insect and disease resistant cultivars has increased and now represents approximately 30 percent of the total plant-breeding budget of companies targeting the European market. Resistant varieties have been used over vast agricultural areas for suppression of disease and insect pests in the United States (Klassen, 1993). However, the synergism and combined benefits of partial host-plant resistance and the use of natural enemies remains to be exploited. In addition, opportunities exist to breed plants that benefit/attract natural enemies and enhance their performance in suppression of pestiferous species (see review by Dicke, 1999). These new strategies will lead to a wider adoption and implementation of host-plant resistance in future IPM.

In addition to host-plant resistance, research is needed to further develop biological control releases and integrate them into appropriate pest management programs. Pest species are generally attacked by an array of natural enemies, from viruses to predatory mammals. Some pest species have well over 100 recorded natural enemy species but quantitatively, insects form the largest group. During the past 100 years of modern biological insect control some 5,000 natural enemy introductions have been made into new areas worldwide. In 1,500 of these, introductions have led to substantial, long-term reductions of insect pest populations, and complete control has been observed in 450 instances (Waage, 1994). Hence, biological insect control has an excellent record of successes. Empirical evidence and a large number of case studies (e.g., Huffaker and Messenger, 1976; Fisher et al., 1999) have demonstrated that biological control practiced by experts is a very cost effective method of pest suppression. Benefit to cost analyses for thirty inoculative biological control programs show ratios between 18 to 1 and 1,500 to 1. Biological control benefit/cost ratios are better than ratios for any other pest control method.

Although some of the benefits of biological control have been quantified, no concerted efforts have been made to measure the global economic value of biological and natural control. These benefits must be reckoned in billions of dollars (Tisdell, 1990). In addition to the economic benefit of reduced pest damage, biological control has highly positive social and ecological benefits. Additionally, once a good natural enemy has been identified it can be used indefinitely. Pest resistance to natural enemies, comparable to resistance to pesticides, has never been documented. Therefore, based on economics (see Table 20.2), success ratio, and the number of natural enemies available, there are certainly no reasons to stifle the investment in research to further develop biological control strategies. Based on potential benefits/costs, biological control should be the first pest control tactic to be explored. However, funding for biological control is still very limited compared to funding for other control tactics, especially when one considers how complex and challenging biological control research can be. In my view, however, the slow increase in the use of biological control is more a matter of politics and policy than of any inherent weakness of the method itself.

CHEMICAL CONTROL IS IRRESPONSIBLY CHEAP AND, AS A RESULT, IMPLEMENTATION OF ALTERNATIVES IS LIMITED

Pest control costs are very low and generally amount to only 3 percent of the total crop production costs. The low cost of pesticides is due, at least in part, to the under-estimation of indirect costs associated with environmental damage (e.g., cost of cleaning up chemical waste and cost of reduced biodiversity and its life-support functions) and human health. Chemical industries are not being held responsible for any of these indirect costs. The availability of inexpensive pesticides has resulted in a decrease of treatment thresholds (pest density at which treatment is economical). Based on the deceptively low cost of pesticides, the implementation of alternatives to chemical control is more difficult to date than before these chemicals were on the market. To recuperate at least partially the indirect costs associated with pesticide use, prices of pesticides should be doubled or even tripled using a system of levies, whereby the most environmentally damaging pesticides pay the highest prices. A re-evaluation of cost to benefit ratios for crop protection methods is also essential to obtain a change in crop protection policy and a more honest appraisal of the usefulness of alternatives.

IMPLEMENTATION OF SUSTAINABLE PEST CONTROL IS NOT STIMULATED

Funding for Research on Biologically Based Pest Control Methods

During the past five decades, little, and usually only public money, has been invested in the development of nonchemical pest control methods. The money has been well spent and considerable progress has been made (Klassen, 1993). Results obtained using nonchemical pest control methods are, of course, dependent on the amount of research and development effort invested in them. Implementation of nonchemical control methods, however, has been limited by many other factors besides financial limitations.

Farmers' Attitudes

Until very recently, only a few farmers had been interested in the development of nonchemical control methods. However, being unable to control a pest using traditional chemicals methods has been a strong force driving farmers to change their ideas about IPM. The attitude of farmers is currently changing. European glasshouse-grown vegetable growers and fruit growers, for example, have experienced the positive aspects of IPM and seriously worry about increasing public concern on pesticide usage. These growers have shifted to IPM methods rather than relying on conventional pesticides (van Lenteren, 1995; 1999). Farmer's perceptions regarding chemical pesticides and their alternatives are changing.

The Viewpoint of the Chemical Industries

In addition to cost considerations, the implementation of alternative pest control methods is more complicated than that of chemical control methods. Biologically based pest control methods generally deviate from simple to use, chemical pest control programs, and these "complications" inhibit their adoption. Alternatives such as biological control not only "complicate" pest management programs but they also seem to be commercially unattractive. This is caused by the combination of the impossibility of patenting natural enemies, the highly time-intensive mass-rearing of natural enemies, the short shelf life of living organisms, their host specificity, and the extra knowledge required for their proper use. Chemical industries will not start the production of host-specific pesticides on their own initiative unless the use of broad spectrum pesticides is prohibited or the target pest populations develop high levels of resistance, but time is on our side! We cannot blame the chemical industry for providing the pesticides and for trying to maximize their returns. The use of broad spectrum pesticides is still lawfully permitted and facilitated by registration policies.

Role of the Government

Regulation by government agencies may alter the current pest control picture. Regulation should be aimed at (1) making conventional chemical control less attractive or impossible (by measures concerning registration, taxation, labeling, etc.), (2) stimulating the adoption of other control methods and, above all, (3) educating the public at all levels regarding impact of conventional pesticides and their alternatives to change public attitudes toward nature. It is bizarre how public money is spent on development of alternatives to chemical control when, at the same time, government agencies often do not encourage their adoption and application.

PRESENT CROP PROTECTION POLICIES STIMULATE A REDUCTION OF PESTICIDE USAGE

Following years of increases, the beginning of the 1990s saw a stabilization and even a reduction in pesticide sales by developed countries. In several northern European countries this reduction, relative to the period between 1980 and 1989, has been more than 50 percent of the active ingredient used (expressed in volume). The governments of Sweden, Denmark, Germany, and the Netherlands have created programs to reduce the use, discharge, and dependence on conventional pesticides. These programs are not merely aimed at banning pesticides but attempt to develop alternative and sustainable means of pest control. The Swedish program was the first to be implemented; it achieved a 50 percent reduction in pesticide use in the period between 1987 and 1990, with and an additional 50 percent reduction implemented at present. The German and Danish programs achieved a 50 percent and 25 percent reduction, respectively, in the early 1990s. The Dutch program started in 1991 and was on schedule to obtain the desired minimum of 65 percent reduction by the year 2000.

Long-term views of agricultural development point at even much stronger reductions in chemical control. A strategy study of the Dutch Scientific Council for Government Policy (Rabbinge, 1992) indicates that the area of cultivated land will be reduced to about half of its present size in the European Union. As a result, the anticipated reduction in the volume of pesticide used will be tremendous over the next decade (i.e., 80 to 90 percent in all Europe). This reduction will be due to:

1. Improvement of pesticide application technology (effectiveness at reaching targets) and the amount of active ingredient needed for effective control will decrease.
2. The application of integrated arthropod and disease management systems will become standard for all crops.
3. The abundant use of pesticides due to quick crop rotations will be eliminated, which will particularly lead to a reduction of soil fumigant use.
4. The area on which pesticides are used will be much smaller.

Such developments will strongly emphasize the role of naturally occurring beneficial insects, and will allow for an increase in the use of other biologically based methods.

OVER-SPECIALIZATION IN AGRICULTURE PREVENTS A REDUCTION IN PESTICIDE INPUTS

Agricultural research saw a large diversification during the second part of the twentieth century. In plant breeding, the search for better-yielding varieties has been the central theme for years. Plant production research has concentrated on finding and overcoming limiting factors in plant growth and crop yields. With the advent of insecticides, research on plant resistance against pests was not considered important, because most pests could be easily controlled with these compounds. This has led to the development of highly productive but extremely susceptible plant varieties. In addition, crop fertilization practices causing highly nutritious plants and high insect reproductive rates have stimulated development of frequent pest outbreaks (e.g., aphids, leafminers, and spider mites).

Pest control has been addressed as an independent area of agriculture, particularly during the past fifty-five years. The common viewpoint was that those working independently in this area could easily develop and present solutions to protect all modern, high-yielding crops. This reductionist approach to agricultural research has failed and resulted in crop production and management practices leading to an ever-increasing reliance on pesticides. A holistic approach to agricultural research, through strong integration of research programs between plant breeding/production and crop protection, will prevent such conflicting developments (Lewis et al., 1997).

INTEGRATED PRODUCTION APPROACHES MAKE CHEMICAL CONTROL PRACTICALLY REDUNDANT

A first step toward the reduction of environmental problems caused by agriculture is to test new research developments within the framework of a holistic farming system approach. The focal point in this approach is farm economics and *maximizing net income,* instead of maximizing gross yield. It argues that top yields can only be obtained with excessively high input of fertilizers and pesticides. Reductions in the input levels may lead to somewhat lower yields but, since direct production costs are also reduced, the net income levels may be equal or even higher. Within the farming systems approach, environmental effects such as pollution of soil and water by pesticides are minimized, while biodiversity and resource conservation (landscape, energy, and nonrenewable resources) are maximized. In general, the integrated farming system approach takes into account the various impacts of technology on ecosystems, as well as sociological considerations (employment, public health, and well-being of the agricultural community) not included in current production-driven farming approaches (Wijnands and Kroonen-Backbier, 1993).

The integrated farming system approach attempts to obtain a better balance between the multiple influences that shape agricultural production by increasing awareness of the problems associated with conventional farming practices. Although this general objective is often expressed, its realization will be difficult to achieve as long as politicians and policy-makers merely pay lip service to integrated farming approaches and do not provide the means for its implementation. Research in integrated farming systems is still very limited but the potential is good. Production and pest management practices that can be manipulated include crop rotation, cultivation, fertilization, cultural control measures, biological control, pesticide use, and such. Practical results of integrated farming systems are those of a fifteen-year-long project in the Netherlands, showing a 90 percent reduction in total volume of active ingredients used. Important reductions in environmental pollution using integrated farming approaches come about by a decreased use of fertilizers and an increased use of alternatives to conventional pesticides including crop rotation, use of resistant varieties, and conservation biological control. In integrated farming, an important reduction of environmental pollution is realized through a decrease in fertilizer use and the replacement of chemical pesticides by an intensified knowledge on nonchemical measures (crop rotation, use of resistant varieties, and natural biological control). In integrated farming, artificial fertilizers tend to be replaced by organic manure, and the total amount of nitrogen is lowered to prevent higher crop susceptibility to pests and diseases. Weed, arthropod, and disease problems are reduced in integrated farming through the combined alternatives including the use of weed-competitive or pest-resistant varieties, a reduction of N-fertilization, adoption of specific planting dates and plant densities, mechanical weed control, natural control, and so forth. Chemical pest control in integrated farming is based on pest population sampling and the use of decision thresholds (supervised control).

The results obtained from demonstration farms have been so encouraging that the implementation of integrated farming is strongly supported by the Dutch Ministry of Agriculture to reduce environmental pollution and to create a firmer basis for the survival of agriculture in the longer term. Integrated farming gives the same economic results as present-day (conventional) farming. The generally lower physical yields for the integrated system were compensated by cost reduction as a result of the lower input of pesticides and fertilizers. Indirect costs of fertilizer and pesticide use are not yet included in comparisons, which would give even better comparative results. The economic aspect of integrated farming is strongly emphasized because economic figures are more convincing for farmers, policy-makers, and politicians on the short term, than are long-term environmental reasons. In my opinion, development and application of sustainable agriculture, even if it were more costly than current farming practices, should have priority in Europe.

CONCLUSIONS

The spin-off of current entomological activities to control crop pests and vectors of human diseases with chemical pesticides are negative for nature. In addition to a drastic reduction in the use of conventional chemical pest control methods, new cost/benefit analyses, taking indirect and social costs of pesticides into account, will lead to a reappraisal of the value of environmentally safer control measures. Many biologically based crop protection methods await implementation. However, in order to achieve a drastic change to use of sustainable control methods, insect management should not be viewed as an isolated activity, but within an ecosystem approach. In such an approach, many more opportunities for long-term environmentally safe control methods may be achieved.

REFERENCES CITED

DeBach, P., and D. Rosen. 1991. *Biological Control by Natural Enemies.* Cambridge: Cambridge University Press.

DeFoliart, G. R. 1989. The Human Use of Insects as Food and as Animal Feed. *Bulletin of the Entomological Society of America* 35: 22–35.

Dicke, M. 1999. Direct and Indirect Effects of Plants on Beneficial Organisms. In *Handbook of Pest Management,* ed. J. R. Ruberson, 105–53. New York: Marcel Dekker.

Fisher, T. W., T. S. Bellows, L. E. Caltagirone, D. L. Dahlstein, C. Huffaker, and G. Gordh, eds. 1999. *Handbook of Biological Control.* San Diego: Academic Press.

Georghiou, G. P., and A. Lagunes-Tejeda. 1991. *The Occurrence of Resistance to Pesticides in Arthropods.* AGPP/MISC/91–1, Food and Agricultural Organization, Rome.

Hogue, C. L. 1987. Cultural Entomology. *Annual Review of Entomology* 32: 181–99.

Huffaker, C. B., and P. S. Messenger. 1976. *Theory and Practice of Biological Control.* New York: Academic Press.

Huffaker, C. B., F. J. Simmonds, and J. E. Laing. 1976. The Theoretical and Empirical Basis of Biological Control. In *Theory and Practice of Biological Control,* ed. C. B. Huffaker and P. S. Messenger, 42–80. New York: Academic Press.

Klassen, W. 1993. Pest Management and Biologically Based Technologies: A Look to the Future. In *Pest Management: Biologically Based Technologies,* ed. R. D. Lumsden and J. L. Vaughn, 410–22. Washington, D.C.: American Chemical Society.

Lewis, W. J., J. C. van Lenteren, S. C. Phatak, and J. H. Tumlinson. 1997. A Total Systems Approach to Sustainable Pest Management. *Proceedings of the National Academy of Sciences USA* 94: 12243–48.

Lumsden, R. D., and J. L. Vaughn, eds. 1993. *Pest Management: Biologically Based Technologies.* Washington, D.C.: American Chemical Society.

Metcalf, R. L. 1980. Changing Role of Insecticides in Crop Protection. *Annual Review of Entomology* 25: 219–56.

Oerke, E. C., H. W. Dehne, F. Schoenbeck, and A. Weber. 1994. *Estimated Losses in Major Food and Cash Crops.* Amsterdam: Elsevier.

Rabbinge, R. 1992. Options for Integrated Agriculture in Europe. In *Biological Control and Integrated Crop Protection: Towards Environmentally Safer Agriculture,* ed. J. C. van Lenteren, A. K. Minks, and O. M. B. de Ponti, 211–18. Wageningen, Netherlands: Pudoc.

Smith, R. A., T. E. Mittler, and C. N. Smith, eds. 1973. *History of Entomology.* Palo Alto, Calif.: Annual Reviews.

Southwood, T. R. E. 1977. *Entomology and Mankind.* Proceedings XVth International Congress of Entomology, Washington, D.C., August 19–27, 1976, 36–51.

Tisdell, C. A. 1990. Economic Impact of Biological Control of Weeds and Insects. In *Critical Issues in Biological Control,* ed. M. Mackauer, L. E. Ehler, and J. Roland, 301–16. Andover, United Kingdom: Intercept.

van Lenteren, J. C. 1993. Parasites and Predators Play a Paramount Role in Pest Management. In *Pest Management: Biologically Based Technologies,* ed. R. D. Lumsden and J. L. Vaughn, 68–81. Washington, D.C.: American Chemical Society.

———. 1995. Integrated Pest Management in Protected Crops. In *Integrated Pest Management,* ed. D. Dent, 311–43. London: Chapman and Hall.

———. 2000. Biological Control of Insects and Mites by Augmentation of Natural Enemies. In *Biological Control: Measures of Success,* ed. S. D. Wratten and G. Gurr. Dordrecht, Netherlands: Kluwer.

Waage, J. K. 1994. *Biodiversity, Business, and Biotechnology: Shaping the Future of Biological Control.* Wageningen, Netherlands: Wageningen Agricultural University.

Wijnands, F. G., and B. M. A. Kroonen-Backbier. 1993. Management of Farming Systems to Reduce Pesticide Inputs: The Integrated Approach. In *Modern Crop Protection: Developments and Perspectives,* ed. J. C. Zadoks, 227–34. Wageningen, Netherlands: Wageningen Pers.

Chapter 21

Sustainable Pest Management for Global Food Security

Examples of Approaches in Africa

HANS R. HERREN
International Centre of Insect Physiology and Ecology (ICIPE)
Duduville, Dasarani
Nairobi, Kenya

Food security issues are a matter of concern to governments of developing and developed countries as they affect, although in very different ways, both groups of countries. The need to produce more and better food at the right place and at an affordable price is paramount to achieving world peace and assuring a future for humanity. There are as many approaches to reach these goals as there are scientists and industrialists in the modern "Life Sciences" of developed nations. The true challenge occurs in sub-Saharan Africa, Asia, and Latin America, where food production needs to make quantum leaps in order to match population growth. These leaps in food production, however, will need to be of a different nature from the ones taken by developed countries during the green revolution. Twenty years from now, developing countries will have to increase their food supply, especially grains, by about 70 percent to feed the 6.5 billion people expected to be living in these countries. Improving food security will be an uphill task, given that pests destroy an average of 40 percent of world food production despite the ever-increasing use of pesticides. Losses due to insects will need to be reduced substantially to accommodate the demand for food of the growing population. In addition to high population growth, developing or newly industrialized countries also harbor the poorest world citizens, who commonly have to live on less than one dollar (U.S.) per day. Reducing poverty, therefore, represents a conflict between monetary and environmental issues, which leaves policy-makers with the choice between today's food security over tomorrow's environmental protection. The result is an overexploitation of production and human resources, and the adoption of "quick fix" approaches that seriously jeopardize long-term solutions.

Our world, therefore, faces an urgent need not only to solve the immediate problems of food availability and health services for the people but also faces an urgent need to develop long-term solutions to achieve sustainable development. Appropriate solutions will need to involve the local communities at all levels in the design and implementation of development programs. One of the challenges of these programs in sub-Saharan Africa is to find new and affordable solutions to pest management programs in order to reduce crop damage and post-harvest losses presently estimated at 42 percent. In addition, the development of local scientific capacity both at the human and infrastructure levels is of great importance. Progress toward food security will come from local policy-makers, scientists, and farmers making the decisions on what needs to be done and doing it within a global network of international scientists and policy-makers that are in tune with the specific needs and means of the target populations.

There is no single approach to complex problems, and the real test of a change will be in the investment patterns in different tactics of Integrated Pest Management (IPM), both by the industry and the public sector. As the Rockefeller Foundation president, Dr. Gordon Conway, recently mentioned in one of his speeches, the world is in need of a double green revolution. This revolution will need to include biotechnological tools, ecological principles, and farmer participation. In this chapter, I would like to emphasize the latter two aspects of this "new revolution," as the first one has been discussed more than proportionally to its actual delivery by many in the field during the past two to three years. There are tremendous challenges ahead in the areas of ecology and community participation. Both areas involve working with systems (plants, insects, humans, etc.), not merely molecules, and also involve going out into the field, the bush, the unknown, and without long-term and secure investment. Very often the solutions may actually be more of the "homegrown" type, and not commercially exploitable, at least not in the short term.

FOOD SECURITY AND THE PRODUCTION AND POST-HARVEST LOSS EQUATION

Food security needs to be examined on two levels, the first one of which is the actual production, from seed to harvest. Seed quality and availability are the basic but not the only conditions for a successful crop. In many parts of Africa in particular, and to a lesser extent in Asia and Latin America, farmers keep a part of their crop for seeds. This has many advantages but also inconveniences. The advantages include the availability of seed of an appropriate genotype for the specific location. Locally produced landraces are extremely important to farmers, but even more so to plant breeders who can use them in crop improvement. One other advantage of local seed is that farmers do not need to spend extra resources in seed purchase and can use the little cash they have to obtain other inputs such as fertilizers. On the down side the landraces are not usually high yielding, green revolution type, but will do well under stress conditions of local environments. The high yielding varieties or hybrids are only available in some countries, to farmers who have more access to cash and who farm near open roads.

Food security also depends on the proper storage, processing, and distribution of agricultural products. One of the main obstacles to increased productivity in many regions is the lack of appropriate infrastructure for the distribution of seeds, other inputs, and the marketing of surplus production. In addition, increased productivity is hindered by the lack of good storage and processing facilities. With the present lack of appropriate distribution, storage, and transportation infrastructure, farmers will always be wary of investing money and time for increased productivity. They have too often watched how high yielding crops can be lost to insects and disease.

Food quantity is of concern but so is food quality. The majority of poor people who have sufficient food (calories) do not have the proper quality of food. It is suggested that land management strategies could not only facilitate sustainable crop yields but also help improve the nutrient content of food crops. In addition, better storage facilities can also conserve food nutrients, and new research efforts are needed to find solutions that are appropriate to the farming conditions of the developing countries. Insect pests are probably one of the limiting factors causing the very narrow range of crops being grown, and in particular the preference for cereals over grain legumes, the latter being very prone to insect damage. New pest management solutions for grain legumes, vegetables, and fruits will help in reintroducing high crop diversity into farms. More research is also necessary on the integration of the different crops into a pest free and sustainable production system.

EXAMPLES OF SUSTAINABLE PLANT HEALTH MANAGEMENT

The Cassava Mealybug Biological Control Program

The cassava mealybug (CM), *Phenacoccus manihoti,* was described as a new species by Matile-Ferrero (Herren and Neuenschwander, 1991). It made its first appearance in Congo (or Congo Brazaville) and then Zaire (now Democratic Republic of Congo [DRC]), in the early 1970s after being accidentally introduced into Africa from the New World, its suggested place of origin. Within three years of its first discovery in Congo, CM had spread to neighboring Angola to the south and to Gabon, Cameroon, and Nigeria to the north and west.

In Africa, over 200 million people depend on cassava (*Manihot esculenta* L. Crantz) for most of their calories and in some countries, most of their proteins. Cassava crops, which usually grow well under the most difficult conditions, were totally destroyed within one or two years after CM invasion. The situation became dramatic and needed immediate attention. Facing this problem, the governments of Congo and DRC requested help from the International Institute of Tropical Agriculture (IITA) to stop the spread of the pest and to bring it under control.

Given the exotic origin of the pest, a classical biological control program was the solution of choice. However, several meetings were held at which resistance breeding and the use of pesticides were also advocated. Eventually, biological control was selected as the primary option, with attention being paid to plant breeding programs to select for resistance and/or tolerance. Starting from scratch on all fronts, the IITA Africa-Wide Cassava Mealybug Biological Control Program took off in the fall of 1979. The results of the program are now history. The program was an unprecedented success in the history of classical biological control (Herren and Neuenschwander, 1991).

Within the first eighteen months of the biological control program, CM and its natural enemies had been discovered in Bolivia, Brazil, and Paraguay. Several natural enemy species were imported into Nigeria, where IITA headquarters are located, for further evaluation, mass-rearing, and release. In the meantime CM had spread to most countries between Senegal to the west and Mozambique to the southeast of the continent. The devastation was immense, and the human populations depending on cassava crops were under great food security risks. Starvation occurred in many areas that could not be reached on time with biological control agents. In many areas, farmers gave up on growing cassava and tried other crops, such as maize and sorghum. Given the high requirements of these crops in terms of soil fertility and management, many farmers were less than successful.

After one year of the first releases of a parasitoid wasp and a predatory beetle, it was clear that natural enemy releases were effective. Following the promising results of early releases, the program quickly expanded across Africa. IITA developed and executed a biological control program that included some twenty-eight countries across the cassava belt. The natural enemies were mostly produced in newly built insect factories and released from an aircraft equipped with appropriate equipment. The operation took about twelve years and had a price tag of U.S. $20 million, including all expenses incurred by the participating countries. The cost to benefit ratio of the program was 1 to 240. What was once termed impossible became one of the biggest successes in classical biological control in terms of development aid and pest management investment. From the farmers' point of view, the biological control program gave them back their cassava crops at no extra expense, and most did not even know whatever happened to the "problem."

The biological control program was also a model for training young scientists in biological control. The results of the training efforts were quite impressive and more than 1,200 people were trained at technical and scientific levels (including both M.S. and Ph.D.). Following this success, policy-makers and plant protection specialists in sub-Saharan Africa have requested biological control programs be developed for several other exotic pests including mango mealybug, larger grain borer, spiraling whiteflies, stemborers, and water weeds. Many of these projects have been successfully completed since then. The CM Biological Control Program has also helped in creating more awareness for IPM within the region. Several IPM projects on the major staple crops have been developed at IITA with the cooperation of regional and national partners including programs for the control of cereal stemborers, and mite and disease problems in cowpeas and cassava.

Locust and Grasshopper Control: Redrawing the Battle Lines for Migrant Pests

Locusts and other herbivorous migrant pests such as gregarious grasshoppers and armyworms are species that have the ability to transform from a solitary lifestyle to a highly social one. In such a state, their behavior and physical appearance changes, and the once-isolated harmless individuals come together, multiply, and develop into pest outbreaks of staggering proportions. As they migrate across vast areas of the earth—covering areas up to 20 percent of the total land area in the case of the desert locust—they strip the countryside of most vegetation, sometimes causing irreversible environmental degradation and total crop loss.

At any given time somewhere in Africa, damaging outbreaks of one or more of these pests occur. The Locust and Migrant Pests Program was established at the International Center for Insect Physiology and Ecology (ICIPE) to seek new solutions to the age-old problem of combating these "biblical plagues." ICIPE entered the scene in 1990, in the wake of catastrophic outbreaks during 1985–89, with a research program aimed at understanding the underlying causes of pest outbreaks. Since then, the center has been searching for alternatives for locust control that do not create the large-scale environmental hazards of the past (broadcast applications of conventional chemical insecticides), and that are economically feasible for some of the world's poorest nations. After using the desert locust as a model, the program is now extending the lessons learned to other migrant pest species. Totally new approaches to locust control are being developed and used to redraw the battle lines in the war against this threat to food security. There are about 200 acridid species in Africa, but only a dozen species are pests and most often only very periodically. Between them, however, they affect virtually the whole African continent. Some species migrate as far as the Near East and western Asia. Damaging populations occur each year in the Sahelian and sub-Saharan Africa, but large and long-lasting outbreaks are not frequent.

Locusts are normally distinguished from grasshoppers by their ability to adapt and transform reversibly from a typical solitary lifestyle to a highly gregarious one characterized by dense, highly mobile and synchronous populations. The transformation involves deep-rooted physiological changes and is associated with improved weather conditions and food availability, generally following periods of drought. Gregarious-phase locusts move together in dense groups or swarms. The distinction between locusts and grasshoppers is somewhat arbitrary since many grasshoppers show varying degrees of gregarious behavior, multiplying rapidly and producing grasshopper bands or adult swarms. Desert locust swarms may contain 20 to 150 million individuals, and cover areas ranging from a few hectares to hundreds of square kilometers. Their mobility allows them access to vegetation and escape from predation and parasitism. While grasshoppers may march a few kilometers, adults can fly over 200 kilometers per day.

Thus it is in their gregarious phase that locusts and grasshoppers represent the biggest threat to agriculture. Individual locusts and grasshoppers can eat their own weight in food (2 grams) every twenty-four hours. One-half million desert locusts, which constitutes only a fraction of an average swarm, weigh as much as one ton, and may consume as much food in a day as approximately 2,500 people. Traditional control methods of digging trenches and herding grasshopper bands to be buried, digging out egg masses, setting fires, and more recently, the use of insecticides, have not brought about sustainable control and can only protect small areas from total devastation. The utilization of insecticides has been an expensive and very unsuccessful tactic. Substantial sums of financial assistance have been utilized for the major insecticide campaigns over the last two decades. A total of 200 million dollars (U.S.) were spent to treat 17 million hectares during the 1986–89 campaigns, and another 45 million dollars to treat about 4 million hectares during 1992–94. In addition, these chemical control programs have left huge amounts of unused hazardous insecticides in some localities, and are now being removed at great cost. It is hard to predict the negative side effects of such large-scale insecticide treatments, as well as the futility of tackling a problem that is already out of hand.

Research at ICIPE has concentrated on the chemical ecology and behavior of locusts and grasshoppers in an attempt to develop environmentally acceptable and sustainable control strategies against these pests. Specifically, progress has been made in the following areas:

1. Five sets of pheromones (chemical messages) involved in regulation of gregarious behavior have been discovered and characterized. Some of these serve to maintain the clustering behavior in the hopper bands and adult swarms, while others ensure that the insects mature, mate, and lay their eggs in synchrony. One chemical signal lures egg-laying females to a common site so that emerging offspring remain together.

2. Field trials at two locations in the Red Sea breeding areas and conducted on large grasshopper bands (greater than 1,000,000 insects) have shown that nymphal "Morse code" can be disrupted by another pheromone sent out by

the adult insects. By jamming their airwaves, so to speak, with this pheromone, gregarious nymphs become confused, reduce their feeding, and start disbanding. This increases their susceptibility to natural enemies and pesticides. This discovery has opened up the possibilities for novel methods of grasshopper control including reduced doses of conventional pesticides and the use of entomopathogens.

3. The key pieces of the puzzle of locust gregarious behavior are still being assembled. Great progress has been made in unraveling the processes and factors associated with the transformation of harmless solitary locusts into gregarious pests. The stage is now set to build on these findings to develop a warning system that would facilitate intervention at earlier stages of swarm development.

Moving the Technologies to the Frontlines

To take advantage of the recent work by partners such as Lutte Biologique contre les Locustes et les Sauteriaux (LUBLOSA) and ICIPE, a path of technology integration is being followed to develop a control strategy based on behavioral modification and entomopathogens. Adult pheromones are being tested for their swarm disruption potential. Trials in multiple locations are in progress in collaboration with LUBLOSA and national locust control organizations on both sides of the Red Sea, and in west Africa and north Africa. These tests are designed to evaluate control tactics including conventional insecticides and new biopesticides developed at ICIPE in partnership with scientists from Virginia Tech University.

The aim of ICIPE's programs is to move from a reactive approach to a proactive/preventive approach for the control of locusts and grasshoppers. Migrant pest control has historically begun at the peak of the pest explosion in the vast swarms. This reactive approach, or "bringing in the fire brigade," requires the application of expensive, environmentally hazardous insecticides, often via indiscriminate aerial sprays. The new proactive approach, or "I smell smoke," began in the last two decades and includes earlier intervention using more sophisticated control tactics in order to prevent plague development. It is based on early warning systems dependent on various monitoring methods, including meteorological reports and ground and aerial surveillance of small outbreaks.

Using Nature's Tools for Keeping the Granaries Full

Maize and sorghum are the most important cereal crops for the people of Africa. Lepidopteran stemborers are ubiquitous pests that attack these crops throughout their growth stages, and the larvae cause damage ranging from 20 to 40 percent loss of yield. Seventeen species of stemborers in two families, Pyralidae and Noctuidae, have been found to attack maize and sorghum in various parts of Africa. *Chilo partellus* (Swinhoe), *Chilo orichalcociliellus* Strand, *Busseola fusca* Fuller, *Sesamia calamistis* Hampson, and *Eldana saccharina* Walker are of significant importance. *Busseola fusca* is generally considered to be the more damaging stemborer at high elevations (> 1,500 m). A related species, *Sesamia calamistis,* also occurs throughout Africa but causes serious damage in West Africa. *Eldana saccharina* is an important pest of sugar cane in southern Africa, but is also found in maize and sorghum throughout Africa. *Chilo orichalcociliellus* has a more limited distribution and has only been reported as a pest in the coastal area of east Africa, Malawi, and in Madagascar. The previously mentioned stemborers are all native to Africa. However, one important stemborer, *Chilo partellus,* invaded Africa from Asia sometime in the early part of the twentieth century, and is now found in all countries in east and southern Africa.

Focusing on the two key staple food grains in eastern and southern Africa (maize and sorghum), ICIPE is developing IPM solutions to stemborer pest problems. These solutions include biological control and habitat management. A study carried out by ICIPE's Social Sciences Department on farmer perceptions of maize growing at the Kenyan coast confirmed that insects are the main constraint to maize production and that stemborers are the farmers' "enemy number one" among insect pests. ICIPE's research activities integrate two environmentally and economically acceptable solutions for developing-country farmers and have resulted in novel and widely tested and adopted solutions.

Habitat Management Tactics for Integrated Control of Stemborers (and Weeds)

Most stemborers are polyphagous and attack several gramineous and other wild plants in families Typhaeceae and Cyperaceae. After the introduction and cultivation of maize and extensive plantings of sorghum in areas of Africa, native stemborers became more widely distributed and gained economic importance. As long as there were only wild plants, stemborers were of no special consequence to agriculture. In Africa, the bulk of maize is grown by subsistence farmers on small plots surrounded by land occupied by wild grasses. These wild hosts may provide refugia to stemborer pests in the off-season when there are no cultivated hosts in the field. On the positive side, wild host plants growing on uncultivated land may provide extremely important refugia and sources of nectar, pollen, and alternate host/prey for natural enemies. Wild hosts may also serve to suppress insect outbreaks by keeping the pests on wild "trap" plants.

As part of a novel integrated program for controlling stemborers in cereal crops, ICIPE has developed habitat management strategies for small- to medium-scale farmers practicing mixed agriculture in resource-poor regions of Africa. One of these strategies is the use of trap and repellent plants. This involves the combined use of trap and deterrent crops into a "pull-push" or stimulo-deterrent diversionary strategy. Using this approach, the insects are attracted into highly susceptible trap crops (pull) while being driven away from the target crop by a repellent intercrop (push), in an effort to effectively separate a target crop from damaging pest populations. A wide range of cultivated and wild plants in the families Gramineae and Leguminacae have been assessed in field trials conducted in Kenya to determine their level of susceptibility and repellency to cereal stemborers. The selected plants were napier grass, *Pennisetum purpureum* Schumacher, the Sudan grass *Sorghum vulgare sudanense* Hitchc., the molasses grass *Melinis minutiflora* Pal. de Beauv., and leguminous silver leaf *Desmondium uncinatum* (Jaqc.). Napier grass and Sudan grass are used as trap crops, whereas the molasses grass and the silver leaf are used as deterrent crops. The integrated push-pull strategy is based on the known susceptibility of trap crops and on knowledge of the impact of deterrent crops. High levels of stemborer parasitism are common in molasses grass while the presence of silver leaf considerably reduces damage to maize by *Striga* spp. weeds. This last aspect is being investigated further to better integrate its usefulness in stemborer control. Furthermore, the "pull-push" approach is highly interesting to many small-scale farmers because all selected plants have some economic value as livestock fodder.

The development of these pest control strategies is based on a participatory, on-farm approach by ICIPE. The trials are conducted in a partnership between small- to medium-scale farmers, who are expected to benefit from ICIPE's IPM technologies, and the local research and extension services. The positive results obtained so far have prompted participating farmers to extend the use of the novel technology to larger areas of their farms. The project helped in the reduction of a 70 percent milk shortage in the district and incorporated about 500 more Kenyan farmers to the on-farm trials by the year 2001, as well as farmers of other African countries including Ethiopia, Tanzania, and Uganda.

Classical Biological Control of Chilo partellus

The spotted stemborer (*C. partellus*) is an exotic species that was accidentally introduced into Africa from Asia in the 1930s. It is a highly pestiferous species and is slowly "eating its way" across the African continent, leaving in its wake crop losses of up to 80 percent. The small parasitic wasp *Cotesia flavipes* Cameron, a natural enemy of stemborers, was imported by ICIPE from Pakistan in 1991 and has proven to be the best biological control agent against this exotic pest. As a bonus, this wasp also attacks pestiferous native stemborer species. After extensive host-specificity tests to confirm that the wasps would only attack cereal stemborer species, ICIPE mass-reared and released *C. flavipes* into areas of Kenya in 1993. The Centre's scientists continue to monitor wasp populations in the field. Searches for additional natural enemies of *C. pratensis* have identified several promising candidates, although none have yet been introduced and evaluated. As long as farmers avoid the use of broad spectrum synthetic pesticides, exotic natural enemies and indigenous predators (ants, spiders, mantids, grasshoppers, and the like) can be conserved and put to use in the battle to control stemborers.

As nature cannot be hurried, it will take several years before the final biological impact of *C. flavipes* on suppressing stemborers is known. However, very promising rates of parasitism (rising to more than 50 percent in some sites) have been observed in Kenya and Uganda. The project is now focusing on understanding the causes behind observed differences in *C. flavipes* impact in different localities. In addition to the impact of natural enemy releases, the project has been important in stimulating regional institutional collaboration. ICIPE has developed collaboration with National Agriculture Research and Extension Systems (NARES) in seven African countries to tackle stemborer problems in the region. Regional training programs have resulted in more than 100 national staff members trained in IPM and biological control since 1991, and eleven students from eight African countries and the Netherlands pursuing graduate degrees (eight Ph.D. students and three M.S. students).

Helping Farmers Find the Right IMP Menu

Recognizing that one of the cornerstones of IPM adoption is farmer involvement, ICIPE biologists and social scientists have worked closely to enhance communication between researchers, extension agents, and farmers. Two projects, the Dwale-Kilifi Adaptive Research Project and the Interactive Socioeconomic Research for Biointensive Pest Management Project (ISERIPM), have assessed the effectiveness of a local strain of the bacterium

Bacillus thuringiensis (a natural enemy) in 150 farms at nine sites along the Kenyan coast. The use of *B. thuringiensis* (*Bt*) within the newly introduced strip-relay intercropping system, and in combination with genetically improved, pest-tolerant varieties of maize, sorghum, and cowpea (developed by ICIPE), has resulted in two crops of maize and sorghum, three crops of cowpea, and one of cassava per year in the coastal region bordering the Indian Ocean. After the adaptive research project ended a few years ago, ICIPE handed over the improved cultivars and technologies to the Kenya Agricultural Research Institute (KARI), which was a full partner in the project.

Eco-Friendly Solutions for the Tsetse Problem

Tsetse is Africa's "special" pet. Found only on this continent, tsetse carry the trypanosome parasites responsible for human sleeping sickness (SS) and several fatal animal diseases such as "nagana" in cattle and "sura" in camels. Farmers in tsetse-infested regions are restrained from tilling the land for fear of contracting the dreaded SS and exposing their livestock to the flies. Unfortunately, the humid and subhumid zones of Africa where tsetse flies thrive are also those that hold the greatest potential for increased agricultural production. Over the past three decades, ICIPE's Tsetse Research Program has been a major player in providing new strategies for controlling these disease vectors. ICIPE's simple yet nonpolluting tsetse control technologies continue to gain credibility as they have been tested and proven effective and appropriate even in some of Africa's worst tsetse-infested regions.

Bait Technology

ICIPE has been recognized as a leader in the development of bait traps. Baits are based on natural or synthetic odors that lure the target insects into a structure (trap), which is also visually attractive to them. Tsetse flies use animal odors for long-range host location and visual cues as they get closer to their target. Trap design often mimics the basic shape of animal host species. New baits are being developed that can be integrated with other control strategies. After years of successfully employing odor baits such as buffalo and cow urine, acetone, and other harmless substances to attract the savanna-dwelling species (including *Glossina pallidipes* Austen and *G. morsitans morsitans* Westwood), ICIPE is now turning its attention to riverine tsetse flies. These groups of flies, many of which carry human sleeping sickness, have thus far avoided control efforts. ICIPE is now collaborating with partners in west and east Africa to do field testing of three new compounds (isolated and identified by ICIPE) that are attractant for riverine flies. These substances were isolated from the monitor lizard, which is the preferred host (for blood meals) of this type of flies.

Repellents to Keep the Flies at Bay

ICIPE has recently identified a natural repellent for the savanna group of tsetse flies that reduces the number of infective bites and the rate of infection. This repellent, and several others being identified, originates from nonhost wild animals that are unattractive to blood-suckling flies. These compounds have shown to repel 80 percent of flies attacking cattle and provide new and promising weapons to use in conjunction with present tsetse and trypanosomiasis control tactics. Work is now focusing on developing ways to use these powerful substances in pest control. A potentially effective way to use the repellents would be a "push-pull" strategy, where the flies are pushed away from one area using the repellents and pulled into another area using attractive odor traps. Such tactics can also be used to construct repellent barriers to prevent the flies from re-invading areas previously cleared with other control methods.

Attractants: A Double-Impact Weapon

Another application of the attractive power of nature's chemicals in controlling tsetse flies is the use of larviposition pheromones. These compounds, isolated and identified by ICIPE scientists, are secreted by the larval stage of the insect prior to its pupation. Research of the fly's behavior has shown that the pheromones exert a powerful attraction to gravid (pregnant) female flies searching for favorable places to deposit their larvae. ICIPE is now involved in field testing in Zimbabwe to determine whether these same substances can be exploited to attract the gravid flies to a control site for double impact—by killing both females and larvae.

A Lethal Path for the Wily Flies

Another nonpolluting tsetse control method under development at ICIPE is the use of a special strain of the fungus *Metarhizium* sp. Attacking only tsetse, this entomopathogen is currently undergoing field-testing at a site near Lake Victoria. The fungus is being applied within traps that will attract and contaminate tsetse adults and allow them to escape and return to the environment where they may spread the pathogen to other flies. Dubbed the "Lethal Insect Technique" (LIT), this new technology may be one of the tsetse population management methods to be employed in the BioVillage Initiative in Ethiopia and in the newly developed "farming in tsetse controlled areas."

A Trap for Biting Flies

ICIPE's work over the past three decades has shown that tsetse fly populations can be managed using odor baits and traps through projects in Ethiopia (ongoing); Nguruman and Lambwe Valley, Kenya (ongoing); and the Dagera River basin of eastern Africa. Building on past successes in research and development of effective traps, including the NGU series (NG2B, NG2G) for tsetse fly control, ICIPE's work is being extended to the design of traps that will attract and kill other kinds of biting flies such as stable flies (*Stomoxys* sp.), tabanids, and horseflies. Other than producing painful bites, these insects may mechanically transfer several parasites, including the trypanosomes that cause the camel disease, "sura." The NZI trap (nzi is Kiswahili for fly) is perhaps the only trap available to control biting flies; this trap has proven promising in field tests in Ethiopia and Kenya, and work is in progress to optimize its design for wider use.

ENTOMOLOGICAL SCIENCES IN SUPPORT OF SUSTAINABLE PLANT HEALTH MANAGEMENT: WHAT ARE THE NEEDS OF THE FUTURE?

IPM and Biotechnology for Food Security: Reality vs. Dream

Over the past fifty years or so, most pests and vector problems have been tackled via the use of pesticides. Pesticides proved to be very effective in the beginning, but development of resistance and the environmental and health problems, which resulted from their use and abuse, are calling for a different set of solutions. Integrated Pest Management was to be this new solution. IPM ideas have been introduced widely in Asia, Latin America, and to a lesser degree, in Africa. They are now common practice in Europe and North America. The problem is that overall consumption of pesticides has increased, leaving water wells contaminated, people sick, and the pests still taking away an even larger proportion of our yields. There is clear need to revisit IPM, and take it back to its original meaning and spirit. To be successful, IPM has to provide self-sustaining, environmentally neutral, and economically affordable solutions to pests and disease vector problems.

Solutions to pest and vector problems in the past, and often in the present, have involved quick fixes. These solutions eventually turn out to be costly, not sustainable in the long term, and deleterious to the environment and human health. Whatever solution we may contemplate, the farmers and rural populations will need cash to procure the necessary inputs. Therefore, all plant health management strategies have to include consideration of farmer income levels, and wherever necessary, income-generating activities in the framework of integrated development.

Present and future food availability for immediate consumption and reserves in Africa and other tropical countries is dependent on increased productivity and loss reduction. Increased productivity in developing countries will probably be tied to broader use of animal traction and will involve dealing with the severe constraints caused by animal diseases. In addition, new solutions will be needed to solve human disease problems associated with the many insect-transmitted diseases affecting rural populations dedicated to agriculture. There is a strong and unavoidable link between biodiversity, arthropod diversity, and plant health management. Past unawareness of the existence of such links has driven farmers to the use and abuse of pesticides. Insects and other arthropods often play key roles in soil fertility maintenance and are frequently eliminated through inappropriate agronomic practices and the use of insecticides. Their contribution needs to be studied in more detail. Recent knowledge (or rediscovered old knowledge), points toward the need to understand the role and function of all ecosystem components in the development of pest management strategies.

Insect populations are important components of ecosystems, and their function is vital to sustainable plant health management. The word sustainable, as used in agricultural production, will only have a meaning when all the ecosystem elements are considered. ICIPE's pioneering work in sustainable food production takes into account these links between ecosystem components.

Pollination

The role and function of insect pollinators in ecosystem balance and agricultural production in the tropics has received little scientific interest. More than two-thirds of all flowering plants depend on insects for pollination. For these plants, pollinators are as critical as light and water. Pollination is a vital link in natural communities, connecting plants and animals in key and essential ways. Cross-pollination by insects has been key to the success of flowering plants throughout the world. Even very rare flowering plants may persist because of the great skill exhibited by insects, which continue to find and identify them. In other words, they do not need to be close together, as is rarely possible in the many forbidding environments in Africa. The wealth of types of pollinators, from butterflies to bees to birds and bats, and the wealth of variety of flowering plants, has stimulated each other's evolution. This has led to a remarkable diversity and beautiful adaptations in both flowers and pollinators. Conserving pollinators in an ecosystem means preserving the finely tuned links between plants and animals that permit the successful reproduction of plants.

Pollination is key to human existence as well. Insect pollinators are essential for many fruit and vegetable crops, and the demand for pollinators grows as agricultural productivity grows. Past experience has shown that we should not rely on a single pollinator species. For instance, different crops may set seed best when pollinated by different species. In the future, it is foreseen that there will be an increased reliance on native or wild pollinators (Kevan et al., 1990). There may not be a better place to learn about conserving diversity of native pollinators than in Africa, where both honeybees and a wide spectrum of other pollinators are native. Unfortunately, comparatively little is known about pollination biology in tropical Africa. Excellent work on pollination guilds (Johnson, 1993) and pollen wasps (Gess and Gess, 1996) has come from South Africa, but the remainder of the continent remains little studied.

Soil Dynamics and Insects

Ecosystem processes occurring underneath the soil surface may be among the most critical to overall ecosystem health and sustainable agricultural production. Key processes are: litter and dung decomposition, nitrogen mineralization, and soil movement. In tropical soils, termites consume large quantities of leaf, litter, and woody materials and thus are a major force in nutrient recycling. Some species possess a gut microflora capable of digesting cellulose, while others (Macrotermitinae) cultivate fungus gardens to increase the digestibility of organic matter. Enzymes capable of digesting plant structural polysaccharides have been recorded from many groups of insects, including fly larvae and beetle larvae. Termites render litter more amenable to microbial decomposition by ingesting and fragmenting it and incorporating it into the soil. The role of fly larvae and dung beetles in the decomposition of dung and its incorporation into the soil is also crucial. The introduction of exotic livestock into Australia and the absence of dung decomposing species resulted in the deterioration of pastures and immobilization of plant nutrients in feces and demonstrated the very critical role of these species in nutrient recycling.

Soil insects, on the one hand, are thought to aid in nitrogen mineralization, as their own requirement for this important plant growth nutrient is relatively small. Thus, in the process of digesting plant material, these species excrete excess nitrogen, which is readily available for primary producers (plants) and decomposers. Termites, on the other hand, may be capable of fixing atmospheric nitrogen.

Soil-dwelling insects such as termites and ants have major influences on ecosystems in simply moving soil around. In Zaire, it was estimated that termite mounds covered 30 percent of the soil surface and contained more than 2,800 tons of soil per hectare. Subsurface soil is preferred to topsoil for mound construction, and thus the effect of termites can be a gradual inversion of the soil profile. While we usually think of termites as pests, in fact the effect of man's activities usually reduces species numbers and overall abundance of termites, including many with highly beneficial roles in nutrient cycling. Some species, however, have adapted to man-modified environments, and have become serious pests in the absence of their competitors (Wood, 1996). The quantities of wood removed by termites have not been carefully assessed, although it has been suggested that in some ecosystems when termites are forced to compete with humans removing dead wood for fuel, they may turn to the consumption of living trees. It is already proving difficult to plant trees in many areas where termites consume the seedlings or cuttings. Farmers are also reporting increased termite problems in crops such as maize and cassava. The proliferation of termites has become a serious issue as it increases the cost of shelters to rural population, and the environmental damage caused by the repeated logging to replace the wooden structures after only two to three years. There is, at present, a nucleus of researchers working on termites and other soil fauna in different sites in Africa. An information management project would bring together managers of termite collections globally, and undertake the specimen databasing of this important group.

SUSTAINABLE DEVELOPMENT AS SEEN AND REQUESTED BY THE FARMING COMMUNITIES OF AFRICA: THE BIOVILLAGE INITIATIVE

Human and animal diseases as well as the lack of food security are among the most important factors limiting social and economic development in Africa. Different individual approaches to tackle these problems have either failed or provided limited success. One of the reasons for limited success or failure is the targeting of individual problems rather than dealing with the complex of multiple problems in a comprehensive way. Work by ICIPE in Ethiopian communities has shown that the goals of improving quality of life and promoting economic and social development can only be achieved through a comprehensive system of health management comprising humans, farm animals, crops, and the environment. Thus, in the development of sustainable rural production systems, health management at all trophic levels should be addressed first. Past failures in achieving these goals by development organizations may have involved the limited participation of local communities in the design of solutions to their own problems, the lack of education and training, and the high dependency on external inputs.

Past projects on disease vector management may be used as examples of failures. For too long, mosquito management programs have focused on control of adult populations and have neglected integrated control concepts including breeding-site management involving community participation. ICIPE's tsetse fly control projects have demonstrated that a community-based approach is as good if not better that a household-based approach. Tsetse fly control projects have been successful because the community, rather than individual farmers, was responsible for mass-trapping operations. Moreover, the social organization of African communities also facilitates participatory-based approaches effective even in the case of crop and crop health management programs.

Another reason for past failures of vector management programs may be their short-term nature. Vector management programs have not adequately addressed sustainability issues such as those emphasized by the United Nations' Brundtland Commission. Environmental sustainability or maintenance of life support systems is a prerequisite for social and economic development and sustainability (Goodland, 1995). The scale at which to look for environmental sustainability may also be an issue. Sustainability should not be perceived at only the individual household level. Environmental sustainability should also be considered. Scale consideration should be incorporated in issues of environmental sustainability. When addressing the spatial scale, neither consumption nor production needs to be sustainable in a single household. On the community level, however, a high level of sustainability may be the basis for economic and social development. Adequate resource management achieves environmental sustainability. Resources to be managed not only include water, nitrogen, and organic matter but also time. There is circumstantial evidence that time constraints do not allow African communities to undertake new activities, such as income-generating ones. Unless existing resource management and health management become more efficient, African communities will not have the time to invest in their economic development. As indicated in the Final Report to the World Heath Organization Director General Transition Team of August 12, 1998, on Health, Health Policy, and Economic Outcomes, and in Bloom and Sachs (1998), there is a positive correlation between health status and economic success. Although classical factors, such as poor economic conditions, climate, soils, and topography are invoked to account for Africa's poor economic performance, health is one of the major factors in economic progress. Poor health limits manpower and drains limited resources from the rural communities.

Background

Historically, the BioVillage Initiative evolved from a successful ICIPE tsetse control project in Ethiopia's Gurage and Sodo Bedesse areas. The program led to a significant reduction in the incidence of farm animal diseases and a dramatic increase in production as demonstrated by the doubling of milk yields. Following detailed discussions at the community level, it became apparent that an approach tackling exclusively the disease vector would not fully address the health problems of rural communities. For example, farmers stressed that both oxen and men have to be healthy to plow the fields. It was agreed that only a comprehensive health management program could provide a solution to the most urgent problems limiting rural development, and that a health management program emphasizing preventive rather than curative methods should be initiated. Thereby, it became apparent that health and well-being also have to rely on a secure supply of food and high quality water, as well as a healthy environment. To achieve the latter, efficient waste management with concomitant reduction of disease vector breeding sites was considered to be important. These aspects broadened the comprehensive health management concept into an integrated resource management system.

The rural community agreed that this system was a prerequisite for further exploitation of available resources, and that it would lead to income generating activities and poverty alleviation. Moreover, capacity building was considered an important element in system development and implementation as well as in information dissemination and technology transfer. National partners, regional government offices, as well as the rural communities involved named the so-conceived community-driven comprehensive health and resource management system for rural development BioVillage Initiative.

However, it soon became apparent that this system only provided a crude empirical framework with limited use for the design of an integrated health and resource management system. To overcome these difficulties ICIPE first proposed a theoretical basis on which system structure and functioning can be optimized and prepared for further discussions with the partners involved. This basic framework should allow objectively defined variables to be measured in monitoring activities, which are considered to be very important to further improve the system's qualities including its sustainability.

The BioVillage System

The theoretical framework in which the BioVillage Initiative was developed can be divided into two interdependent subsystems which differ in, among others things, the time scales at which they function.

Subsystem 1: Arthropods and Comprehensive Health Management

Populations of arthropods interact with humans, farm animals, stored products, plants, decaying organic matter, and soils. An adequate knowledge of the spatio-temporal dynamics of these populations is a prerequisite for the design and evaluation of health management strategies. For example, this knowledge is essential in the development of adequate strategies to separate disease vectors from their human and farm animal hosts in time and space. Once identified, potential health management tactics are discussed with the members of the community. The participation of the community ensures social acceptability and technology adoption. Strategic planning addresses economic constraints, food security, and prevention of crop, animal and human diseases, and pests.

Subsystem 2: Integrated Resource Management

Components of this subsystem include soil, water, crops, farm animals, stored products, organic waste, and humans. These components are connected by a flow of nutrients including nitrogen, phosphorous, and potassium. An adequate knowledge of nutrient cycles, including nutrient uptake and deposition by crops and organic matter decomposition, is prerequisite for the design of sustainable resource management systems. For example, adequate water and waste management practices improve sanitation and reduce the number of mosquito breeding sites. Once identified, potential resource management practices are discussed with the members of the community.

The participation of the community ensures social acceptability of practices. Strategic planning addresses economic constraints, food security, and prevention of crop, animal and human diseases, and pests.

Design and Implementation of the System

The design and implementation of the BioVillage system relies on both ecological principles and applied social science through interactions among ICIPE scientists, partner organizations, and local communities.

Design Principles

The system's components are presented to the community, and the community decides on both the components and the management operations to be implemented. The community also decides on management responsibilities, which ideally should follow principles identified by Becker and Ostrom (1995) for sustainable use of common resources. Monitoring work is organized by a BioVillage program leader and carried out by a BioVillage manager and by collaborating institutions. The information obtained enables specialists from ICIPE and scientists from collaborating institutions to make recommendations to the community on how to improve both the structure and the functioning of the system.

Demonstration Sites

The communities, ICIPE, and Ethiopian partner institutions establish demonstration sites adjacent to existing villages and inhabited by villagers. These sites allow the testing of technologies under realistic conditions and facilitate the transfer of knowledge and experiences to the neighboring communities. This enables the communities to select appropriate technologies and to identify needs to ICIPE and its partners. The management of the demonstration sites enables the partners to find a common terminology and to plan as well as to carry out participatory work. Moreover, it serves as a basis for training activities.

The construction of nearly insect-proof tukuls (local name for a dwelling) for humans and farm animals is expected to provide protection during the time of vector activity. Using the villagers' expertise and locally available building materials, the traditional tukul can easily be modified. The flow of organic matter is organized as to meet sanitation requirements and to minimize the necessity for external inputs of resources such as energy and fertilizers. Using a bio-digester, energy (methane gas) is extracted from organic waste and the leftover slurry is utilized as organic fertilizers. This minimizes loss of nutrients such as nitrogen. High quality organic fertilizers can be produced by integrating an organic converter into the system. High quality organic fertilizers are important in maintaining soil quality. As demonstrated in China, for example, energy (in the form of methane gas) can be extracted from the bio-digester for cooking purposes. The availability of energy reduces the need to collect firewood and conserves trees that in turn contribute to a healthy environment and may supply fruits and botanical insecticides for pest control. Moreover, the time usually invested into firewood collection can be used for income generating activities. In addition, waste management also reduces the breeding sites of disease vectors.

The BioVillage Initiative also involves the design of cropping systems to produce a constant and secure supply of high quality food to humans and animals. This can be achieved by an adequate selection of plant species to be grown at the proper place and time and by the adoption of sustainable crop management practices. These include the use of available organic fertilizers, botanical pesticides, and native natural enemies to reduce the need for external inputs. Food security also involves safe storage of food and feed surplus. Food storage is improved through the use of insect-proof tukuls, and through food processing facilitated by energy available from bio-digesters.

Economic and Social Sustainability

In collaboration with local communities and regional governments, ICIPE is currently recommending several income generating activities. Among these, bee keeping, silk production, natural enemy production (*Trichogramma* spp.), and leather processing are the most promising ones. In addition, local communities have proposed community activities to promote social sustainability of BioVillage. Environmental and economic sustainability will be ensured by attaching a price to valuable resources such as biogas and water, for re-investment into the system. Results will enable the design of a virtual BioVillage system as a tool for information dissemination, technology transfer, and adoption.

CONCLUDING REMARKS

Arthropods are essential components of life on earth. It is the duty of the entomologists and scientists of related disciplines to take a holistic approach in the design and execution of their research programs. The tendency of increasing specialization of the different scientific disciplines into the smaller components of life is a natural result of increased amount of accumulated knowledge. Specialization is positive from the strict point of view of knowledge accumulation but should be complemented by a corresponding intensification in interdisciplinary and multi-trophic research and training activities. Such interdisciplinary partnerships, from genes to landscapes, will assure that solutions to our health, agriculture, and environmental problems are addressed with some chance of success. New technologies developed through molecular biology and genetic engineering cannot by themselves solve the world's food security problems. These promising new technologies can only be successfully and economically deployed, and in an environmentally sound way, through their integration with other disciplines. It is therefore our duty to promote partnerships in research to assure that all relevant insect-science disciplines are continuously developed and strengthened.

There is much talk about food security and how to assure it. The solutions, however, are less apparent. Unsubstantiated claims and statements have been made regarding the need to invest large sums of money in genetic engineering and the creation of new models of our major staple crops. However, I believe that our food security lies more in diversity of food crops and of the environment. The saying "do not put all your eggs in the same basket" holds true in assuring food security for all, and we all would be well advised to remember it. Therefore, we must urge our fellow scientists, our policy-makers, and the public at large to support broad research efforts covering a thorough understanding of the ecosystems we use to produce our crops and the ones we live in. We also need open partnerships between the North and the South, which will in the near future wean the developing countries from the need to beg for research, money, and food.

As Professor Robert van den Bosch, one of the most outspoken promoters of biological control, wrote, "the insects shall eat the last piece of flesh off the last human bone." This means we will never be able to fully control insects, by whatever means we may devise. It may then be appropriate to look for ways and means to live in harmony with our smaller, but mighty, neighbors. Instead of further sharpening the knife, we should use our intellectual power to modify the way in which we live, and practice agriculture and forestry to prevent conflict with arthropods. The cheap and quick fixes have left behind many known and yet unknown time bombs and have no future. Biodiversity and systems research is time consuming and therefore expensive. The key to succeeding in this endeavor is certainly good science, but even more important is public opinion. Given the poor image of insects in public opinion, it is easy to understand the difficulty in getting the attention of the policy-makers and their constituency to support research for control measures, which do not "kill on the spot." Therefore, in addition to our research job, we will have to spend or invest more time, and some of our funds, in public education.

REFERENCES CITED

Becker, C. D., and E. Ostrom. 1995. Human Ecology and Resource Sustainability: The Importance of Institutional Diversity. *Annual Review of Ecology and Systematics* 26: 113–33.

Bloom, D. E., and J. D. Sachs. 1998. Geography, Demography, and Economic Growth in Africa. *Brookings Papers on Economic Activity* 2: 1998.

Gess, F. W., and S. K. Gess. 1993. Effects of Increasing Land Utilization on Species Representation and Diversity of Aculeate Wasps and Bees in the Semi-Arid Areas of Southern Africa. In *Hymenoptera and Biodiversity,* ed. J. LaSalle and I. D. Gauld, 83–114. Oxon, United Kingdom: CAB International.

Goodland, R. 1995. The Concept of Environmental Sustainability. *Annual Review of Ecology and Systematics* 26: 1–24.

Herren, H. R., and P. Neuenschwander. 1991. Biological Control of Cassava Pests in Africa. *Annual Review of Entomology* 36: 257–83.

Johnson, S. D. 1993. Red Flowers and Butterfly Pollination in the Fynbos. In *Plant-Animal Interactions in Mediterranean-Type Ecosystems,* ed. M Arianoutsou and R. H. Groves, 137–48. Dordrecht, Netherlands: Kluwer.

Kevan, P. G., E. A. Clark, and V. G. Thomas. 1990. Insect Pollinators and Sustainable Agriculture. *American Journal of Alternative Agriculture* 5: 13–22.

Wood, T. G. 1996. The Agricultural Importance of Termites in the Tropics. *Agricultural Zoology Reviews* 7: 117–55.

Part VII

The Future of Entomological Research

Chapter 22

Evolutionary Pest Management

An Approach to the Twenty-First Century

KEVIN M. HEINZ
Department of Entomology
Texas A&M University–College Station

The concept of pattern or regularity is basic to science, nature, and agriculture. Since pattern implies some sort of repetition, its existence means that some prediction or explanation is possible. This explanation often takes the form of a mechanism that is consistent with the pattern. However, within the continuum of nature, repetition is usually imperfect. This imperfection of repetition may provide the means of making comparisons on the behavior of a system with a variety of conditions in order to possibly identify the most basic, unifying pattern (MacArthur, 1972). Benefits to science and agriculture may be derived from studies of patterns because direct experimentation at appropriate spatial or temporal scales is often not feasible. Pattern analysis provides a solution to the problem of studying large-scale events such as detailing ecological and agricultural landscapes or species dynamics through prolonged periods of time.

While the concept of pattern may be ubiquitous to most of science and its application, a major drawback to the analyses of patterns is the inability of such analyses to elucidate "cause-and-effect" relationships that generate the observed patterns. To understand cause-and-effect relationships, one must develop a critical attitude about the problem and construct a procedure for assessment. Without guidance, this attitude development and procedure construction step often results in a random assortment of experiments to be performed with little relatedness to the problem that is attempting to be solved. This often-used yet scattered research paradigm has lead to substantial criticism of ecological research and the applicability of results to agricultural problems. From where does guidance for ecological research reside?

Fundamental and applied ecology is a quantitative study of systems at various levels of integration (Macfadyen, 1975), from the autecology of individuals to the structure within and among landscapes. At its core is a body of theory linking the various levels of integration (Watt, 1971). Yet attacks on the durability and applicability of ecological theory have been numerous (McIntosh, 1985; Loehle, 1988; Peters, 1991; Pickett et al., 1994; Weiner 1995). In terms of entomological research, such attacks may be a result of ecological entomologists focusing on questions that address "how" systems operate rather than "why" they operate. A pattern of "how" oriented research may be explained by a positive correlation that frequently exists between the status of an insect pest and the amount of research effort expended. However, the proximate "how" approach provides little predictive power relative to the ultimate "why" approach as to how systems will respond to environmental perturbations.

Predicting and charting an appropriate course for the future of agriculture and the research supporting this technology is difficult. There are many conflicting trajectories among various research paradigms that encompass a wide range of issues: international economics, biotechnology, environmental safety, societal consciousness, education, and transportation services. If such diversity is analogous to the genetic diversity observed within species, then events in the future may result from processes similar to evolution by natural selection. Various trajectories will be tested within the context of their current environment. Only imagination and technological capability are likely to limit the number and types of trajectories. The environment within which the fitness of each trajectory will be tested is delineated by current economic, socio-environmental, and regulatory constraints. Those trajectories that fail the test of selection will be eliminated from the agricultural population and the scientific research supporting the practice.

Rather than lay out a roadmap of the future, which will have an ever-expanding error term as we move from today's point of origin, a better route might be for me to discuss a method that should be successful in advancing fundamental and applied ecological research through the twenty-first century and those that follow. The approach intermingles evolutionary theory with pest management research. This approach provides binomial, testable hypotheses that may be most easily studied experimentally in agricultural systems. I will develop the theory and tests of sex allocation behavior in ectoparasitoids as an example whereby evolutionary thought has contributed to predictable pest management strategies applicable to production agriculture. I will conclude by constructing a framework whereby evolutionary thought may be applied to pest management research. This framework will address the roles of fundamental and applied research in developing robust solutions to pest management problems.

AN EXAMPLE OF EVOLUTIONARY THEORY

A characteristic common to sexually reproducing, dioecious organisms is the division of reproductive investment into sons and daughters. The proportion of male to female progeny produced by the parents of a given genotype can influence the parents' inclusive fitness (Fisher, 1930; Maynard Smith, 1978). Therefore, it is of great interest to examine the way natural selection determines the allocation of resources to male versus female function, and hence offspring sex ratios (Charnov, 1982). The morphology and physiology of parasitic Hymenoptera permit ovipositing females to control fertilization, and hence the sex of each egg laid. Unfertilized eggs develop into haploid sons whereas fertilized eggs develop into diploid daughters. Host quality, which is frequently measured in terms of host size, is well known to influence the sex allocation decisions made by parasitic wasps (Charnov, 1982; King, 1987; Godfray, 1994). When host quality differentially affects the fitness of sons and daughters, selection will favor mothers with the ability to manipulate sex ratio in response to current host quality distributions (Charnov, 1979; Charnov et al., 1981; Bull, 1981).

Charnov et al. (1981) were the first to explain how this common sex allocation pattern in parasitoids may be adaptive. This model was developed specifically for solitary species where at most one adult offspring develops per host and the quantity of resources available to a developing wasp is expected to correlate closely with the size of the host (King, 1987, 1993; Godfray, 1994). As with observations made by early natural historians (Clausen, 1939) the first prediction of the model is that only sons should be produced in low quality or small hosts and only daughters in high quality or large hosts. Charnov et al. (1981) argued that host size affects female fitness more strongly than male fitness; therefore, an ovipositing mother will realize greater reproductive success by allocating daughters to larger hosts and sons to smaller hosts. Second, the model predicts relative, rather than absolute, host size to be important.

For instance, if a female wasp encounters a population of small-sized and medium-sized hosts, then she should lay more sons in the small-sized hosts and more daughters in the medium-sized hosts. If that same female were to subsequently encounter a population of medium-sized and large-sized hosts, she should now lay sons in the medium-sized hosts and daughters in the large hosts.

This body of theory establishes a set of binomial hypothesis (see Figure 22.1), each of which is experimentally tractable. Development of simplistic binomial hypotheses creates a methodology whereby complex processes may be elucidated in a hierarchical reductionist manner. As one proceeds through the chain, each hypothesis is rejected or accepted, thereby eliminating a particular factor. In the discussion following, an example of how this methodology operates is revealed using sex allocation studies on *Diglyphus begini* (Ashmead) (Hymenoptera: Eulophidae), a parasitoid of *Liriomyza* leafminers, as a model system.

Population-level studies conducted within greenhouses (Heinz and Parrella, 1990a) and studies with individuals conducted in the laboratory (Heinz and Parrella, 1990b; P. Ode and K. Heinz, 2002) demonstrated that female progeny emerged from significantly larger hosts than male progeny, and the pattern was independent of the range of host sizes encountered by female parasitoids. These data support the concept that female progeny are overproduced on large hosts and male progeny are overproduced on small hosts, and a host is only large or small relative to the hosts encountered.

A precaution frequently overlooked in sex ratio studies (King, 1993) is whether inference derived from secondary sex ratios correctly reflects primary sex ratios. Although several authors have noted that shifts in parasitoid sex ratios may be attributed to differential mortality (Wilkes, 1963; Meyers, 1978; Wellings et al., 1986), few studies have addressed the influence of differential mortality on the relationship between sex ratio and host size. Female wasps tend to be larger than males and hence may require more food (larger hosts) than males. Host sizes sufficiently large for the complete development of male eggs may be inadequate for female development. A direct method for testing this hypothesis is to conduct transfers of parasitoid eggs among hosts of varying size as first described by Luck and Podoler (1985). If small hosts are inadequate for female development, parasitoid eggs originally oviposited on small hosts and transferred to large hosts should produce a sex ratio similar to eggs originally oviposited on large hosts and transferred to another host of equal size.

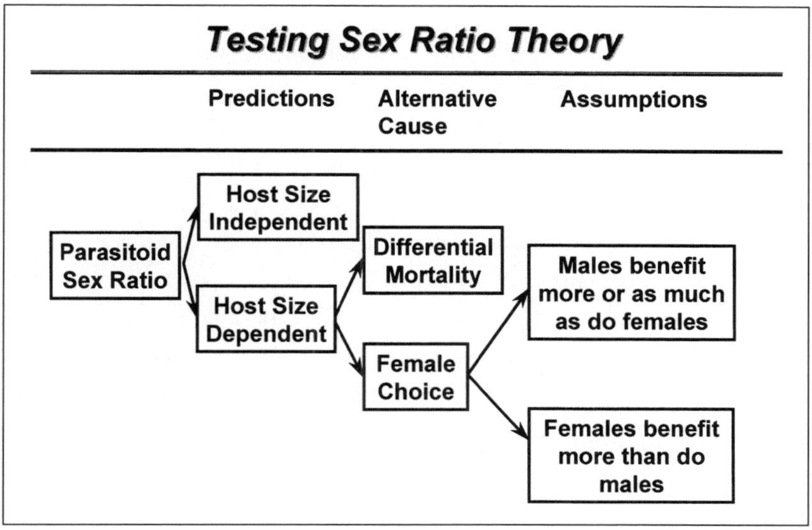

Figure 22.1: Use of sex ratio theory generates an experimental design of relatively simple experiments that test a hierarchy of binomial hypotheses. A relatively complex set of evolutionary theory is rigorously evaluated upon completion of all segments of the tree.

In separate studies involving *D. begini* and *Liriomyza trifolii* (Burgess) (Heinz and Parrella, 1990a; 1990b), and parasitoid-host combination of *Diglyphus isaea* (Walker) (Hymenoptera: Eulophidae) and *Liriomyza huidobrensis* (Blanchard) (Diptera: Agromyzidae) (P. Ode and K. Heinz, 2002), parasitoid eggs originally oviposited on small hosts and subsequently transferred to large hosts yielded predominantly male offspring whereas eggs oviposited on large hosts and transferred to other large hosts (as an experimental control) yielded predominantly female offspring. Hence the host-size-specific sex ratios observed for *Diglyphus* spp. were not due to sex-specific differential mortality between the sex of progeny due to a difference in host size; the secondary sex ratios observed were indicative of the primary sex ratios. Females appear to be choosing the sex of their offspring based upon the relative sizes of hosts utilized.

The handful of studies that compare male and female reproductive success have generally supported the assumption that female reproductive success is more positively affected as host size increases than is male reproductive success (Charnov et al., 1981; Jones, 1982; van den Assem et al., 1989; Heinz, 1991; Ode et al., 1996; Ueno, 1999). Comparative measures of reproductive success require a common unit of fitness be determined. While measures of lifetime fecundity are relatively easy to obtain for females, a comparative measure for male reproductive success must reflect both the relative ability of males to inseminate females successfully and the ability of these mated females to oviposit female eggs on suitable hosts.

Heinz (1991) compared progeny production by males relative to females under low and high levels of female *D. begini* accessibility to potential male mates and low and high *L. trifolii* host densities available for female oviposition. The assumption central to the Charnov (1979; 1982) sex allocation model, that females gain more in fitness by being larger than do males, was confirmed only under a specific set of environmental conditions. Regardless of host density for ovipositing females, when male accessibility to females was low, there was a significant positive correlation between wasp size and female reproductive success but no correlation between wasp size and male reproductive success (Heinz, 1991). Wasp sizes were positively correlated with male and female reproductive success at high and low host densities when male accessibility to females was high; however, there were no sex-specific differences in fitness gains. The observed increase in male reproductive success with wasp size was probably due to an increase in mating success in large wasps rather than an increase in insemination quality with large body size. While the assumption of the Charnov (1979; 1982) sex allocation model was satisfied only under specific environmental conditions, large male wasps never gained more in relative fitness than large female wasps (Heinz, 1991).

APPLICATION OF SEX ALLOCATION THEORY TO PEST MANAGEMENT SOLUTIONS

In many agricultural and urban landscapes, tolerance for arthropod damage is very low. One solution for pest control in these situations is the repeated release of natural enemies through augmentative or inundative biological control. While these approaches can provide desirable levels of pest suppression (van Lenteren, 1986; van Lenteren et al., 1997), they are often prohibitively expensive relative to chemical control programs in terms of production and implementation costs (Parrella et al., 1992; Heinz, 1998). As a result, many otherwise very effective biological control agents are not reared by commercial insectaries. Modification of mass-rearing procedures to reduce the production costs of natural enemies is one strategy that could make available otherwise unaffordable natural enemies for use in biological control. If purchase costs are substantially reduced, biological control is more likely to be perceived as a viable alternative to traditional chemical control.

In augmentative releases, females are more valuable than males because only they are responsible directly for killing pests by oviposition and/or host feeding. One important cost associated with many arrhenotokous parasitic Hymenoptera (parasitoid) mass-rearing programs is the production of more males than is necessary to mate females that are produced. Excess males represent a surplus that can make the commercial production costs of parasitoids excessive in terms of the cost-per-female-released. Mass-rearing programs that incorporate techniques to increase the proportion of females produced may allow insectaries to produce a higher quality product with little or no increase in production costs. While the price per natural enemy shipment charged to clientele may not change, biological control practitioners may realize a significant savings if they are able to attain the same degree of pest control with fewer releases of natural enemies.

Diglyphus isaea and its congener *D. begini* have been shown to be effective augmentative biological control agents against *Liriomyza* leafminers infesting a wide range of greenhouse and field-grown crops and ornamentals (Minkenberg, 1989; Heinz and Parrella, 1990b; Heinz et al., 1993). Large-scale mass-production programs for *D. isaea* (Hendrickson, 1975; Hunter, 1997) and its host (Parrella et al., 1989; Heinz, 1996) have been developed to support biological control efforts using this species. However, current rearing procedures for *D. isaea* remain expensive (van Lenteren, 1986), in part because of the production of male-biased sex ratios (0.6 to 0.7 proportion male; Hendrickson and Barth, 1978). As a consequence, adoption of this species as a biocontrol agent is hindered. Commercial prices charged per individual wasp range from $0.065 to $0.349 making this a relatively expensive biocontrol agent (van Lenteren et al., 1997). Costs of biocontrol using *Diglyphus* spp. is compounded by the need for frequent large releases to achieve acceptable levels of leafminer control (Heinz et al., 1988). Development of a rearing system that maximizes female wasp production may reduce overall mass-rearing costs and thereby increase the attractiveness of using *D. isaea* to control agromyzid leafminers in commercial settings.

P. Ode and K. Heinz (2002) developed a mass-rearing technique that increased the proportion of female *D. isaea* produced and thus reduce the need for frequent releases in biological control programs. By presenting females groups of sequentially larger hosts to attack, they were able to generate progressively more female-biased sex ratios. After three days of providing increasingly larger hosts, the sex ratios produced by individual females declined from 57 percent male to 36 percent male; sex ratios produced by groups of females dropped from 64 percent male to 45 percent male. Attributes discussed here for *Diglyphus* spp. permitted manipulation of sex allocation behavior. Rearing modifications at the insectary scale to induce such sex ratios shifts may be relatively inexpensive, particularly if they primarily involve more stringent controls to produce evenly aged host distributions and a standardized schedule to expose each wasp culture to increasingly large sets of hosts. If protocol changes for the insectary are relatively inexpensive, most of the cost savings resulting from the change in sex ratio can be directly translated into cost reduction for the biological control practitioner. Ideally, this will encourage the more widespread use of *D. isaea* as a viable pest control strategy.

Using a similar host size manipulation procedure, Heinz (1998) was also able to generate substantially more female-biased sex ratios in mass-reared cultures of the boll weevil parasitoid, *Catolaccus grandis* (Hymenoptera: Pteromalidae). This suggests that the ability to artificially induce more female-biased sex allocation in mass-reared cultures of parasitoids has widespread potential. This technique holds the greatest promise of success in solitary parasitoids of hosts that cease to develop after parasitism (idiobionts) (Wellings et al., 1986; King, 1987; Godfray, 1994). Fitness of gregarious parasitoids (those where more than one offspring routinely complete development in a single host) is determined largely by the number of offspring per host rather than a direct effect of host size. Females attacking hosts that continue to grow after parasitism (koinobionts) are less likely to exhibit host-size dependent sex allocation behavior because they are less likely to be able to predict the final amount of host resources available for their offspring. Many currently available natural enemies from insectaries are solitary (e.g., Hunter, 1997) and are potential candidates for improved sex ratios through modified mass-reared protocols.

Use of sex allocation theory has also shown to be useful in predicting the outcome of natural enemy introduction schemes (Luck and Podoler, 1985), whereby the species best able to utilize small hosts for the production of female progeny tends to be the most successful invader. Further, application of sex allocation theory is useful in estimating natural enemy release rates (Heinz et al., 1993) and in the timing of natural enemy releases (in terms of the relative availability of various host sizes). Other opportunities yet explored are also likely to exist.

A 1996 report filed by the National Research Council (U.S.A.) suggested that pest management has strayed from its ecological underpinnings. As demonstrated in this set of examples, evolutionary theory provides clear sets of testable hypotheses capable of crosscutting existing ideologies and technologies. Use of this approach permits the detection of patterns within and among systems, but also provides a means to understand reasons "why" the patterns are as observed. The structure to "why" questions typically lays in theory, and through validation of theoretical predictions can be used to form the foundation of current pest management practices and those of the future. Answers to the "why" questions provide scientific rigor to making predictions about systems as they evolve through time and space.

ACKNOWLEDGMENTS

The research discussed here has been supported by grants and gifts-in-kind from Yoder Bros., Inc.; the American Floral Endowment; the National Biological Control Institute; and the USDA–National Research Initiative Competitive Grants Program.

REFERENCES CITED

Bull, J. J. 1981. Sex Ratio Evolution When Fitness Varies. *Heredity* 46: 9–26.

Charnov, E. L. 1979. The Genetical Evolution of Patterns of Sexuality: Darwinian Fitness. *American Naturalist* 113: 465–80.

——— 1982. *The Theory of Sex Allocation.* Princeton, N.J.: Princeton University Press.

Charnov, E. L., R. L. los-den Hartogh, W. T. Jones, and J. van den Assem. 1981. Sex Ratio Evolution in a Variable Environment. *Nature* 289: 27–33.

Clausen, C. P. 1939. The Effect of Host Size upon the Sex Ratio of Hymenopterous Parasites and Its Relation to Methods of Rearing and Colonization. *Journal of the New York Entomological Society* 47: 1–9.

Godfray, H. C. J. 1994. *Parasitoids: Behavioral and Evolutionary Ecology.* Princeton, N.J.: Princeton University Press.

Heinz, K. M. 1991. Sex-Specific Reproductive Consequences of Body Size in the Solitary Ectoparasitoid, *Diglyphus begini. Evolution* 45: 1511–15.

———. 1996. Space- and Cohort-Dependent Longevity in Adult *Liriomyza trifolii* (Burgess) (Diptera: Agromyzidae) Mass-Rearing Cultures. *Canadian Entomologist* 128: 1225–37.

———. 1998. Host Size-Dependent Sex Allocation Behaviour in a Parasitoid: Implications for *Catolaccus grandis* (Hymenoptera: Pteromalidae) Mass-Rearing Programmes. *Bulletin of Entomological Research* 88: 37–45.

Heinz, K. M., and M. P. Parrella. 1990a. The Influence of Host Size on Sex Ratios in the Parasitoid *Diglyphus begini* (Hymenoptera: Eulophidae). *Ecological Entomology* 15: 391–99.

———. 1990b. Holarctic Distribution of the Leafminer Parasitoid *Diglyphus begini* (Hymenoptera: Eulophidae) and Notes on Its Life History Attacking *Liriomyza trifolii* (Diptera: Agromyzidae) in Chrysanthemum. *Annals of the Entomological Society of America* 83: 916–24.

Heinz, K. M., J. P. Newman, and M. P. Parrella. 1988. Biological Control of Leafminers on Greenhouse Marigolds. *California Agriculture* 42: 10–12.

Heinz, K. M., L. Nunney, and M. P. Parrella. 1993. Toward Predictable Biological Control of *Liriomyza trifolii* (Diptera: Agromyzidae) Infesting Greenhouse Cut Chrysanthemums. *Environmental Entomology* 22, 1217–33.

Hendrickson, R. M. 1975. Mass-Rearing of *Diglyphus isaea* Walker (Hymenoptera: Eulophidae) on *Liriomyza trifoliearum* Spencer (Diptera: Agromyzidae). *Journal of the New York Entomological Society* 83: 243–44.
Hendrickson, R. M., and S. E. Barth. 1978. Notes on the Biology of *Diglyphus intermedius* (Hymenoptera: Eulophidae), a Parasite of the Alfalfa Blotch Leafminer, *Agromyza frontella* (Diptera: Agromyzidae). *Proceedings of the Entomological Society of Washington* 80: 210–15.
Hunter, C. D. 1997. *Suppliers of Beneficial Organisms in North America.* Sacramento: California Department of Pesticide Regulation.
Fisher, R. A. 1930. *The Genetical Theory of Natural Selection.* Oxford: Oxford University Press.
Jones, W. T. 1982. Sex Ratio and Host Size in a Parasitoid Wasp. *Behavioral Ecology and Sociobiology* 10: 207–10.
King, B. H. 1987. Offspring Sex Ratios in Parasitoid Wasps. *Quarterly Review of Biology* 62: 367–96.
———. 1993. Sex Ratio Manipulation by Parasitoid Wasps. In *Evolution and Diversity of Sex Ratio in Insects and Mites,* ed. D. L. Wrensch and M. M. Ebbert, 418–41. New York: Chapman and Hall.
Loehle, C. 1988. Philosophical Tools: Potential Contributions to Ecology. *Oikos* 51: 97–104.
Luck, R. F., and H. Podoler. 1985. Competitive Exclusion of *Aphytis lingnanensis* by *A. melinus:* Potential Role of Host Size. *Ecology* 66: 904–13.
Macfadyen, A. 1975. Some Thoughts on the Behaviour of Ecologists. *Journal of Animal Ecology* 44: 351–63.
MacArthur, R. H. 1972. *Geographical Ecology: Patterns in the Distribution of Species.* New York: Harper and Row.
McIntosh, R. P. 1985. *The Background of Ecology.* Cambridge: Cambridge University Press.
Meyers, J. H. 1978. Sex Ratio Adjustment under Food Stress: Maximization of Quality or Numbers of Offspring? *American Naturalist* 112: 381–88.
Minkenberg, O. P. J. M. 1989. Temperature Effects on the Life History of the Eulophid Wasp *Diglyphus isaea,* an Ectoparasitoid of Leafminers (*Liriomyza* spp.), on Tomatoes. *Annals of Applied Biology* 115: 381–97.
National Research Council. 1996. *Ecologically Based Pest Management: New Solutions for a New Century.* Washington, D.C.: National Academic Press.
Ode, P. J., M. F. Antolin, and M. R. Strand. 1996. Sex Allocation and Sexual Asymmetries in Intra-Brood Competition in the Parasitic Wasp *Bracon hebetor. Journal of Animal Ecology* 65: 690–700.
Ode, P. J., and K. M. Heinz. 2002. Host-Size Dependent Sex Ratio Theory and Improving Mass-Reared Parasitoid Sex Ratios. *Biological Control* 24(1): 31–41.
Parrella, M. P., K. M. Heinz, and L. Nunney. 1992. Biological Control through Augmentative Releases of Natural Enemies: A Strategy Whose Time Has Come. *American Entomologist* 38: 172–79.
Parrella, M. P., J. T. Yost, K. M. Heinz, and G. W. Ferrentino. 1989. Mass-Rearing of *Diglyphus begini* (Hymenoptera: Eulophidae) for Biological Control of *Liriomyza trifolii* (Diptera: Agromyzidae). *Journal of Economic Entomology* 82: 420–25.
Peters, R. H. 1991. *A Critique for Ecology.* Cambridge: Cambridge University Press.
Pickett, S. T. A., J. Kolsa, and C. G. Jones. 1994. *Ecological Understanding.* San Diego: Academic Press.
Smith, J. Maynard. 1978. *The Evolution of Sex.* Cambridge: Cambridge University Press.
Ueno, T. 1999. Host-Size-Dependent Sex Ratio in a Parasitoid Wasp. *Researches in Population Ecology* 41: 47–57.
van den Assem, J., J. J. A. van Iersel, and R. L. Los-Den Hartogh. 1989. Is Being Large More Important for Female Than Male Parasitic Wasps? *Behaviour* 108: 160–95.
van Lenteren, J. C. 1986. Parasitoids in the Greenhouse: Successes with Seasonal Inoculative Release Systems. In *Insect Parasitoids,* ed. J. Waage and D. Greathead, 341–74. Orlando, Fla.: Academic Press.

van Lenteren, J. C., M. M. Roskam, and R. Timmer. 1997. Commercial Mass-Production and Pricing of Organisms for Biological Control of Pests in Europe. *Biological Control* 10: 143–49.

Watt, K. E. F. 1971. Dynamics of Populations: A Synthesis. In *Dynamics of Populations: A Synthesis,* ed. P. J. den Boer and G. R. Gradwell, 568–80. Wageningen, Netherlands: Centre for Agricultural Publishing and Documentation.

Weiner, J. 1995. On the Practice of Ecology. *Journal of Ecology* 83: 153–58.

Wellings, P. W., R. Morton, and P. J. Hart. 1986. Primary Sex Ratio and Differential Progeny Survivorship in Solitary Haplo-Diploid Parasitoids. *Ecological Entomology* 11: 341–48.

Wilkes, A. 1963. Environmental Causes of Variation in the Sex Ratio of an Arrhenotokous Insect, *Dahlbominus fuliginosus* Nees (Hymenoptera: Eulophidae). *Canadian Entomologist* 95: 183–202.

CHAPTER 23

Genetic Engineering of Insects

Current and Future Applications

CRAIG COATES
Department of Entomology
Texas A&M University–College Station

The ability to genetically engineer and manipulate insect species holds great promise for the fields of agriculture and medicine. Genetic transformation technology will dramatically enhance our ability to investigate the basic biology of a number of insect species. This new information will undoubtedly lead to the identification of novel targets and control mechanisms for insects of agricultural and medical importance. It is also possible that genetically manipulated insects will be used directly to help control or modify insect pest populations. The goals of this chapter are to describe the current state of insect transformation and to predict some of the future applications of genetic transformation technology.

GENETIC TRANSFORMATION OF *DROSOPHILA MELANOGASTER* AND ADVANCES FOR OTHER INSECTS

The vinegar fly *Drosophila melanogaster* Meigen was first genetically transformed almost twenty years ago (Spradling and Rubin, 1982). Since that time, an immense volume of biological information has been produced as a result of genetic transformation studies performed on this model genetic organism. Studies in *D. melanogaster* that utilize genetic transformation technology continue to answer a vast array of biological questions pertaining to *Drosophila* development, behavior, physiology, and sex determination. In many cases the information obtained in the *Drosophila* system has been extrapolated directly to humans and resulted in the identification of a similar process, or lesion, in a human function or disease state.

The successful development of a transformation system for *D. melanogaster* inspired several efforts to develop similar systems for insects of agricultural and medical importance. Unfortunately, for years this was a tale of many defeats and few triumphs, the reasons for which are discussed here. Thankfully, the number of insect species that have been successfully transformed is currently increasing at a rapid rate as shown in table 23.1.

A number of scientific advances have been primarily responsible for advancing the field of insect transformation. These include (1) an increase in the number and variety of transposable elements that are available; (2) the development of a nondestructive, autonomous, transformation marker gene; and (3) the use of plasmid-based mobility assays to assess transposable element function in target insects. It is likely that the number of researchers using transformation technology for investigations of insect biology will continue to increase as more improvements are made to existing systems.

TRANSPOSABLE ELEMENTS AS GENETIC TRANSFORMATION VECTORS

All current examples of insect transformation have occurred through the use of transposable elements (TEs) of the type II class. These TEs are broadly characterized by the presence of short inverted terminal repeat sequences and an internally encoded transposase protein that can act in *cis* or *trans* to mobilize elements through a DNA only mechanism. The *P* element is one of the best-studied class II TEs and is the TE most commonly used for *D. melanogaster* transformation (Spradling and Rubin, 1982). However, when this TE was used for similar experiments in other insect species, no germline transformants were recovered, with one exception in the human malaria vector mosquito, *Anopheles gambiae* (Giles) (Miller et al., 1997). However, this was the result of a nonhomologous recombination event, occurred at a low frequency, and has not been replicated in other laboratories.

Subsequent research demonstrated that the *P* element is not active in several insect species outside of *Drosophila,* presumably as a result of a requirement for host factors such as the inverted terminal repeat binding protein (Handler et al., 1993; Beall et al., 1994). The transient plasmid-based assays used to determine *P* element inactivity were adapted such that they could be

Table 23.1. **Timeline of Insect Transformation.**

Year	Species	Reference
1982	*Drosophila melanogaster*	Rubin and Spradling, 1982
1987	*Anopheles gambiae*	Miller et al., 1987
	Bactrocera tryonii	Wyhard S., pers. comm.
1995	*Ceratitus capitata*	Loukeris et al., 1995
1998	*Aedes aegypti*	Coates et al., 1998 Jasinskiene et al., 1998
1999	*Anastrepha suspensa*	Handler et al., 2000
	Bombyx mori	Toshiki et al., 2000
	Bactrocera dorsalis	Handler A., pers. comm.
	Pectinophora gossypiella	Peloquin J., pers. comm.
	Stomoxys calcitrans	Lehane M., pers. comm.
	Tribolium castaneum	Berghammer et al., 1999
	Culex quinifaciatus	Atkinson P., pers. comm.

used to assess the ability of numerous TEs to excise and transpose in the embryonic soma of a number of non-drosophilid insects (Coates et al., 1997; Sarkar et al., 1997; Lobo et al., 1999; Catteruccia et al., 2000). At least four different transposable element systems are now available for non-drosophilid insect transformation.

GENETIC TRANSFORMATION MARKER GENES

The availability of a suitable genetic transformation marker gene is an essential component of a successful transformation system. The nature of transformation events dictates that injected individuals must be allowed to develop, mate, and produce progeny that can be identified as transgenic. The injected individuals themselves are not transgenic and those that will produce transgenic progeny cannot be identified at this stage are almost always in the minority and in some cases may be rare. This problem is further compounded as transgenic individuals tend to be the minority of the total siblings produced. The end result of these effects is that the number of transgenic individuals produced from each experiment may be very small and will almost certainly be a small proportion of the total individuals.

The identification of transgenic individuals is achieved through the use of screenable or selectable marker genes. Screenable marker genes are detected by screening all progeny for a change in phenotype such as eye color, or an enzymatic activity such as *lacZ*. Ideally, the screening procedure is nondestructive such that the identified individuals can be used to produce the following generation. Alternatively, a selectable marker gene, such as an insecticide resistance allele, can be used such that only transgenic individuals survive the selection regime. However, this approach can be problematic as insects carrying naturally occurring resistance alleles will be recovered as false positives and some transgenic individuals may be lost as a result of low expression levels.

The marker gene of choice for *D. melanogaster* is the *white*$^+$ (w^+) eye color gene (Klemenz et al., 1987). This is used in conjunction with a w^- mutant strain, allowing the simple identification of transgenic individuals by the partial or complete restoration of eye color. A similar scheme was used to produce transgenic *Aedes aegypti* (L.), except in this case the *cinnabar* gene from *D. melanogaster* was used as the marker gene, rescuing a lesion in the homologous gene from the mosquito (Cornel et al., 1997; Coates et al., 1998; Jasinskiene et al., 1998). While very effective for these insects, analogous systems will be difficult to develop in species that lack mutant strains and well-established genetics.

Perhaps the most significant advance in the development of transformation systems for non-drosophilid insects has been the use of the Green Fluorescent Protein (GFP) as a transformation marker (Berghammer et al., 1999; Toshiki et al., 2000; Pinkerton et al., 2000). The GFP is autonomous and does not require any host factors. Screening for GFP expression is nondestructive and expression is easily detected by excitation with UV and detection of green fluorescent light. The GFP is nontoxic, and perhaps most importantly, it can be used with any strain of a given species. To further the utility of the GFP, alternative forms have been developed with shifted spectra and enhanced emission energies (Zhang et al., 1996; Beavis and Kalejta, 1999).

BASIC RESEARCH APPLICATIONS

Model Genetic Systems

That *D. melanogaster* is a model genetic organism is undisputed. A vast array and amount of biological information has been generated through studies on this species, many of them through the direct use of the *P* element genetic transformation system. The results of these studies have impacted not only other insects but also vertebrate systems, including aspects of human health. The fields of research that have been spanned are extremely broad and some of the more significant include those of development, physiology, sex determination, and behavior. However, with all this wonderful research and completed studies, significant gaps in knowledge remain, even within insects. Many of the biological processes that are important for agricultural pest species, such as host preference and diapause, are unlikely to be revealed by studies conducted in *D. melanogaster.* Similarly, important molecules and processes involved in blood feeding and pathogen transmission by medically important insects are unlikely to exist in *D. melanogaster.*

Therefore, there is a great need and desire to develop similar model genetic systems in other insects. It is unlikely that this will be necessary for every species, as in many cases similar systems can be analyzed and extrapolations applied. However, it is expected that there will be a need to develop systems for at least one insect from each major group. The availability of these model genetic systems will enhance research in all fields and in many insect species. The most likely candidates for which model genetic systems will be developed are those in which a large amount of genetic information already exists. Examples include *Aedes aegypti, Bombyx mori* (L.), and *Tribolium castaneum* (Herbst). For these species, researchers have already generated physical and genetic chromosome maps, mutant strains, and characterized a significant number of gene sequences. These three insects would at least provide a representative amongst blood feeding insects, lepidopterans, and beetles. Clearly, there may be processes in other insects that will not be completely elucidated from studies in these species, requiring the development of complementary model genetic systems. Once developed, model genetic systems can be used to study gene function and regulation in a target species.

Gene Function

One of the applications of the *P* element system in *D. melanogaster* has been the use of insertional mutagenesis. This is a technique whereby a TE is randomly inserted into a genome, where it can inactivate, or mutate genes. Lines that are homozygous for an insertion are generated and the phenotype examined. If an insertion has affected a biological process of interest, the inserted TE provides a molecular tag such that the researcher can identify and isolate the functional gene from an unmodified genome.

To determine or confirm the function of genes that have been cloned through traditional means, transformation systems can be used to insert cloned genes back into genomes to observe an additive effect, or the rescue of a mutation or deficiency. Similarly, genes from one species can be inserted into the genome of another species to determine if gene function is conserved. As an example, these experiments are particularly important for the determination of the genetic basis of an insecticide resistance phenotype. A candidate gene may have been isolated through traditional mapping and cloning techniques and a particular mutation or allele may appear to be associated with the resistance phenotype. To confirm this hypothesis with certainty, the putative resistant allele can be transformed into an insecticide sensitive genetic background to determine if this results in a gain of resistance.

Control of Gene Expression

Many biological processes that are important in insect pest species are controlled, or induced, by a combination of environmental, behavioral, and developmental conditions. An understanding of the molecular mechanisms, or switches, that control these processes would be extremely valuable and may result in the identification of novel targets for control strategies. Genetic transformation technology can be used to identify and characterize DNA elements involved in the control of gene expression. These so-called enhancer traps utilize a TE vector that carries a reporter gene such as *lacZ* or GFP, under the control of a weak promoter. If this construct is inserted near an enhancer sequence, *lacZ* or GFP expression will be controlled by this enhancer. Developmental stages or tissues of interest can be monitored for *lacZ* or GFP expression to identify suitable enhancer sequences.

The use of enhancer traps in non-drosophilid insects will result in a greater understanding of the control of gene expression in specific tissues, in each sex, at different stages of development, and under certain environmental conditions such as those that induce diapause. Furthermore, this creates the potential to specifically express recombinant gene products in a desired tissue, sex, developmental stage, and in response to environmental cues. The ability to manipulate an organism in such a manner will have a dramatic impact on laboratory studies of the basic biology of a species and may also be applied to a field release of a genetically modified species for control purposes.

APPLIED RESEARCH APPLICATIONS

Improvements to Sterile Insect Technique Strains

There are several avenues whereby genetic transformation technology could improve current Sterile Insect Technique (SIT) strains and strategies. The simplest of these is the incorporation of a molecular marker into release strains. Current identification and tracking methods involve the use of dusting, or the presence of a phenotypic marker, which may be associated with deleterious effects. A molecular marker would facilitate the discrimination of released individuals versus wild type insects in the field. The GFP shows excellent potential for this purpose, as it can be expressed at all life stages if desired and has a long half-life, even after the death of the insect (A. Handler, personal communication).

In general, the SIT strategy relies on the release of sterile male insects only, causing mating disruption as the sterile males mate with wild-type females and prevent the production of progeny. Furthermore, in some cases, such as those involving tephritid fruit flies, release of sterile females would be detrimental, as the females will still oviposit and thus cause damage to the fruit, even if they do not produce viable progeny. The release of sterile females would also dilute out the sterile males that are available for mating to wild-type females.

Successful application of the SIT for insect control has been used for the screw-worm fly, the Mediterranean fruit fly, and tsetse fly. In many cases, the main reason for SIT program failure is a breakdown of the release strain during mass-rearing and also that released males are not as fit as wild-type males and thus are not competitive for females in the wild. The application of genetic transformation technology to the production of SIT strains may result in the availability of more robust strains. Furthermore, the technology can be utilized to produce male-only strains, thus reducing production and labor costs, as SIT strains currently require sorting to remove females. A proof-of-principle experiment has been completed in which a repressible female-lethal genetic strain of *D. melanogaster* was created (Thomas et al., 2000). In this case the removal of tetracycline from the rearing media results in female lethality. One of the main reasons for the poor fitness level of SIT strains is that they must be irradiated, or chemically treated, to induce sterility. These treatments presumably induce somatic effects that result in a reduced fitness level. In a similar manner to altering the sex ratio, genetic engineering technology could be used to produce a sterile male strain.

Generation of Genetic Control Release Strains

The major difference between a genetic control strain and a SIT strain is that a genetic control strain would not be sterilized. This should remove a significant portion of the fitness disadvantage that SIT strains currently suffer from. A successful genetic control strain would rely on the presence of a gene whose expression results in the death of all progeny produced from a mating involving a released individual. This would allow for the possibility of female release. Furthermore, the system could be manipulated such that the lethal effect is sex-specific. That is, all the female progeny die, whereas the male progeny survive and are then able to spread the lethal gene throughout the population and into the next generation. These systems could also be manipulated such that transgene expression is triggered by a chemical, or a protein produced in a transgenic crop, thus the lethal effect would only be observed in the growing area.

Improved Beneficial Insects

The potential for genetic transformation technologies to improve beneficial insects should not be overlooked, particularly for economically important social insects such as honeybees and silk moths. Like most social species, these insects are susceptible to disease and large population losses when a pathogen spreads throughout social groups. The field of insect immunology is advancing rapidly, with the recent identification of several molecules and mechanisms involved in insect immune responses and protection (Vilmos and Kurucz, 1998). Genes encoding for these molecules could be isolated from a resistant insect species and genetically transformed into a susceptible insect species, thus providing protection from a number of pathogens. Similarly, recombinant molecules from other systems could be expressed in transgenic insects, for example, single chain antibodies with antipathogen activity. In the case of viral pathogens, an antisense RNA strategy can be considered.

There is also some potential for the technology to be used to improve natural enemies and other species used for biological control purposes. These strategies might involve the introduction of insecticide resistance genes such that the beneficial would be protected during periods of insecticide application. This might be particularly useful in greenhouse environments where concerns over the accidental release of the transgenic organism, or escape of the resistance gene, might be alleviated.

Modifications to Human Disease Vectors

One of the more controversial applications of genetic transformation technology is the modification of insect vectors of human disease. Proof-of-principle experiments have been completed that demonstrate that vectors can be modified transiently such that they are no longer able to transmit a particular virus or parasite (Higgs et al., 1998; James et al., 1999). The molecules that were used in these experiments are ideal candidates for inclusion in genetic transformation constructs and experiments are underway along these lines. While these modified strains may not be released into the field in the short term, they nonetheless provide a powerful laboratory tool for further investigations of vector-pathogen interactions.

Scientific and Public Concerns

The potential release of genetically modified insects raises a number of scientific and public concerns. Given the current social climate and discussions generated by the production and application of genetically modified plants, extreme caution must be followed in the production and use of genetically modified insects. Much of the concern focuses on the use of TEs and their potential movement. That is, what is the likelihood that the TE will be re-mobilized within the insect genome and alter gene expression, thus causing the accidental production of a more deleterious pest species? Furthermore, certain TEs appear to have moved horizontally across species barriers in the past (Robertson and Lampe, 1995). If an element were to move into the genome of another organism, what would be the effect? Would gene inactivation or changes in gene regulation occur and could this cause a new pest species to arise? To minimize these potential effects, the TE must be chosen carefully and issues of re-mobilization and cross-mobilization carefully evaluated for each species.

While TE mobility is a concern, the potential deleterious effects must be put in perspective. The process of gene inactivation and mutation as a result of TE movement constantly occurs in nature, and the potential cost must be weighed against the gain obtained from the production and use of a transgenic organism. Furthermore, all the examples mentioned in this chapter involve the use of an inactive TE that carries a transgene. This means that the TE is no longer capable of catalyzing its own mobility. For such a TE to be mobilized, an active, homologous TE would have to be present in that genome. If this is the case, it is unlikely that a TE of that family would be chosen for the original transformation system. If the inactive TE were to be horizontally transferred to a species containing active elements, it is unlikely that the introduction of a single additional copy will have a significant effect, as there will already be multiple elements present in that genome. However, to minimize these particular concerns, efforts are underway to create "suicide vectors." These TE vectors are further inactivated once inserted in the genome, such that they cannot be mobilized further, even in the presence of a homologous, active TE.

The author would like to suggest that consideration of the transgene itself should be of much greater concern than the TE. Let us assume that even with the best intentions and use of sophisticated molecular technology, the transgene "escapes" and is expressed in a nontarget insect. What are the consequences of transgene escape into a nontarget organism? As an example, it is this concern that will most likely prevent the incorporation of an insecticide resistance gene into a transgenic organism that is to be released into the environment. It seems more sensible that the transgene itself be the focus of risk analysis when considering release of a transgenic organism, not simply the fact that the organism is transgenic and contains an extra inactive TE. Above all, discussions and concerns over the potential release of transgenic insects should not be allowed to hinder the use of genetic transformation technology for furthering the study of insect biology. It is likely that this application will reap the greatest and most immediate rewards for the fields of agriculture and medicine.

SUMMARY

The development of genetic transformation for non-drosophilid insects is a relatively new technology and one that is growing rapidly to include an increasing number of diverse insect species. It is expected that this will result in the development of a number of model genetic systems approaching the level of sophistication of *D. melanogaster.* This would be a significant advance for researchers working on all aspects of insect biology, as it is likely that many molecular processes and mechanisms will be highly conserved within groups of insects. In many cases this will result in as yet unimagined forms of control for insect pests of agricultural and medical significance. Whether the new control strategies occur directly, or indirectly, through the use and release of transgenic insects remains to be determined.

REFERENCES CITED

Beall, E. L., A. Admon, and D. C. Rio. 1994. A *Drosophila* Protein Homologous to the Human p70 Ku Autoimmune Antigen Interacts with the *P* Transposable Element Inverted Repeats. *Proceedings of the National Academy of Sciences USA* 91: 12681–85.

Beavis, A. J., and R. F. Kalejta. 1999. Simultaneous Analysis of the Cyan, Yellow, and Green Fluorescent Proteins by Flow Cytometry Using Single-Laser Excitation at 458 nm. *Cytometry* 37: 68–73.

Berghammer, A. J., M. Klingler, and E. A. Wimmer. 1999. A Universal Marker for Transgenic Insects. *Nature* 402: 370–71.

Catteruccia, F., T. Nolan, C. Blass, H. M. Muller, A. Crisanti, F. C. Kafatos, and T. G. Loukeris. 2000. Toward *Anopheles* Transformation: *Minos* Element Activity in Anopheline Cells and Embryos. *Proceedings of the National Academy of Sciences USA* 97: 2157–62.

Coates, C. J., C. L. Turney, M. Frommer, D. A. O'Brochta, and P. W. Atkinson. 1997. Interplasmid Transposition of the *mariner* Transposable Element in Non-Drosophilid Insects. *Molecular Genetics and Genomics* 253: 728–33.

Coates, C. J., N. Jasinskiene, L. Miyashiro, and A. A. James. 1998. *Mariner* Transposition and Transformation of the Yellow Fever Mosquito, *Aedes aegypti. Proceedings of the National Academy of Sciences USA* 95: 3748–51.

Cornel, A. J., M. Q. Benedict, C. S. Rafferty, A. J. Howells, and F. H. Collins. 1997. Transient Expression of the *Drosophila melanogaster cinnabar* Gene Rescues Eye Color in the White Eye (WE) Strain of *Aedes aegypti. Insect Biochemistry and Molecular Biology* 27: 993–97.

Handler, A. M., S. P. Gomez, and D. A. O'Brochta. 1993. A Functional Analysis of the *P*-Element Gene-Transfer Vector in Insects. *Archives of Insect Biochemistry and Physiology* 22: 373–84.

Handler, A. M., and R. A. Harrell. 2001. Transformation of the Caribbean Fruit Fly, *Anastrepha suspensa,* with a *piggyBac* Vector Marked with Polyubiquitin-Regulated GFP. *Insect Biochemistry and Molecular Biology* 31: 199–205

Higgs, S., J. O. Rayner, K. E. Olson, B. S. Davis, B. J. Beaty, and C. D. Blair. 1998. Engineered Resistance in *Aedes aegypti* to a West African and a South American Strain of Yellow Fever Virus. *American Journal of Tropical Medicine and Hygiene* 58: 663–70.

James, A. A., B. T. Beerntsen, M. de L. Capurro, C. J. Coates, J. Coleman, N. Jasinskiene, and A. U. Krettli. 1999. Controlling Malaria Transmission with Genetically Engineered, *Plasmodium*-Resistant Mosquitoes: Milestones in a Model System. *Parassitologia* 41: 461–71.

Jasinskiene, N., C. J. Coates, M. Q. Benedict, A. J. Cornel, C. S. Rafferty, A. A. James, and F. H. Collins. 1998. Stable Transformation of the Yellow Fever Mosquito, *Aedes aegypti,* with the *Hermes* Element from the Housefly. *Proceedings of the National Academy of Sciences USA* 95: 3743–47.

Klemenz, R., U. Weber and W. J. Gehring. 1987. The White Gene as a Marker in a New *P*-Element Vector for Gene transfer in *Drosophila. Nucleic Acids Research* 15: 3947–59.

Lobo, N., X. Li, and M. J. Fraser. 1999. Transposition of the *piggyBac* Element in Embryos of *Drosophila melanogaster, Aedes aegypti,* and *Trichoplusia ni. Molecular Genetics and Genomics* 261: 803–10.

Loukeris, T. G., I. Livadaras, B. Arca, S. Zabalou, and C. Savakis. 1995. Gene Transfer into the Medfly, *Ceratitis capitata,* with a *Drosophila hydei* Transposable Element. *Science* 270: 2002–05.

Miller, L. H., R. K. Sakai, P. Romans, R. W. Gwadz, P. Kantoff, and H. G. Coon. 1987. Stable Integration and Expression of a Bacterial Gene in the Mosquito *Anopheles gambiae*. *Science* 237: 779–81.

Pinkerton, A. C., K. Michel, D. A. O'Brochta, and P. W. Atkinson. 2000. Green Fluorescent Protein as a Genetic Marker in Transgenic *Aedes aegypti*. *Insect Molecular Biology* 9: 1–10.

Robertson, H. M., and D. J. Lampe. 1995. Distribution of Transposable Elements in Arthropods. *Annual Review of Entomology* 40: 333–57.

Sarkar, A., C. J. Coates, S. Whyard, U. Willhoeft, P. W. Atkinson, and D. A. O'Brochta. 1997. The *Hermes* Element from *Musca domestica* Can Transpose in Four Families of Cyclorrhaphan Flies. *Genetica* 99: 15–29.

Spradling, A. C., and G. M. Rubin. 1982. Transposition of Cloned *P*-Elements into *Drosophila* Germ Line Chromosomes. *Science* 218: 341–47.

Thomas, D. D., C. A. Donnelly, R. J. Wood, and L. S. Alphey. 2000. Insect Population Control Using a Dominant, Repressible, Lethal Genetic System. *Science* 287: 2474–76.

Toshiki, T., T. Chantal, R. Corinne, K. Toshio, A. Eappen, K. Mari, K. Natuo, T. Jean-Luc, M. Bernard, C. Gerard, S. Paul, F. Malcolm, P. Jean-Claude, and C. Pierre. 2000. Germline Transformation of the Silkworm *Bombyx mori L.* Using a *piggyBac* Transposon-Derived Vector. *Nature Biotechnology* 18: 81–84.

Vilmos, P., and E. Kurucz. 1998. Insect Immunity: Evolutionary Roots of the Mammalian Innate Immune System. *Immunology Letters* 62: 59–66.

Zhang, G., V. Gurtu, and S. R. Kain. 1996. An Enhanced Green Fluorescent Protein Allows Sensitive Detection of Gene Transfer in Mammalian Cells. *Biochemical and Biophysical Research Communications* 227: 707–11.

Chapter 24

Driving Forces for Target-Based Insecticide Design in the Twenty-First Century

PATRICIA PIETRANTONIO
Department of Entomology
Texas A&M University–College Station

As we look to the future of entomological research, one valid question is to ask whether synthetic insecticides are going to be used or replaced by other means of pest control in the next century. If we agree that synthetic insecticides are key for the success of modern agriculture and to preserve public health, then, to what extent will insecticides be used? How will new, safer insecticides be developed? This chapter focuses on some factors that will influence insecticide discovery and use in the twenty-first century.

THE NEED FOR SAFE SYNTHETIC INSECTICIDES

Synthetic organic insecticides currently represent 89 percent of the active ingredients in the market (Casida and Quistad, 1998). At the center of our concern as entomologists is the management of insect pest populations in agricultural and urban settings. From the urban perspective, it appears that pesticide use will not only be maintained but also might actually increase at times as pests continue to expand their previously occupied geographical areas. Many of these arthropod pests transmit human diseases, for example, ticks transmit Lyme disease on the East Coast and in the southeastern states of the United States (Linden et al., 2000). *Culex quinquefasciatus* Say mosquitoes carrying the West Nile virus are now a recurrent problem in New York and the East Coast of the United States. Malathion is being applied against the vector because the disease can be deadly, and there is no vaccine to prevent it. In Texas *C. quinquefasciatus* transmit St. Louis encephalitis, a viral disease for which there is also no vaccine. Nuisance arthropods such as the imported red fire ant, *Solenopsis invicta* Buren, also continue to expand their range in the United States (Mackay and Fagerlund, 1997). Insecticides are needed against these pests; however, not all types of insecticides are accepted by the public. The public is intolerant of arthropod presence in homes and neighborhoods, and intolerant of insect damage to produce and grains. Paradoxically, the public is also intolerant of the risks associated with insecticide use (Coats, 1994). The public's attitude, as reflected in consumer reports, is a driving force in insecticide development.

THE LOSS OF INSECTICIDES

Many pesticides, and particularly many insecticides, are being taken off the market because of environmental risk concerns and problems associated with insecticide resistance. This loss will most significantly affect public health and small-market commodities for which insecticide choices are limited. Insecticide development for these commodities is virtually nonexistent due to the small market share of these crops reducing product economic feasibility (Casida and Quistad, 1998).

The organochlorinated insecticides DDT, cyclodienes, and related compounds have been eliminated from western markets due to undesirable toxicological effects on nontarget organisms and their high persistence in the environment. Organochlorinated insecticides accumulate in food chains and negatively affect wildlife and human health as endocrine disruptors (Kamrin, 1997). Compounds such as Mirex have been banned in the United States and most western countries (Kamrin, 1997).

Organophosphates and carbamates are nerve poisons that inhibit the enzyme acetylcholinesterase, and many of these compounds have high mammalian toxicity (Fukuto, 1990). Several of them are being revised or going to be revised by the Environmental Protection Agency (EPA) and their uses will be further regulated or restricted. The revision by EPA has been prompted by the awareness within the scientific community and the public of the risk of insecticide use to human health, especially for children, the elderly, and agricultural workers, and the potential damage to the environment through ground water contamination and concurrent effects on nontarget species.

In addition to regulation, many insecticides are being taken out of agricultural markets because of problems with insecticide resistance. Effective synthetic insecticides are in short supply and need to be used with moderation to prevent development of resistant pest populations. Reliance on specific active ingredients often leads to increases in both frequency and rate of application, which in turn lead to development of insecticide resistance. Thus, when a new compound is registered, resistance management plans are necessary to keep it on the market. In most cases, however, scientists have failed to "manage" resistance and can only delay or mitigate it (Hoy, 1998). This has been exemplified with pyrethroids. Pyrethroids are generally cost-effective products with low mammalian toxicity. However, their excessive use and abuse has developed resistance in major pests including pests of agricultural crops (Luttrell et al., 1994) and disease vectors. Excessive use of pyrethroids in cotton has led to their current inefficacy to control the budworm *Heliothis virescens* (F.), and their decreasing efficacy to control the bollworm, *Helicoverpa zea* (Boddie), in Texas (Pietrantonio et al., 2000a) and the U.S. cotton belt (Martin et al., 1999). A comparable situation with heliothine resistance to pyrethroids has been documented in Australia where moderate pyrethroid use continues despite resistance (Fitt, 1994).

Safe and effective insecticides to control pests in urban areas are also a limited resource. These insecticides are most needed for disease control and prevention when viruses are detected in mosquito vectors. Preserving the effectiveness of these insecticides including malathion, resmethrin, and others will be one of the main challenges of pest protection in urban settings. Entomologists will have to work closely with public health officials to delay or manage insecticide resistance. Education of the public will be critical for the long-term management of populations of mosquitoes in densely populated areas. We have recently documented the loss of efficacy of malathion for the control of *C. quinquefasciatus* mosquitoes in Harris County, Texas. This is significant because this species is the vector of St. Louis encephalitis, which is endemic in the area, and for which there is no vaccine. In addition, in one of the study areas in Houston, female mosquitoes were not only resistant to malathion but also showed tolerance to resmethrin, the only effective insecticide currently on the market for control of this species. This is a precarious situation considering mosquito control is the only disease prevention method currently available against St. Louis encephalitis (Pietrantonio et al., 2000b). New, safe, and inexpensive insecticides are needed to control resistant populations of several mosquito species worldwide. Safe insecticides targeting agricultural pests in urban areas may also be needed. For example, malathion has been applied in a broad scale in urban areas of California and Florida for fruit fly eradication (Myers, 1998).

BIOTECHNOLOGY CONTRIBUTES TO THE DISCOVERY AND/OR DEPLOYMENT OF SAFER INSECTICIDES

Most of the leading insecticide manufacturers in the world are now including a strong biotechnological component in their discovery programs. This has been prompted by the high profitability of biotechnology and by recent advances in several fields such as plant, microbe, and insect genomics (sequencing), robotics, combinatorial chemistry, and new technology for a variety of cellular assays including fluorescent technology, immunological methods, and others. In addition, awareness of the environmental and health risks of synthetic insecticides has caused more stringent regulations for pesticide development and registration. The trend of the market is an emphasis on green chemistry or "softer chemistries" such as the natural toxin derivative Spinosad (Tracer, insecticide) for which its manufacturer Dow-Elanco received the Green Chemistry Presidential Award in 1999. Spinosad controls budworm and bollworm in cotton and its dried residues are less toxic to beneficial insects than other insecticides registered for this use (Pietrantonio and Benedict, 1999).

Impact of Transgenic Crop Technology to Pest Control

The potential impact of biotechnology to crop protection cannot be minimized. Transgenic corn and cotton represented 39 percent of the total acreage planted with transgenic crops in 1998 (28 percent corresponded to insect resistant cultivars and 11 percent to herbicide resistant cultivars). Countries (on the American continent) adopting transgenic crops as a significant portion of their cultivated acreage include the United States, Canada, and Argentina. Most of transgenic cotton and corn express Cry1Ac and Cry1Ab protoxins, respectively, from the bacterium *Bacillus thuringiensis* (*B.t.*). The use of transgenic crops has resulted in a reduction of the total volume of pesticides applied worldwide.

The early detection of resistant individuals in the field is one of the current challenges for preserving the usefulness of transgenic crop technology. Immediately after the deployment of transgenic cotton, for example, sampling or scouting of heliothines was complicated by the fact that there was no information on the spatial distribution of larvae on these crops. The distribution of larvae may vary according to the level of toxin expression in plants, which is not uniform in different plant structures and throughout the growing season. Knowledge of the spatial distribution of heliothines may facilitate the identification of surviving individuals in the field and may aid in the detection of initial stages of resistance development (Pietrantonio and Heinz, 1999). Biotechnology will continue to affect crop protection because of its use in the development of resistant cultivars. It will also change how synthetic insecticide discovery will be achieved.

PROGRESS IN TARGET-BASED INSECTICIDE DESIGN

The pesticide industry, aware of environmental contamination problems and nontarget effects, has shifted toward softer chemistries and a biotechnological approach to insecticide discovery. In addition to transgenic crops, the insecticide industry continues to search for highly potent and selective synthetic chemicals. Emphasis is now given to chemistries that are highly effective yet safe to nontarget species (especially mammals), chemistries that decompose relatively rapidly into nontoxic metabolites in the environment and do not bio-accumulate or contaminate ground water. An overall reduction in the total amount of pesticide applied per acre is the goal for the next century. In the 2nd Pan Pacific Conference on Pesticide Science held in Hawaii in October of 1999, the target rate of 1 g of active ingredient per hectare per application was posed as the challenge to the pesticide industry to be achieved within the next ten years.

In addition to the loss of conventional insecticides from the market, some of the driving forces for target-based insecticide development are: (1) the need for safer and more selective insecticides, driven by more stringent registration regulations; (2) the availability of improved instrumentation technology; (3) the availability of genomic sequence information; (4) advances in combinatorial chemistry; and (5) technology owned by pharmaceutical companies with agrochemical divisions.

Target-based insecticide discovery process begins with knowledge of a specific system in the insect that could be disrupted by a novel insecticide. There has to be negative consequences or what we call a biochemical lesion to the organism when the target is disrupted. For example, the most exploited insect targets are ion channels and enzymes. Among channels, most insecticides target the voltage-gated sodium channel, the gamma amino butyric acid (GABA) receptor, a chloride channel, and the nicotinic acetylcholine receptor, a cation channel (Bloomquist, 1996). In order to identify compounds selective for the desired targets, the corresponding receptor cDNAs are expressed in cell lines in culture. Once that protein is expressed, the screening of libraries of thousands of compounds can begin. Knowledge of the sequence and the tri-dimensional structure of the protein target can be used to improve the structure of the compound or lead that showed activity (Blundell, 1996). Worldwide, all of the major companies dedicated to insecticide discovery have formed partnerships with companies in the life sciences or producers of combinatorial libraries with the purpose of target-based pesticide discovery. Most of the technologies associated with these partnerships are for genomics and functional genomics. The companies include Monsanto (Pharmacia), FMC, Novartis, Dupont, Bayer, and others. Bayer is currently screening compounds based on in vitro assays of suitable molecular targets for development of new insecticides. This current trend indicates that entomology professionals working in crop protection need to be knowledgeable in genetics, biochemistry, and molecular biology to best understand the application of novel crop protection strategies and to contribute to their success.

Technologies

Target-based discovery screening and development of pesticides has been facilitated by five major technological advances. These are advances in combinatorial chemistry (Verdine, 1996), high-throughput screening that reflects the speed provided by robotization and miniaturization of in vitro assays (Broach and Thorner, 1996), automatic gene sequencing, the availability of thousands of gene sequences, and public access to general and specialized databases that allow comparison of gene sequences. Knowledge of gene sequences allows the selection of molecular targets and screening of compounds that ultimately may cause a "biochemical lesion" or result in a therapeutic effect in the case of pharmaceuticals. The combinatorial libraries of random compounds can be attached to solid beads, which facilitates the rapid deduction of the structure of an active compound because the bead carries the "history of the synthesis process" (Hogan, 1996). The screening of compounds by means of cellular assays in vitro is quickly performed by robots using 384 well-plates. The pharmaceutical industry has identified four groups of protein targets for the development of therapeutics for human and veterinary medicine: nuclear receptors, ion channels, enzymes, and G protein-coupled receptors (GPCRs). Not surprisingly, these are the same four groups in which most of the commercial insecticides act upon (Fukuto, 1990; Bloomquist, 1996; Dhadialla et al., 1998; Hiripi et al., 1999). Further, the pharmaceutical industry has identified that the G protein-coupled receptor superfamily has the highest potential with about 5000 targets identified. However, only one group of commercial insecticides that target a GPCR has been developed for pest control. These are the formamidines that act on the octopamine receptor (Vanden Broeck, 1996).

G PROTEIN-COUPLED RECEPTORS (GPCRS) AS TARGETS FOR INSECTICIDE DEVELOPMENT

One of the purposes of insecticide toxicology is the discovery of compounds for the disruption of insect intercellular communication. For any organism to exist, cells must communicate with each other through surface or adhesion proteins, by sending chemical signals through gap junction channels, by sending neurotransmitters to neighboring cells at synapses, or by releasing long-distance primary messengers in the form of hormones into the hemolymph. Therefore, toxicants disrupt the systems involved in the organism internal communication system. These molecular targets are the same identified as therapeutic targets by the pharmaceutical industry (GlaxoWellcome, 1996). Many insect hormones and neurotransmitters act through GPCRs present in the cell plasma membrane. GPCRs are seven transmembrane region receptors involved in signal transduction. These receptors are called G protein-coupled because they are associated intracellularly with GTP and GDP binding proteins (G proteins). Receptor activation by the endogenous ligand (hormone, neuromodulator, or neurotransmitter) transduces the signal from the ligand to other cellular effectors such as enzymes or channels that perform the cellular functions regulated by the endogenous receptor ligand. GPCRs control or regulate critical functions in insects such as diuresis, olfaction, feeding behavior, aggressiveness, and rythmicity, to name a few, and are currently underexploited as targets for insecticidal action. The *Drosophila* genome project has revealed 163 transcripts that potentially belong to GPCRs (Adams et al., 2000). This knowledge and the availability of cDNA clones will help speed up research on these new targets. The elusive insect odorant and taste receptors have been identified; their identification will likely be followed by expression experiments and encourage research for the development of synthetic pheromone mimics, repellents, and attractants. Two projects on GPCRs in my laboratory contribute to target-based insecticide design. One focuses on the excretory system of the mosquito *Aedes aegypti* (L.), and the second involves the identification and characterization of a leucokinin-like receptor from the southern cattle tick, *Boophilus microplus* (Canestrini) (Acari: Ixodidae) (Holmes et al., 2000).

The female of *A. aegypti* is the vector of dengue and dengue hemorrhagic fever. In the Americas, dengue is present from Florida and Texas (Gill et al., 2000), to Argentina (Avilés et al., 1999), and is globally a reemerging disease (Gratz, 1999). We are studying the Malpighian tubule as a system to understand how the female mosquito is able to eliminate the high water load during and after a blood meal. Diuresis is a complex physiological process and the effectors involved can potentially be exploited for mosquito control. The Malpighian tubule is also a major detoxifying organ, and the molecular identification of its detoxifying enzymes provides probes with which to identify resistance mechanisms in field populations. We have recently cloned a cytochrome P450 cDNA from the mosquito *A. aegypti* (Pietrantonio, unpublished), the first one known in this species of mosquitoes that is most related to other insect cytochrome P450 enzymes known to be inducible by xenobiotics. This clone will serve as a probe to investigate *A. aegypti* resistant populations worldwide. We have cloned and biochemically invested several GPCRs involved in Malpighian tubule function. We showed that a photoaffinity leucokinin peptide analogue was suitable to identify the putative leucokinin receptor because it was biologically active in the transepithelial voltage assay in vitro, thus behaving functionally as the endogenous hormone (Pietrantonio et al., 2000c). We used the analogue to characterize a protein of 54 kDa that binds this ligand saturably and with estimated parameter values of binding affinity and density consistent with receptor function (Pietrantonio et al., 2000c). Further, we used a molecular approach selective for this type of receptor sequence. We successfully obtained a PCR product that was sequenced. On the basis of the deduced amino acid sequence we developed antibodies and confirmed by western blot that this receptor was indeed expressed in the Malpighian tubule (Pietrantonio, unpublished). A full-length serotonin receptor cDNA has been fully sequenced (Pietrantonio and Jagge, unpublished). In addition, we have cloned a leucokinin-like receptor from the southern cattle tick that has helped identify by sequence similarity the

Drosophila leucokinin receptor, the first identified insect leucokinin receptor, the sequence of which had not been previously annotated by the *Drosophila* genome project (Holmes et al., 2000).

CONCLUDING REMARKS

One of the roles of insect toxicologists in academia is the discovery and validation of molecular targets for insecticide design. Some of the G protein-coupled hormone receptors characterized could be involved in the regulation of feeding; this has implications for the control of disease transmission by the female mosquito. It is my hope that our research on GPCRs can contribute to new developments for mosquito control that will be applied in the field to improve public health worldwide.

REFERENCES CITED

Adams, M. D., S. E. Celniker, and R. A. Holt. 2000. The Genome Sequence of *Drosophila melanogaster. Science* 287: 2185–95.
Avilés, G., G. Rangeón, V. Vorndam, A. Briones, P. Baroni, D. Enria, and M. S. Sabattini. 1999. Dengue Reemergence in Argentina. *Emerging Infectious Diseases* 5: 575–78.
Bloomquist, J. R. 1996. Ion Channels as Targets for Insecticides. *Annual Review of Entomology* 41: 163–90.
Blundell, T. L. 1996. Structure-Based Drug Design. *Nature* (supplement) 384: 23–26.
Broach, J. R., and J. Thorner. 1996. High-Throughput Screening for Drug Discovery. *Nature* (supplement) 384: 14–16.
Casida, J. E., and G. B. Quistad. 1998. Golden Age of Insecticide Research: Past, Present, or future? *Annual Review of Entomology* 43: 1–16.
Coats, J. R. 1994. Risks from Natural Versus Synthetic Insecticides. *Annual Review of Entomology* 39: 489–515.
Dhadialla, T. S., G. R. Carlson, and D. P. Le. 1998. New Insecticides with Ecdysteroidal and Juvenile Hormone Activity. *Annual Review of Entomology* 43: 545–69.
Fitt, G. P. 1994. Cotton Pest Management: Part 3. An Australian Perspective. *Annual Review of Entomology* 39: 543–62.
Fukuto, T. R. 1990. Mechanism of Action of Organophosphorous and Carbamate Insecticides. *Environmental Health Perspectives* 87: 245–54.
Gill, J., M. L. Stark, and G. G. Clark. 2000. Dengue Surveillance in Florida. *Emerging Infectious Diseases* 6: 30–35.
GlaxoWellcome. 1996. Redesigning Drug Discovery. *Nature* (supplement) 384: 1–5.
Gratz, N. G. 1999. Emerging and Resurging Vector-Borne Diseases. *Annual Review of Entomology* 44: 51–75.
Hiripi, L., L. Nagy, and R. M. Hollingworth. 1999. *In vitro* and *in vivo* Effects of Formamidines in Locust (*Locusta migratoria migratorioides*). *Acta Biologica Hungarica* 50: 81–87.
Hogan, J. C., Jr. 1996. Directed Combinatorial Chemistry. *Nature* (supplement) 384: 17–19.
Holmes, S. P., H. He, A. C. Chen, G. W. Ivie, and P. V. Pietrantonio. 2000. Cloning and Tran-

scriptional Expression of a Leucokinin-Like Peptide Receptor from the Southern Cattle Tick, *Boophilus microplus* (Acari: Ixodidae). *Insect Molecular Biology* 9: 457–65.

Hoy, M. A. 1998. Myths, Models, and Mitigation of Resistance to Pesticides. *Philosophical Transactions of the Royal Society of London, Series B* 353: 1787–95.

Kamrin, M. A. 1997. Chlorinated Hydrocarbons. In *Pesticide Profiles: Toxicity, Environmental Impact, and Fate,* ed. M. A. Kamrin, 239–98. Boca Raton, Fla.: CRC Press, Lewis Publishers.

Linden, J. V., S. J. Wong, F. K. Chu, G. B. Schmidt, and C. Bianco. 2000. Transfusion-Associated Transmission of Babesiosis in New York State. *Transfusion* 40: 285–89.

Luttrell, R. G., G. P. Fitt, F. S. Ramalho, and E. S. Sugonyaev. 1994. Cotton Pest Management: Part 1. A Worldwide Perspective. *Annual Review of Entomology* 39: 517–26.

Mackay, W. P., and R. Fagerlund. 1997. Range Expansion of the Red Imported Fire Ant, *Solenopsis invicta* Buren (Hymenoptera: Formicidae), into New Mexico and Extreme Western Texas. *Proceedings of the Entomological Society of Washington* 99: 757–58.

Martin, S. H., R. D. Bagwell, M. L. Boyd, B. L. Freeman, G. A. Herzog, D. R. Johnson, M. B. Layton, B. R. Leonard, N. Liu, G. T. Payne, P. V. Pietrantonio, M. E. Roof, M. J. Sullivan, J. W. Van Duyn, and J. R. Weeks. 1999. Status of Bollworm, *Helicoverpa zea,* Susceptibility to Pyrethroids: IRAC-US 1998 Update. In *Proceedings. Beltwide Cotton Conferences,* vol. 2, ed. P. Dugger and D. Richter, 867–71. Memphis, Tenn.: National Cotton Council.

Myers, J. H. 1998. Eradication and Pest Management. *Annual Review of Entomology* 43: 471–91.

Pietrantonio, P. V., and J. H. Benedict. 1999. Effect of New Cotton Insecticide Chemistries, Tebufenozide, Spinosad, and Chlorfenapyr, on *Orius insidiosus* and Two *Cotesia* Species. *Southwestern Entomologist* 24: 21–29.

Pietrantonio, P. V., and K. Heinz. 1999. Distribution of Heliothine Larvae in *Bt* and Non-*Bt* Cotton in Texas. In *Proceedings. Beltwide Cotton Conferences,* vol. 2, ed. P. Dugger and D. Richter, 945–48. Memphis, Tenn.: National Cotton Council.

Pietrantonio, P. V., L. Nemec, and J. Sronce. 2000a. Monitoring for Resistance to Cypermethrin in Budworm (*H. virescens*) and Bollworm (*H. zea*) and to Malathion in Boll Weevil in the Brazos River Bottom, Texas. In *Proceedings. Beltwide Cotton Conferences,* vol. 2, ed. P. Dugger and D. Richter, 1236–39. Memphis, Tenn.: National Cotton Council.

Pietrantonio, P. V., G. Gibson, S. Nawrocki, F. Carrier, and W. P. Knight Jr. 2000b. Insecticide Resistance Status, Esterase Activity, and Electromorphs from Mosquito Populations of *Culex quinquefasciatus* Say (Diptera: Culicidae), in Houston (Harris County), Texas. *Journal of Vector Ecology* 25: 74–89.

Pietrantonio, P. V., G. Gibson, D. Petzel, A. Strey, and T. K. Hayes. 2000c. Characterization of a Leucokinin Binding Protein in *Aedes aegypti* (Diptera: Culicidae) Malpighian Tubule. *Insect Biochemistry and Molecular Biology* 30: 1147–59.

Vanden Broeck, J. J. M. 1996. G Protein-Coupled Receptors in Insect Cells. *International Review of Cytology* 164: 189–268.

Verdine, G. L. 1996. The Combinatorial Chemistry of Nature. *Nature* (supplement) 384: 11–13.

CONTRIBUTORS

Allen E. Knutson. Texas Cooperative Extension, Texas A&M Research and Extension Center. Dallas, Texas.

A. R. Kraaijeveld. Centre for Population Biology, Department of Biology, Imperial College at Silwood Park. Ascot, Berkshire SL5 7PY. United Kingdom.

Barry Beaty. Arthropod-Borne and Infectious Diseases Laboratory, Department of Microbiology, Colorado State University. Fort Collins, Colorado 80523.

Carrie Koplinka-Loehr. New York State IPM Program, Cornell University. Ithaca, New York 14853.

Craig Coates. Department of Entomology, Texas A&M University. College Station, Texas 77843-2475.

Curtis H. Petzoldt. New York State IPM Program, Cornell University. Ithaca, New York 14853.

Daniel Simberloff. Department of Ecology and Evolutionary Biology, University of Tennessee. Knoxville, Tennessee 37996.

Donald W. Hall. Department of Entomology and Nematology, University of Florida. Gainesville, Florida 32611-0620.

Donald A. Rutz. Department of Entomology, Cornell University. Ithaca, New York.

Edward A. Hiler. College of Agriculture and Life Sciences, Texas Agricultural Experiment Station and Texas Cooperative Extension. The Texas A&M University System. College Station, Texas 77843.

Frank G. Zalom. Department of Entomology, University of California. Davis, California 95616.

Hans R. Herren. International Centre of Insect Physiology and Ecology (ICIPE), Duduville, Dasarani, Nairobi. Kenya.

H. C. J. Godfray. Centre for Population Biology, Department of Biology, Imperial College at Silwood Park. Ascot, Berkshire SL5 7PY. United Kingdom.

J. Keith Waldron. New York State IPM Program, Cornell University. Ithaca, New York 14853.

J. C. van Lenteren. Laboratory of Entomology, Department of Plant Sciences, Wageningen University and Research Centre. 6700 EH. Wageningen, Netherlands.

James P. Tette. New York State IPM Program, Cornell University. Ithaca, New York 14853.

John L. Capinera. Department of Entomology and Nematology, University of Florida. Gainesville, Florida 32611-0620.

Joseph Kovach. IPM Program, Ohio State University, OARDC. Wooster, Ohio 44691.

Karen English. New York State IPM Program, Cornell University. Ithaca, New York 14853.

Kevin M. Heinz. Department of Entomology, Texas A&M University. College Station, Texas 77843-2475.

Margaret Haining Cowles. Law office of Margaret Haining Cowles. Geneva, New York 14456.

Marshall W. Johnson. UC Kearney Agricultural Center, University of California. Riverside, California 92521.

May R. Berenbaum. Department of Entomology, University of Illinois. Urbana, Illinois 61801-3795.

M. D. E. Fellowes. School of Animal and Microbial Sciences, University of Reading, Whiteknights, Reading, Berkshire, RG6 6AJ. United Kingdom.

Michael J. Samways. Department of Entomology and Nematology, University of Stellenbosch, Matieland 7602. South Africa.

Nick Toscano. Department of Entomology, University of California. Riverside, California 92521.

Patricia Pietrantonio. Department of Entomology, Texas A&M University. College Station, Texas 77843-2475.

Raymond E. Frisbie. Department of Entomology, Texas A&M University. College Station, Texas 77843-2475.

Richard J. Leskosky. Unit for Cinema Studies, University of Illinois. Urbana, Illinois 61801.

Richard L. Jones. Agricultural Experiment Station, University of Florida–IFAS. Gainesville, Florida 32611-0200.

S. Bradleigh Vinson. Department of Entomology, Texas A&M University. College Station, Texas 77843-2475.

Seymour D. Van Gundy. College of Natural and Agricultural Sciences and Department of Nematology, University of California. Riverside, California 92521.

Thomas W. Fuchs. Texas Cooperative Extension. San Angelo, Texas 76901.

William W. Murdoch. Department of Ecology, Evolution, and Marine Biology, University of California. Santa Barbara, California 93106-9610.

ISBN 1-58544-432-4
54500
9 781585 444328